Couplings and Joints

MECHANICAL ENGINEERING

A Series of Textbooks and Reference Books

EDITORS

1. Spring Designer's Handbook, *by Harold Carlson*
2. Computer-Aided Graphics and Design, *by Daniel L. Ryan*
3. Lubrication Fundamentals, *by J. George Wills*
4. Solar Engineering for Domestic Buildings, *by William A. Himmelman*
5. Applied Engineering Mechanics: Statics and Dynamics, *by G. Boothroyd and C. Poli*
6. Centrifugal Pump Clinic, *by Igor J. Karassik*
7. Computer-Aided Kinetics for Machine Design, *by Daniel L. Ryan*
8. Plastics Products Design Handbook, Part A: Materials and Components; Part B: Processes and Design for Processes, *edited by Edward Miller*
9. Turbomachinery: Basic Theory and Applications, *by Earl Logan, Jr.*
10. Vibrations of Shells and Plates, *by Werner Soedel*
11. Flat and Corrugated Diaphragm Design Handbook, *by Mario Di Giovanni*
12. Practical Stress Analysis in Engineering Design, *by Alexander Blake*
13. An Introduction to the Design and Behavior of Bolted Joints, *by John H. Bickford*
14. Optimal Engineering Design: Principles and Applications, *by James N. Siddall*
15. Spring Manufacturing Handbook, *by Harold Carlson*
16. Industrial Noise Control: Fundamentals and Applications, *edited by Lewis H. Bell*
17. Gears and Their Vibration: A Basic Approach to Understanding Gear Noise, *by J. Derek Smith*

18. Chains for Power Transmission and Material Handling: Design and Applications Handbook, *by the American Chain Association*
19. Corrosion and Corrosion Protection Handbook, *edited by Philip A. Schweitzer*
20. Gear Drive Systems: Design and Application, *by Peter Lynwander*
21. Controlling In-Plant Airborne Contaminants: Systems Design and Calculations, *by John D. Constance*
22. CAD/CAM Systems Planning and Implementation, *by Charles S. Knox*
23. Probabilistic Engineering Design: Principles and Applications, *by James N. Siddall*
24. Traction Drives: Selection and Application, *by Frederick W. Heilich III and Eugene E. Shube*
25. Finite Element Methods: An Introduction, *by Ronald L. Huston and Chris E. Passerello*
26. Mechanical Fastening of Plastics: An Engineering Handbook, *by Brayton Lincoln, Kenneth J. Gomes, and James F. Braden*
27. Lubrication in Practice, Second Edition, *edited by W. S. Robertson*
28. Principles of Automated Drafting, *by Daniel L. Ryan*
29. Practical Seal Design, *edited by Leonard J. Martini*
30. Engineering Documentation for CAD/CAM Applications, *by Charles S. Knox*
31. Design Dimensioning with Computer Graphics Applications, *by Jerome C. Lange*
32. Mechanism Analysis: Simplified Graphical and Analytical Techniques, *by Lyndon O. Barton*
33. CAD/CAM Systems: Justification, Implementation, Productivity Measurement, *by Edward J. Preston, George W. Crawford, and Mark E. Coticchia*
34. Steam Plant Calculations Manual, *by V. Ganapathy*
35. Design Assurance for Engineers and Managers, *by John A. Burgess*
36. Heat Transfer Fluids and Systems for Process and Energy Applications, *by Jasbir Singh*
37. Potential Flows: Computer Graphic Solutions, *by Robert H. Kirchhoff*
38. Computer-Aided Graphics and Design, Second Edition, *by Daniel L. Ryan*
39. Electronically Controlled Proportional Valves: Selection and Application, *by Michael J. Tonyan, edited by Tobi Goldoftas*
40. Pressure Gauge Handbook, *by AMETEK, U.S. Gauge Division, edited by Philip W. Harland*
41. Fabric Filtration for Combustion Sources: Fundamentals and Basic Technology, *by R. P. Donovan*
42. Design of Mechanical Joints, *by Alexander Blake*
43. CAD/CAM Dictionary, *by Edward J. Preston, George W. Crawford, and Mark E. Coticchia*

44. Machinery Adhesives for Locking, Retaining, and Sealing, *by Girard S. Haviland*
45. Couplings and Joints: Design, Selection, and Application, *by Jon R. Mancuso*
46. Shaft Alignment Handbook, *by John Piotrowski*
47. BASIC Programs for Steam Plant Engineers: Boilers, Combustion, Fluid Flow, and Heat Transfer, *by V. Ganapathy*
48. Solving Mechanical Design Problems with Computer Graphics, *by Jerome C. Lange*
49. Plastics Gearing: Selection and Application, *by Clifford E. Adams*
50. Clutches and Brakes: Design and Selection, *by William C. Orthwein*
51. Transducers in Mechanical and Electronic Design, *by Harry L. Trietley*

ADDITIONAL VOLUMES IN PREPARATION

Mechanical Engineering Software

1. Spring Design with an IBM PC, *by Al Dietrich*

Couplings and Joints

Design, Selection, and Application

JON R. MANCUSO
Zurn Industries, Inc.
Mechanical Drives Division
Erie, Pennsylvania

MARCEL DEKKER, INC. New York and Basel

Library of Congress Cataloging-in-Publication Data

Mancuso, Jon R., [date]
Couplings and joints.

(Mechanical engineering ; 45)
Bibliography: p.
Includes index.
1. Couplings. 2. Joints (Engineering) I. Title.
II. Series.
TJ183.M36 1986 621.8'25 85-27433
ISBN 0-8247-7400-0

MARCEL DEKKER, INC.
270 Madison Avenue, New York, New York 10016

Current printing (last digit):
10 9 8 7 6 5 4 3 2 1

PRINTED IN THE UNITED STATES OF AMERICA

This book is dedicated to my family

to my parents, Mr. and Mrs. Samuel A. Mancuso,
for their influence on my decision to become an engineer

to my wife Rose,
for her understanding of my profession

and to my children,
Samuel, Lynn, Jonette, and Kristie,
for the inspiration they gave me throughout the creation of this work

Preface

The flexible coupling method of connecting rotating equipment is a vital and necessary technique. Large shafts in loosely mounted bearings, bolted together by flanged rigid couplings, do not provide for efficient and reliable mechanical power transmission. This is especially true in today's industrial environment, where equipment system designers are demanding higher speeds, higher torques, greater flexibility, additional misalignment, and lighter weights for flexible couplings. The need for flexible couplings is becoming more acute as is the need for technological improvements in them.

The basic function of a coupling is to transmit torque from the driver to the driven piece of rotating equipment. Clutches (which constitute another subject for a book and are not covered here) are couplings designed to cease transmitting torque at a certain load or speed. Flexible couplings expand upon the basic function by also accommodating misalignment and end movement. During the initial assembly and installation of rotating equipment, precise alignment of the equipment shaft axes is not only difficult to achieve, but in most cases it is economically prohibitive. In addition, misalignment during equipment operation is even more difficult to control. Misalignment in operation can be caused by flexure of structures under load, settling of foundations, thermal expansion of equipment and their supporting structures, piping expansion or contraction, and many other factors. A flexible coupling serves as a means to compensate for, or minimize the effects of, misalignment. Flexible couplings, however, have their own limitations. Therefore, calculations and predictions are required to know what the maximum excursions can be. Only then can the correct coupling be selected.

A system designer or coupling user cannot just put any flexible coupling into a system and hope it will work. It is the responsibility of the designer or user to select a compatible coupling for the system. The designer must also be the coupling selector. Flexible coupling manufacturers are not system designers and should not be expected to assume the role of coupling selectors. Since they are not system experts, they can only take the information given them and size, design, and manufacture a coupling to fulfill the requirements specified. Some coupling manufacturers can offer some assistance but they usually will not accept responsibility for a system's successful operation.

The purpose of this book is to aid the coupling selector (system designer or user) in understanding flexible couplings so the best coupling can be selected for each system. In order to help coupling selectors do their jobs properly, four chapters are included that cover the basics of couplings. These chapters provide answers to often-asked questions, such as:

Why a flexible coupling?
What are this coupling's limitations and capabilities?
How much unbalance will this coupling produce?
How can I compare this coupling with that coupling?
How do I install this coupling?
How is it disassembled?
How is it aligned?
How is it lubricated?

There are two basic classes of couplings: the rigid coupling and the flexible coupling. A chapter on rigid couplings has been included. Rigid couplings should only be used when the connecting structures and equipment are rigid enough so that very little misalignment can occur and the equipment is strong enough to accept the generated moments and forces.

There are hundreds of different types of flexible couplings. In this book, many types are covered in general and the more common ones in detail. The eight most common types of couplings in use today are:

gear couplings
grid couplings
chain couplings
universal joints
elastomeric shear couplings
elastomeric compression couplings

disk couplings
diaphragm couplings

Each one of these couplings is covered in detail. Included in each section are variations available, the principles of operation, coupling constructions, design criteria, failure modes, and other important information.

It is hoped that this book will achieve two things: generally, to stress to the reader the importance of couplings in power transmission systems; and, more specifically, to provide the coupling selector with the basic tools required for the successful application of couplings to particular needs.

This book could not have been written without the help and cooperation of many individuals and coupling manufacturers. In particular, acknowledgment is given to the Mechanical Drives Division of Zurn Industries, Inc. and especially to the following personnel: Norman Anderson, Sam Steiner, Jim Paluh, Bill Herbstritt, and my secretary Marty Keim. Special thanks are extended to: Edward Heubel and Bill Herbstritt, for many of the sketches; Michael M. Calistrat of Boyce Engineering, Howard Schwerdlin of Lovejoy, and Q. W. Hein of the Falk Corporation, for their extra help.

Jon R. Mancuso

Contents

Couplings and Joints

1
History of Couplings

I. EARLY HISTORY

The flexible coupling is an outgrowth of the wheel. In fact, without the wheel and its development there would have been no need for flexible couplings.

It has been reported that the first wheel was made by an unknown Sumerian more than 5000 years ago in the region of the Tigris and Euphrates Rivers. The earliest record we have dates to 3500 B.C. History records that the first flexible coupling was the universal joint (see Figure 1.1), used by the Greeks in or around 300 B.C. History also indicates that the Chinese were using this concept sometime around A.D. 25. The father of the flexible coupling was Jerome Cardan who invented what was described as a simple device consisting of two yokes, a cross, and four bearings. This joint, the common ancestor of all flexible couplings, is still in use today and is continually being upgraded with the latest technology. Cardan did not design the Cardan shaft for rotating shaft applications, only as a suspension member. The Cardan joint is also known as the Hooke joint. In approximately A.D. 1650, Robert Hooke made the first application of this joint to a rotating shaft in a clock drive. Hooke wrote the equation for fluctuations in angular velocity caused by a single Cardan joint.

From 1700 to 1800, history records very little in the way of further developments in flexible couplings. The advent of the industrial revolution and especially, later, the automobile revolution precipitated the development of many flexible couplings.

In 1886, F. Roots theorized that if he thinned down the flange section of a rigid coupling it would flex and prevent the equipment and shaft from failing. This is the forerunner of today's diaghragm coupling (see Figure 1.2A). The Davis compression coupling (Figure 1.2B) was developed to eliminate keys by compressing hubs onto the shaft. It was thought to be safer than other coupling devices since no protruding screws were required. What is believed to be the first chain coupling (Figure 1.2C) was described in the May 1914 issue of *Scientific American*.

II. THE PERIOD 1900—1930

The coupling industry developed rapidly in the 1920's as a direct result of the invention of the automobile. Some of the coupling manufacturers that were established in this period included:

Original name	Date started	Today's name	Figure number
Thomas Flexible Coupling Company	1916	Rexnord, Inc., Coupling Division	1.3A
Fast Couplings, The Barlette Hayward Co.	1919	Koppers Company, Inc., Power Transmission Division	1.3B
Lovejoy Couplings	1927	Lovejoy, Inc.	1.3C
Poole Foundry and Machine	1920	Poole Company	1.3D
American Flexible Coupling	1928	Zurn Industries, Inc. Mechanical Drives Division	1.3E
Ajax Flexible Coupling Company, Inc.	1920	Renold, Inc., Engineered Products	1.3F
T. B. Wood's & Sons, Inc.	1920	T. B. Wood's & Sons Company	1.3G
Bibby	1919	Bibby	1.3H

III. THE PERIOD 1930–1945

During this period many general-purpose flexible couplings were introduced into the industrial market. The most commonly used couplings during this period were the following:

Chain coupling (Figure 1.4A)
Grid coupling (Figure 1.4B)
Jaw coupling (Figure 1.4C)
Gear coupling (Figure 1.4D)
Disk coupling (Figure 1.4E)
Slider block coupling (Figure 1.4F)
Universal joint (Figure 1.4G)

IV. THE PERIOD 1945–1960

The late 1940s to the 1960s saw rapid technological advancement and use of rotating equipment. Larger and higher-horsepower equipment was used. This brought about the need for more "power dense" flexible coupling with greater misalignment.

Around this time the fully crowned gear spindle (Figure 1.5A) was developed and introduced in the steel industry. Also, in this period the use of the gas turbine in industrial applications (generators, compressors) was becoming popular. With increased use of the gas turbine came the requirement for higher-speed couplings. Therefore, the gear coupling and disk coupling were upgraded and improved to handle those higher-speed requirements (see figures 1.5B and 1.5C).

One of the ever-increasing demands of rotating equipment is for higher and higher operating speeds. With the increased speed of operation came system problems that required lighter-weight couplings, and the torsional characteristics of couplings became more important. This necessitated the need for improvements in resilent couplings (see Figure 1.5D), as these couplings not only had to help tune a system but in many cases had to be able to absorb (dampen) anticipated peak loads caused by torsional excitations.

Up to now we have been discussing relatively large couplings for industrial applications, generally these with a bore larger then 1/2 in. But there is another type of flexible coupling in widespread use—the miniature coupling—used to drive servomechanisms, office equipment, and other small mechanisms.

V. THE PERIOD 1960–1980

Higher horsepower and higher speed continue to be increasing requirements of rotating equipment. The 1960s saw the introduction of many new types of couplings. Many coupling manufacturers introduced a standard line of crowned tooth gear couplings (see Figure 1.6A). Today, the gear coupling is probably the most widely used coupling on the market. There are about six major manufacturers that make a product line which is virtually interchangeable.

The grid coupling and chain coupling are also very popular couplings for general-purpose applications (see Figures 1.6B and 1.6C). The rubber tire coupling (Figure 1.6D) is widely used, with many companies offering one. More sophisticated resilent couplings (Figures 1.6E and 1.6F) were introduced in this period to help with the ever-increasing system problems, as were several flexible membrane couplings (disk or diaphragm) (Figures 1.6G, 1.6H, and 1.6I). The application of nonlubricated couplings has grown rapidly in the past decade and indications are that their use will continue to grow. The gear coupling has again been upgraded to meet the challenge of higher speeds (Figures 1.6J and 1.6K).

In the next chapters we describe and discuss the types of couplings available today. Many of the flexible couplings used today will be around for years. As with the ancestor of flexible couplings, the universal joint (Figure 1.6L), technological improvements in material, design, and manufacturing will help upgrade couplings so they can handle the ever-increasing needs and demands of rotating equipment.

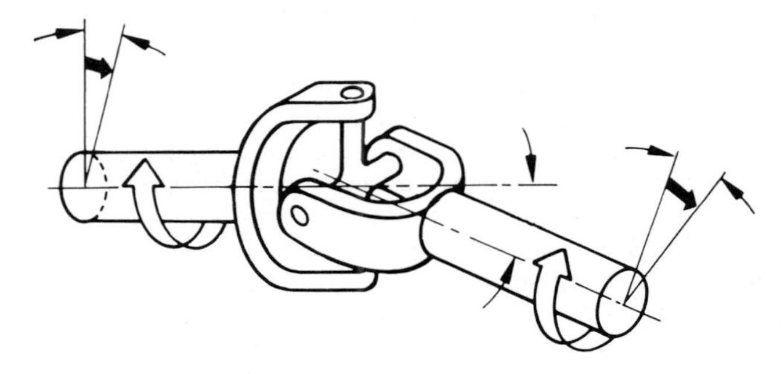

Figure 1.1 The ancestor of the flexible coupling—the universal joint.

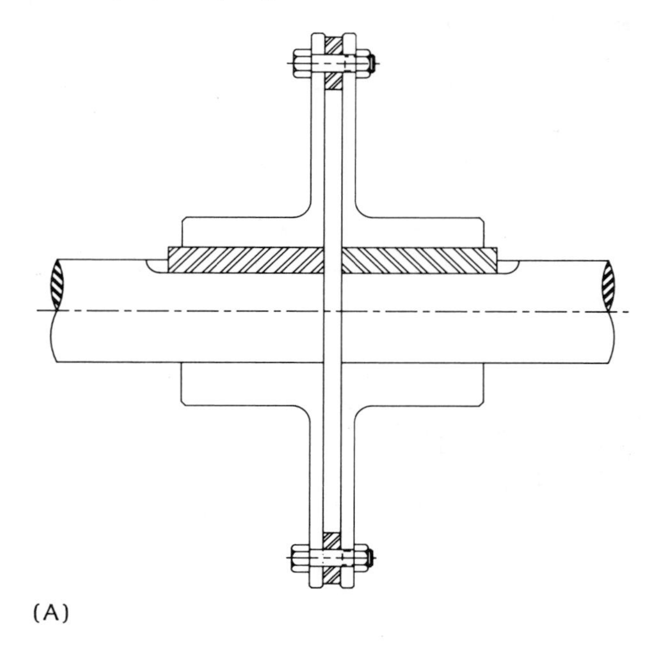

(A)

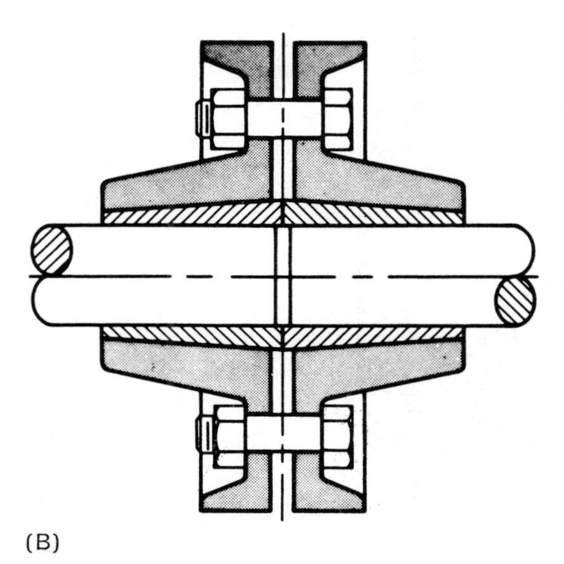

(B)

Figure 1.2 Early couplings: (A) Diaphragm coupling patented in 1886. (B) Davis compression coupling. (C) Early chain coupling.

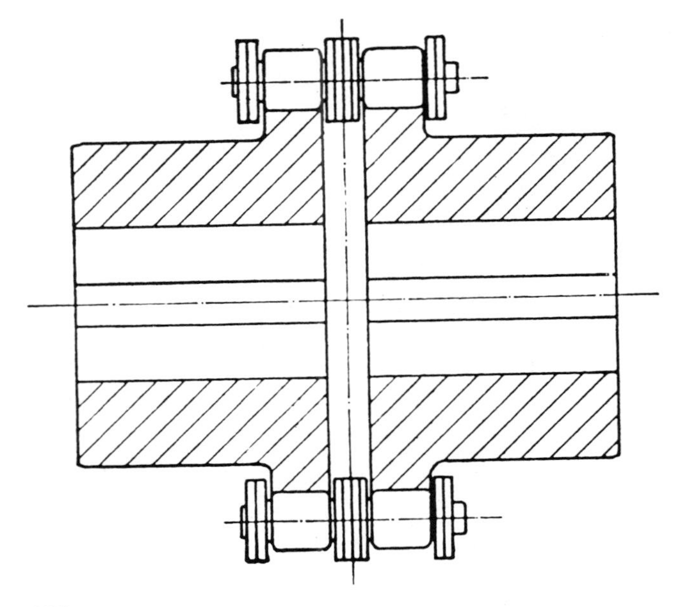

(C)

Figure 1.2 (Continued)

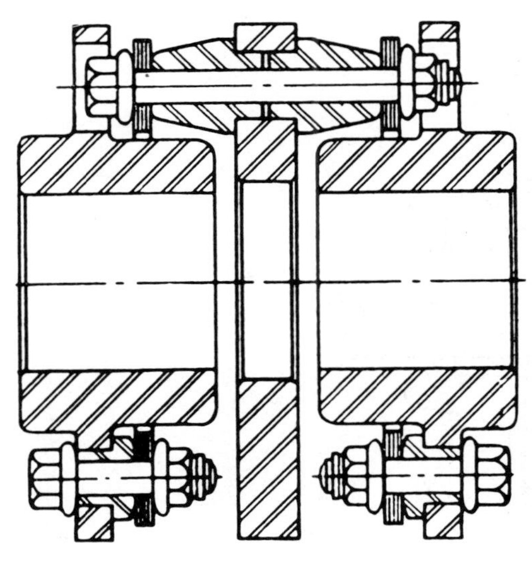

(A)

Figure 1.3 Couplings of the period 1900–1930: (A) Thomas disk coupling. (B) Fast gear coupling.

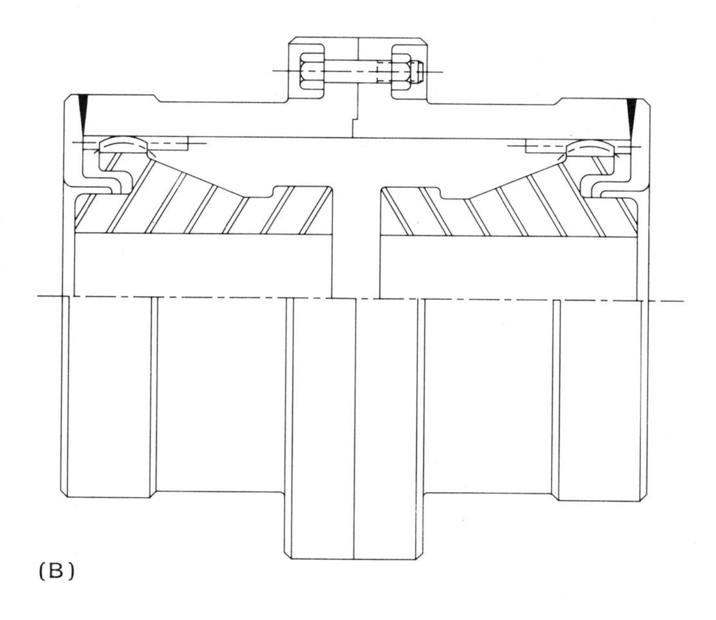

(B)

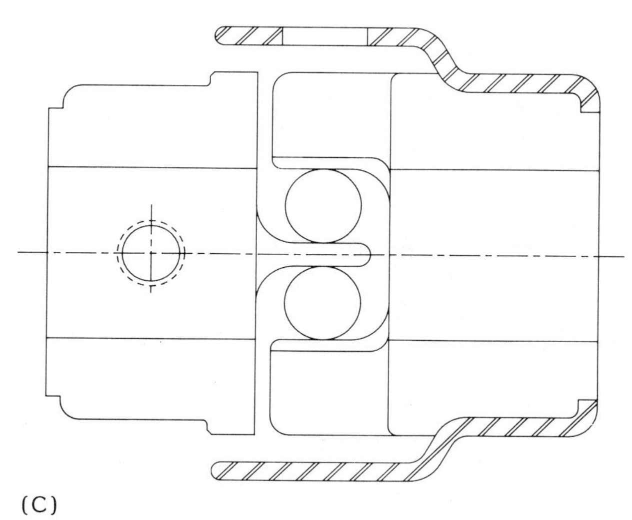

(C)

Figure 1.3 (Continued)

(C) Lovejoy jaw coupling. (D) Poole gear coupling. (E) American slider block coupling. (F) Ajax pin and bushing coupling. (G) T. B. Wood's Resilient coupling. (H) Bibby grid coupling.

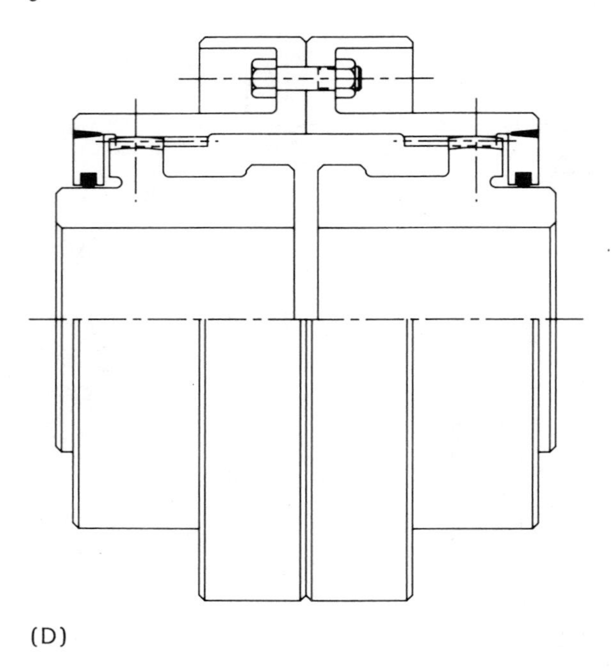

(D)

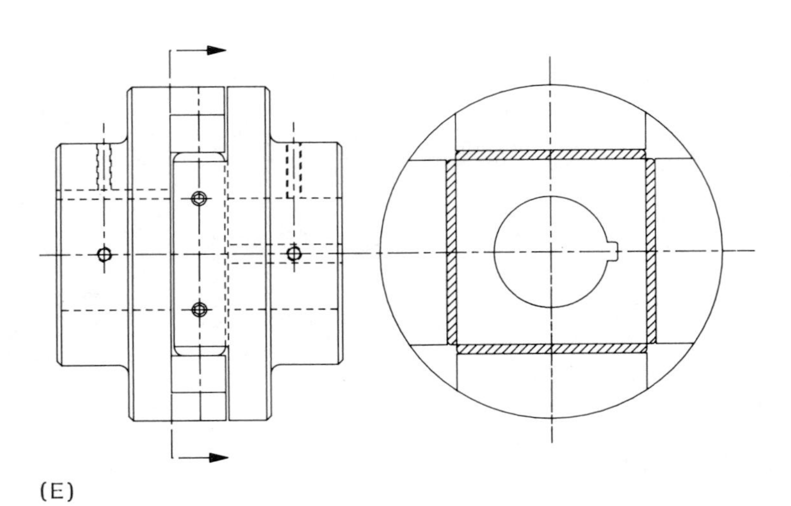

(E)

Figure 1.3 (Continued)

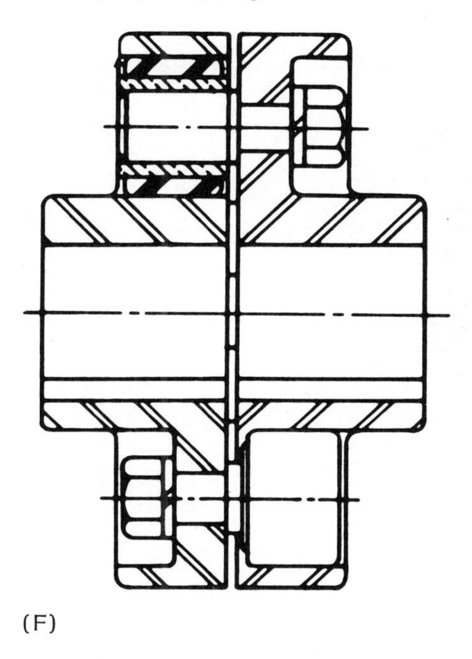

(F)

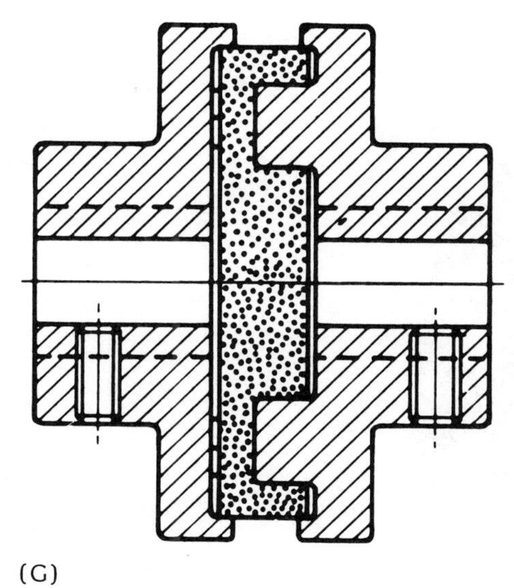

(G)

Figure 1.3 (Continued)

(H)

Figure 1.3 (Continued)

(A)

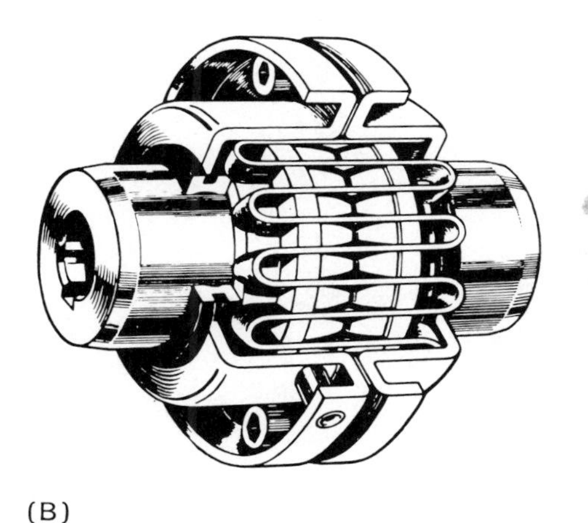

(B)

(C)

Figure 1.4 Couplings of the period 1930–1945: (A) Chain coupling (courtesy of Morse Industrial Corporation). (B) Grid coupling (courtesy of Falk Corporation). (C) Jaw coupling. (D) Gear coupling (courtesy of Poole Company). (E) Disk coupling (courtesy of Coupling Division of Rexnord, Thomas Couplings). (F) Slider block coupling (courtesy of Zurn Industries, Inc., Mechanical Drives Division). (G) Universal joint (courtesy of Spicer Universal Joint Division of Dana Corporation).

(D)

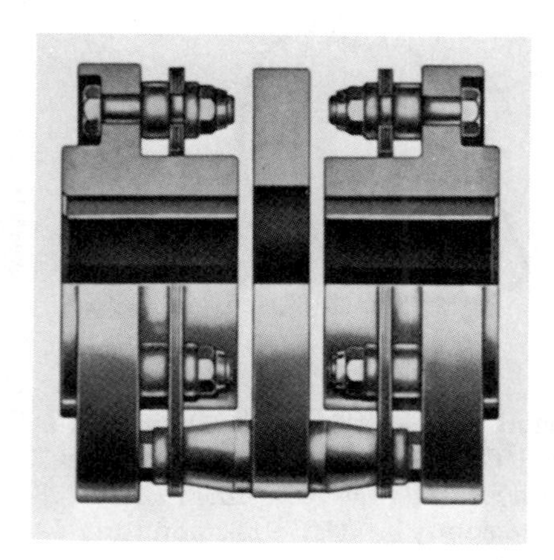

(E)

Figure 1.4 (Continued)

(F)

(G)

Figure 1.4 (Continued)

(A)

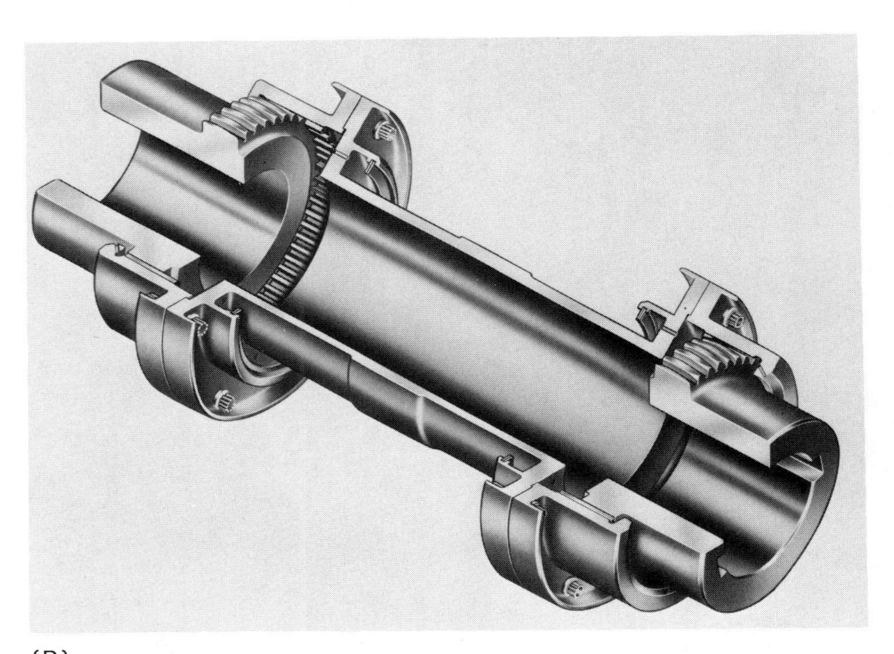

(B)

Figure 1.5 Couplings of the period 1945–1960: (A) Gear spindle coupling (courtesy of Zurn Industries, Inc., Mechanical Drives Division). (B) High-speed gear coupling (courtesy of Zurn Industries, Inc., Mechanical Drives Division). (C) High-speed disk coupling (courtesy of Coupling Division of Rexnord, Thomas Couplings). (D) Resilent coupling (courtesy of Koppers Company, Inc., Engineered Metal Products Group).

(C)

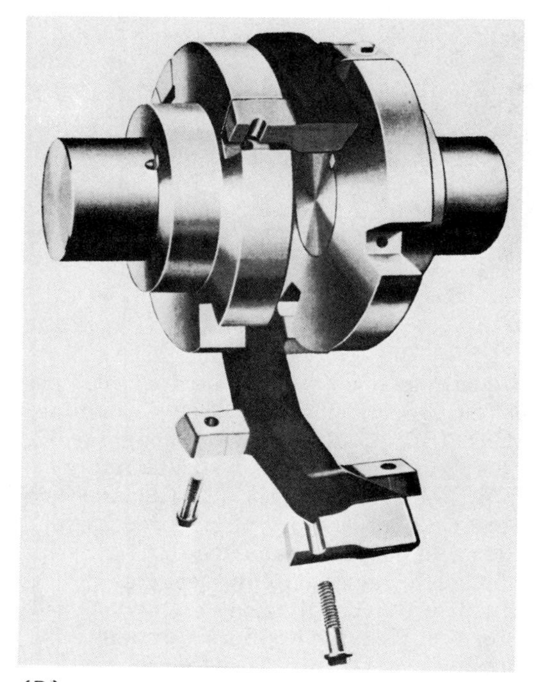

(D)

Figure 1.5 (Continued)

(A)

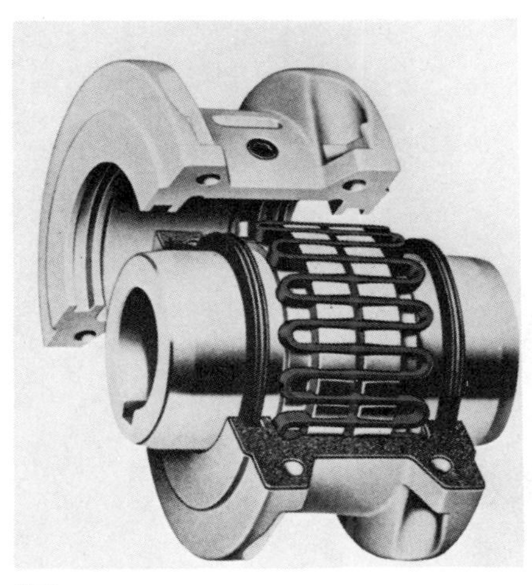

(B)

Figure 1.6 (A) Crowned Tooth gear coupling (courtesy of Zurn Industries, Inc., Mechanical Drives Division). (B) Grid coupling (courtesy of Falk Corporation). (C) Chain coupling (courtesy of Dodge Division of Reliance Electric). (D) Rubber tire coupling (courtesy of Dodge Division of Reliance Electric). (E) Elastomeric coupling (courtesy of Koppers Company, Inc., Engineered Metal Products Group). (F) Elastomeric coupling (courtesy of Lovejoy, Inc). (G) Multiple convoluted diaphragm coupling (courtesy of Zurn Industries, Inc., Mechanical Drives Division). (H) Tapered contoured diaphragm coupling (courtesy of Koppers Company, Inc., Engineered Metal Products Group). (I) Multiple diaphragm coupling (courtesy of Flexibox International, Metastream Couplings). (J) High-speed sealed lubed gear coupling (courtesy of Koppers Company, Inc., Engineered Metal Products Group). (K) High-speed

(C)

(D)

Figure 1.6 (Continued)

---

lightweight gear coupling (courtesy of Zurn Industries, Inc., Mechanical Drives Division). (L) Modern industrial universal joint (Courtesy of Voithe Transmit GmbH).

(E)

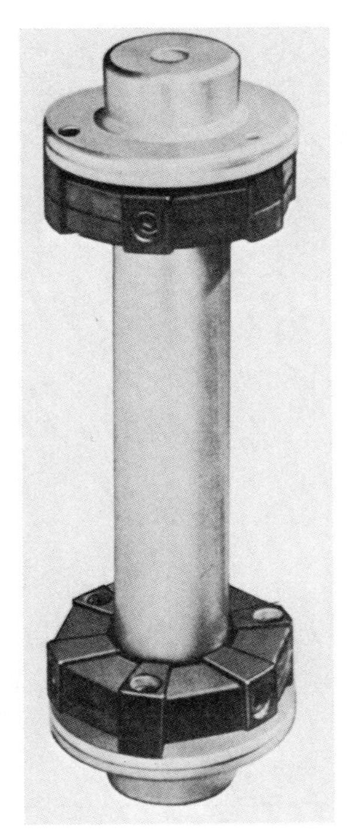

(F)

Figure 1.6 (Continued)

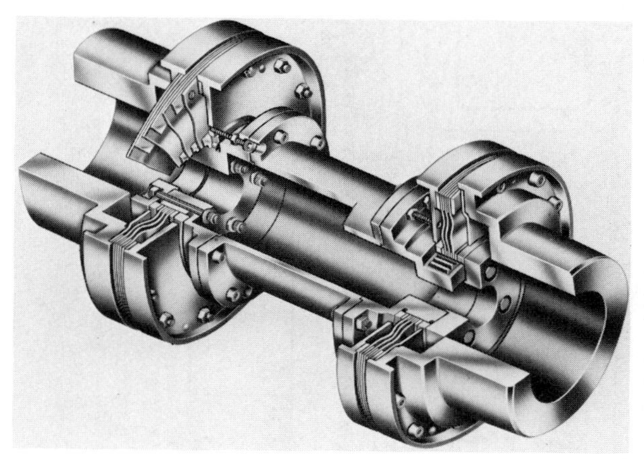

(G)

(H)

Figure 1.6 (Continued)

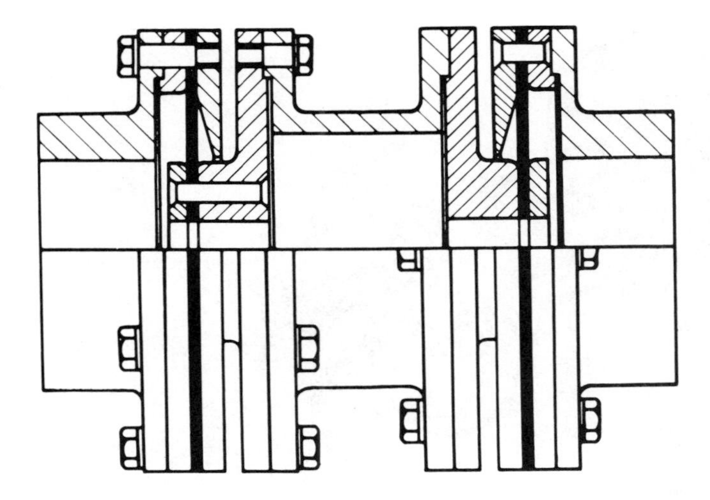

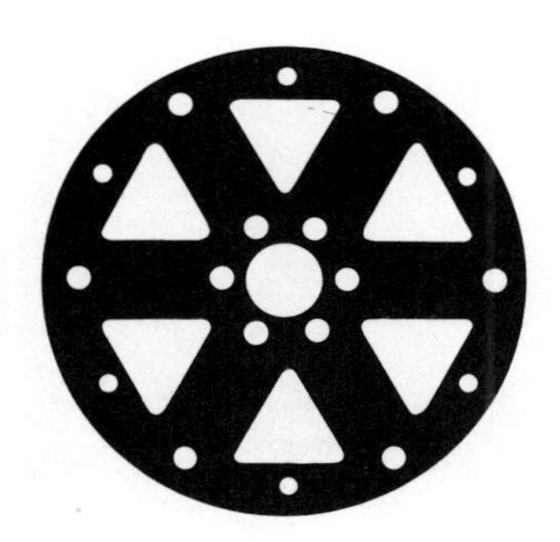

(l)

Figure 1.6 (Continued)

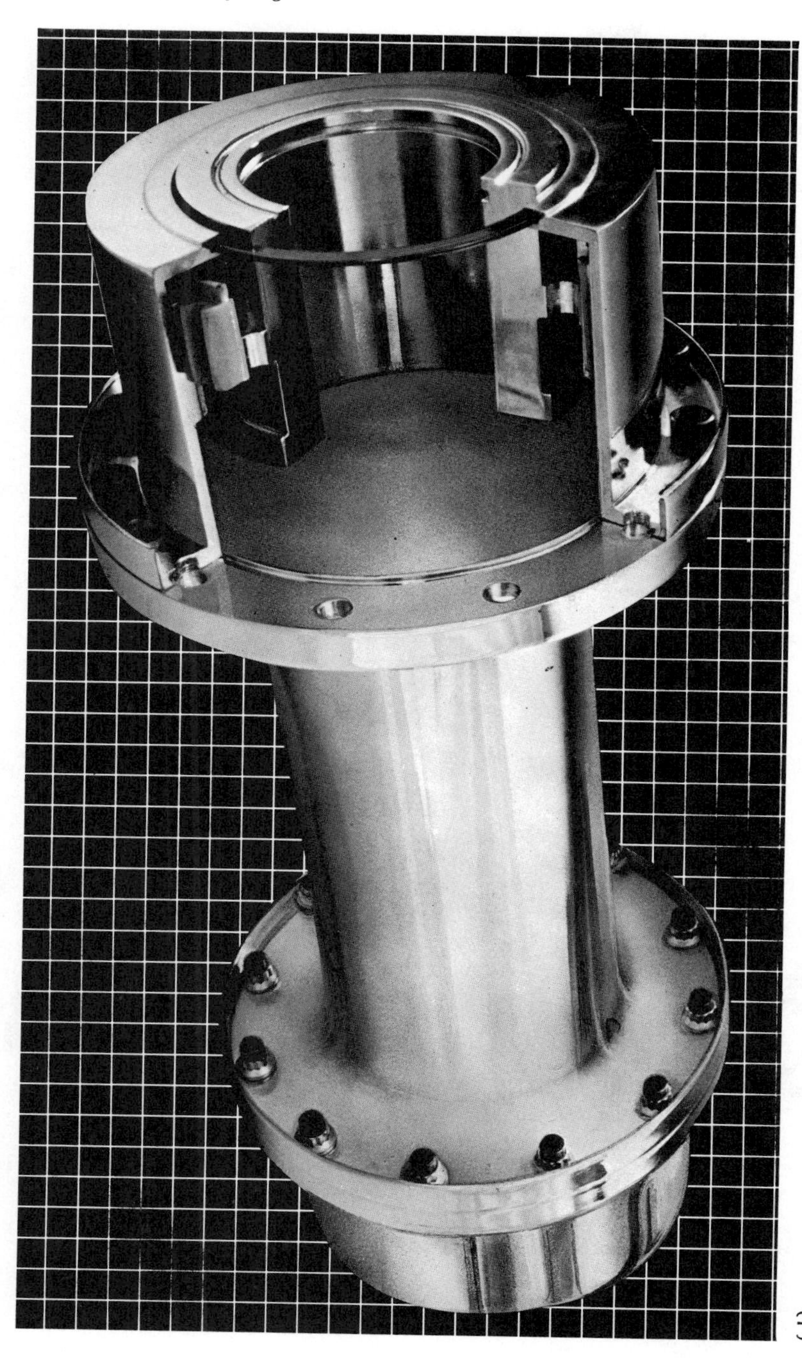

(J)

Figure 1.6 (Continued)

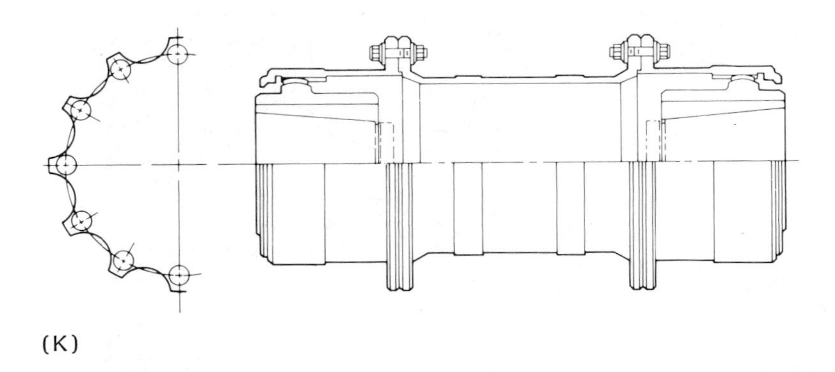

(K)

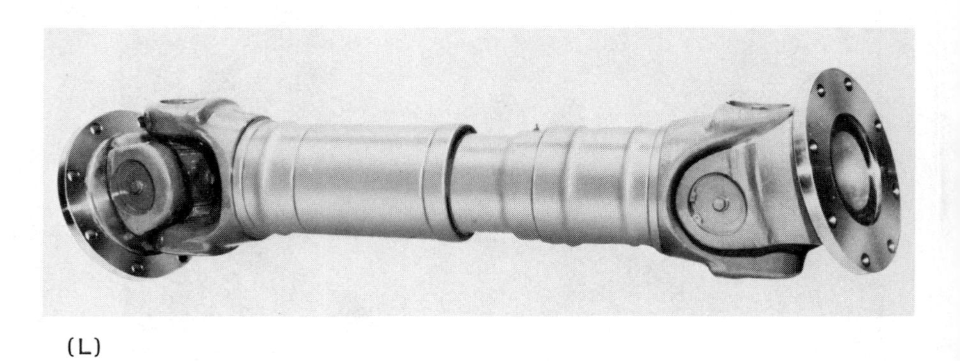

(L)

Figure 1.6 (Continued)

2
Overview of Couplings

I. ADVANTAGES OF USING FLEXIBLE COUPLINGS

Historically, rotating equipment was first connected by means of rigid flanges (Figure 2.1). Experience indicated that this method did not accommodate the motions and excursions experienced by the equipment. As discussed in Chapter 1, F. Roots was the first to thin these flanges and allow them to flex. Rigid couplings are used to connect equipment that experiences very small shaft excursions or with shafts made long and slender enough so that they can accept the forces and moments produced from the flexing flanges and shafts.

The three basic functions of a coupling are:

1. To transmit power (Figure 2.2A)
2. To accommodate misalignment (Figure 2.2B)
3. To compensate for end movement (Figure 2.3C)

A. Power Transmission

Flexible couplings must couple two pieces of rotating equipment, equipment with shafts, flanges, or both. These interface connections are numerous and are covered in Section 3.IV.

Flexible couplings must also transmit power efficiently. Usually, the power lost by a flexible coupling is small, although some couplings are more efficient than others. Power is lost in friction heat from the sliding and rolling of flexing parts and at high speed, windage and frictional losses indirectly cause lost efficiency. Most flexible couplings are better than 99% efficient.

B. Accommodating Misalignment and End Movement

Flexible couplings must accommodate three types of misalignment:

1. *Parallel offset*: Axes of connected shafts are parallel but not in the same straight line (see Figure 2.3A).
2. *Angular*: Axes of shafts intersect at the center point of the coupling, but not in the same straight line (see Figure 2.3B).
3. *Combined angular and offset*: Axes of shafts do not intersect at the center point of the coupling and are not parallel (see Figure 2.3C).

Most flexible couplings are designed to accommodate axial movement of equipment or shaft ends. In some cases (e.g., motors) couplings are required to limit axial float of the equipment shaft to prevent internal rubbing of a rotating part within its case.

Accommodation of misalignment and end movement must be done without inducing abnormal loads in the connecting equipment. Generally, machines are set up at installation quite accurately. There are many things that force equipment to run out of alignment. The thermal effects of handling hot and cold fluids cause some movement in the vertical and axial direction, together with differentials of temperature in driver media such as gas and steam. The vertical motions could be a result of support structure expansions due to temperature differences, distortion due to solar heating, axial growth or a combination of these. Horizontal motions are usually caused by piping forces or other structural movements, temperature differentials caused by poor installation practices, and expansions or contractions caused by changes in temperature or pressure differentials of the media in the system.

It is a fact of life that machinery appears to live and breathe, move, grow, and change form and position; this is the reason for using flexible couplings. A flexible coupling is not the solution to all movement problems that can or could exist in a sloppy system. Using a flexible coupling in the hope that it will compensate for any and all motion is naive. Flexible couplings have their limitations. The equipment or system designer must make calculations that will give a reasonable estimate of the outer boundaries of the anticipated gyrations. Unless those boundaries are defined, the equipment and system designer may just be transferring equipment failure into coupling failure (see Figure 2.4).

One thing to remember is that when subjected to misalignment and torque, *all* couplings react on connected equipment components. Some produce greater reactionary forces than others and if overlooked, can cause shaft failures, bearing failures, and other failures of equipment components (see Figure 2.5). Rigid couplings produce the greatest reactions. Mechanically flexible couplings such as gear, chain, and grid couplings produce high to moderate moments and forces on equipment that are a function of torque and misalignment. Elastomeric couplings produce moderate to low moments and forces that are slightly dependent on torque. Metallic membrane couplings produce relatively low moments and forces which are relatively independent of torque. The most commonly used flexible couplings today are those that produce the greatest flexibility (misalignment and axial capacity) while producing the lowest external loads on equipment.

II. TYPES OF COUPLINGS

There are many types of couplings. They can virtually all be put into two classes, two disciplines, and four categories (Figure 2.6). The two classes of couplings are:

1. The rigid coupling (Figure 2.7)
2. The flexible coupling

The two disciplines for the application of flexible couplings are:

1. The miniature discipline, which covers couplings used for office machines, servomechanisms, instrumentation, light machinery, and so on (see Figure 2.8).
2. The industrial discipline, which covers couplings used in the steel industry, the petrochemical industry, utilities, off-road vehicles, heavy machinery, and so on.

This book covers industrial couplings.

The four basic categories of flexible industrial couplings are:

1. Mechanically flexible couplings
2. Elastomeric couplings
3. Metallic membrane couplings
4. Miscellaneous couplings

The general operating principles of the four basic categories of industrial couplings are as follows:

1. *Mechanically flexible couplings*: In general, these couplings obtain their flexibility from loose-fitting parts and/or rolling or sliding of mating parts (see Figures 2.9 to 2.12). Therefore, they usually require lubrication unless one moving part is made of a material that supplies its own lubrication needs (e.g., a nylon gear coupling). Also included in this category are couplings that uses a combination of loose-fitting parts and/or rolling or sliding, with some flexure of material.
2. *Elastomeric couplings*: In general, these couplings obtain their flexibility from stretching or compressing a resilent material (rubber, plastic, etc.) (see Figures 2.13 and 2.14). Some sliding or rolling may take place, but it is usually minimal.
3. *Metallic membrane couplings*: In general, the flexibility of these couplings is obtained from the flexing of thin metallic disks or diaphragms (see Figures 2.15 and 2.16).
4. *Miscellaneous couplings*: These couplings obtain their flexibility from a combination of the mechanisms described above or through a unique mechanism (see Figure 2.17).

III. COUPLING FUNCTIONAL CHARACTERISTICS AND CAPACITIES

The *coupling selector* (equipment designer or system designer must decide what coupling is best for the system. The designer must review the possible candidates for a flexible coupling and make a selection. The person responsible for the selection of couplings should build a file of the most recent coupling catalogs. (See Appendix A for the names and addresses of coupling manufacturers.) This file should be reviewed at regular intervals because designs, models, and materials are constantly being updated and improved.

Couplings are usually selected based on their characteristics and capacities. The two most important capabilities relate to torque and speed. Figure 2.18 is a graph that defines the torque and speed envelopes of the most common types of couplings. In addition the functional capacities listed in Figure 2.19 should be considered. Finally, the designer must take into account coupling characteristics (Figure 2.20), which tell basically how the coupling will affect the system.

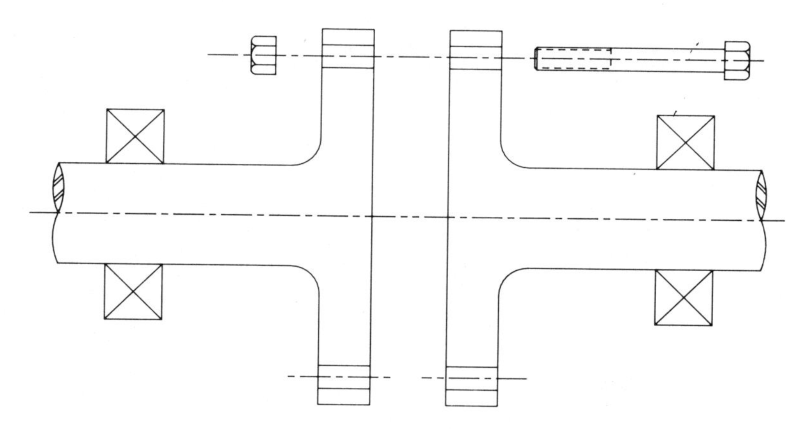

Figure 2.1 Rigid flanged connection.

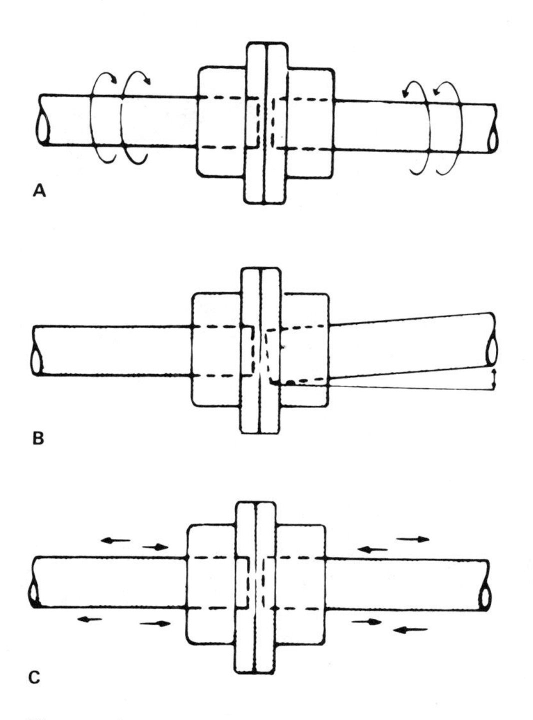

Figure 2.2 Functions of a flexible coupling: (A) transmit torque; (B) accommodate misalignment; (C) compensate for end movement.

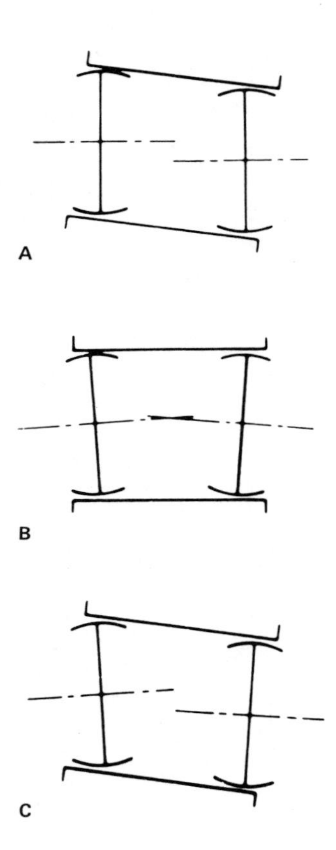

Figure 2.3 Types of misalignment: (A) parallel offset; (B) angular; (C) combined angular and offset.

Figure 2.4 Coupling failure.

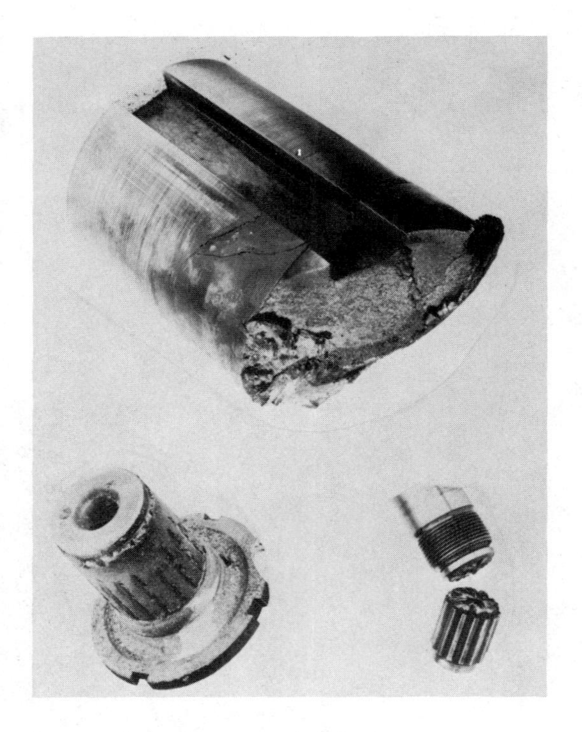

Figure 2.5 Equipment failure.

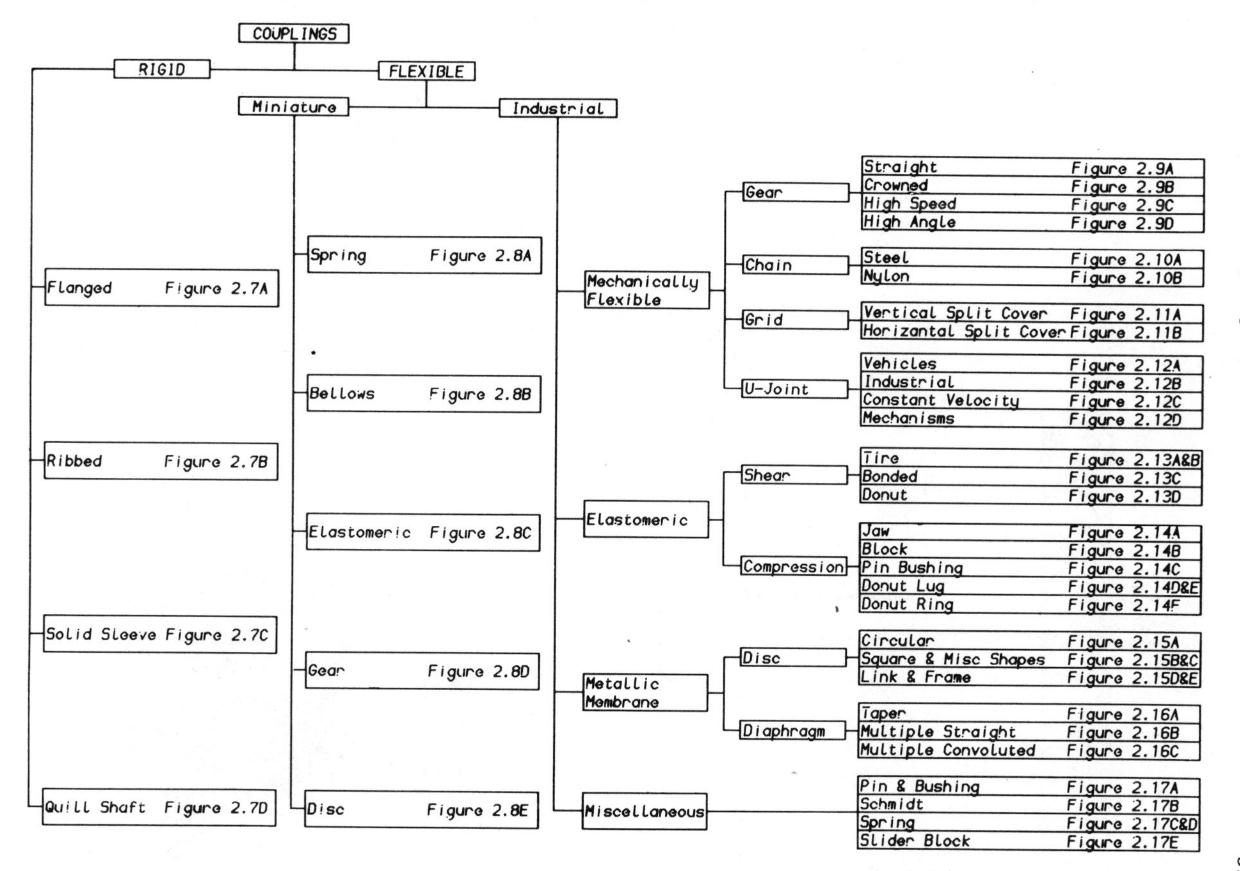

Figure 2.6 Types of couplings.

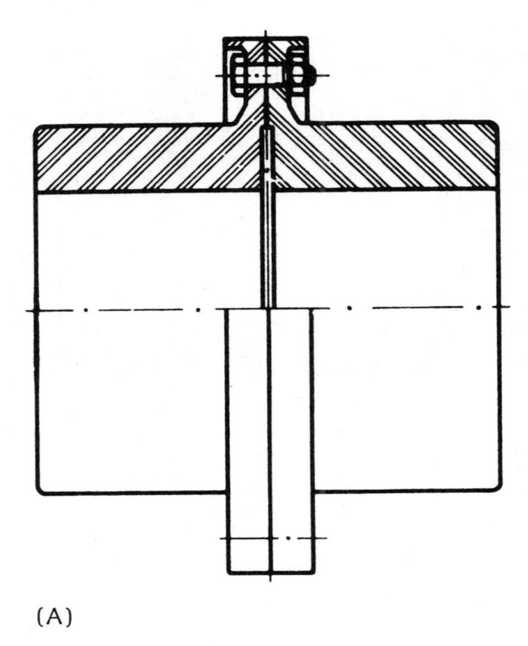

(A)

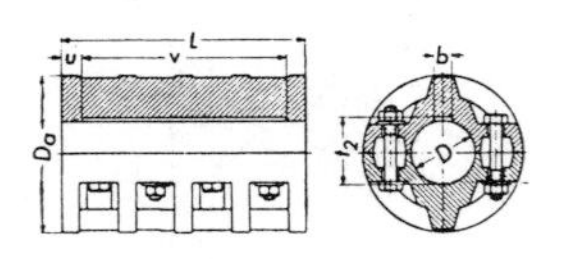

(B)

Figure 2.7 Types of rigid couplings: (A) rigid flanged coupling; (B) rigid ribbed coupling; (C) rigid sleeve coupling (courtesy of SKF Steel, Coupling Division); (D) quill shaft coupling.

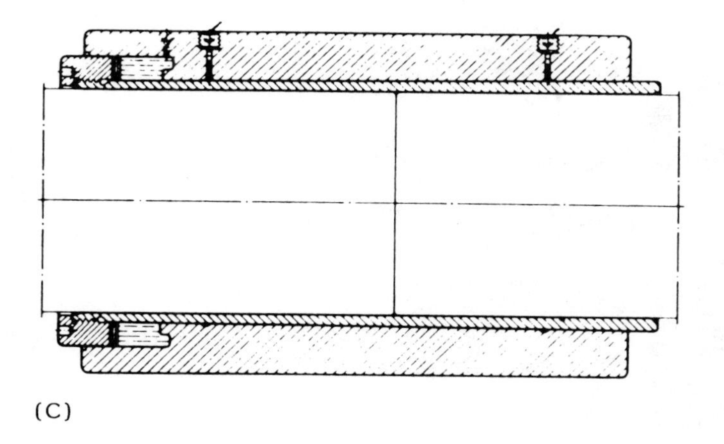

(C)

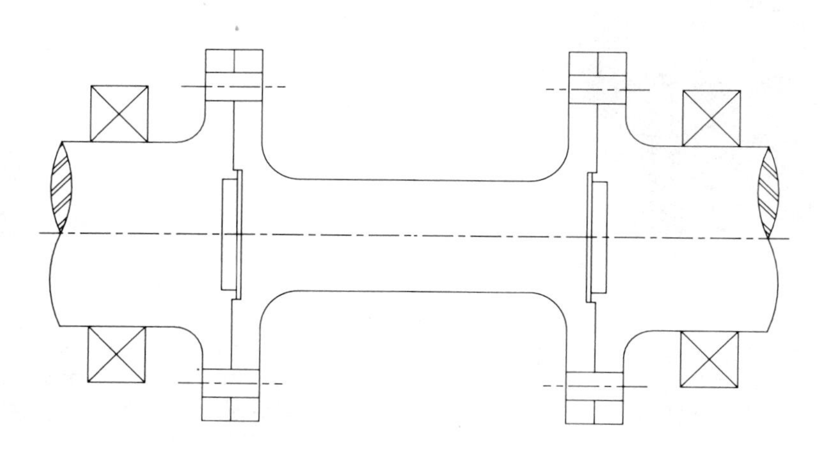

(D)

Figure 2.7 (Continued)

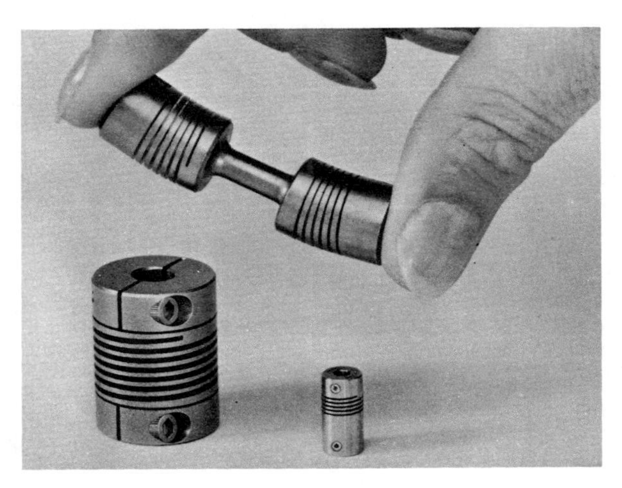

(A)

(B)

Figure 2.8 Types of miniature couplings: (A) miniature spring coupling (courtesy of Helical Products Company, Inc.); (B) miniature bellows coupling (courtesy of Metal Bellows Corporation); (C) miniature elastomeric coupling (courtesy of Acushnet Company); (D) miniature gear coupling (courtesy of Guardian Industries, Inc.); (E) miniature disk (courtesy of Coupling Division of Rexnord, Thomas Couplings).

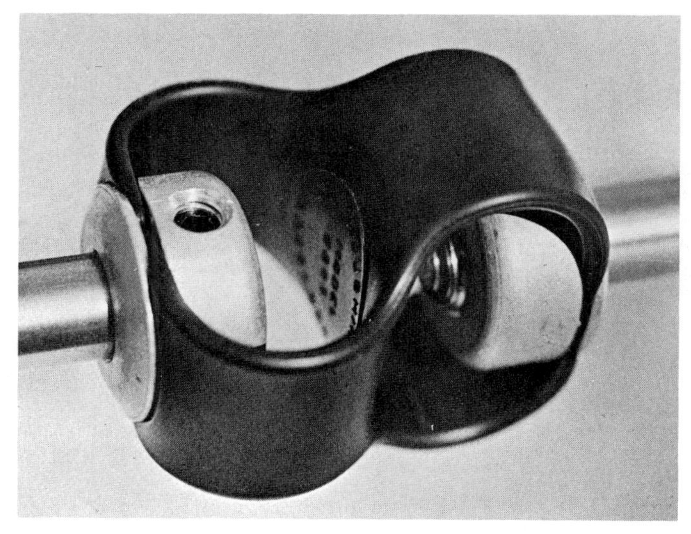

(C)

(D)

Figure 2.8 (Continued)

(E)

Figure 2.8 (Continued)

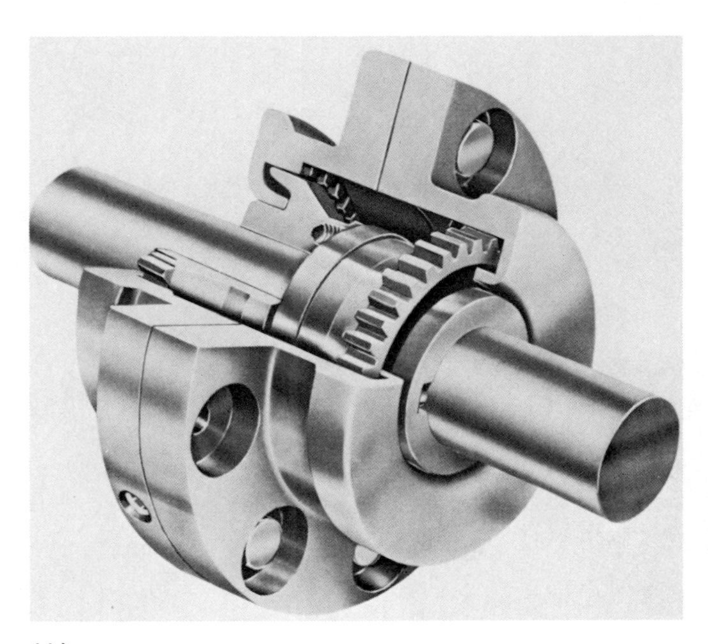

(A)

Figure 2.9 Types of gear couplings: (A) straight tooth gear coupling (courtesy of Koppers Company, Inc., the Engineered Metal Products Group); (B) crowned tooth gear coupling.

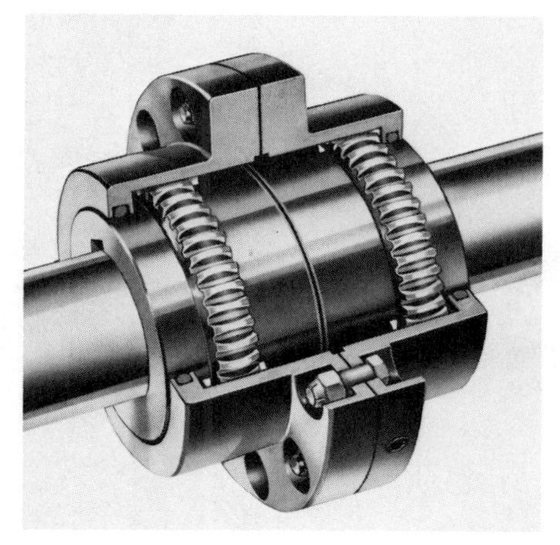

(B)

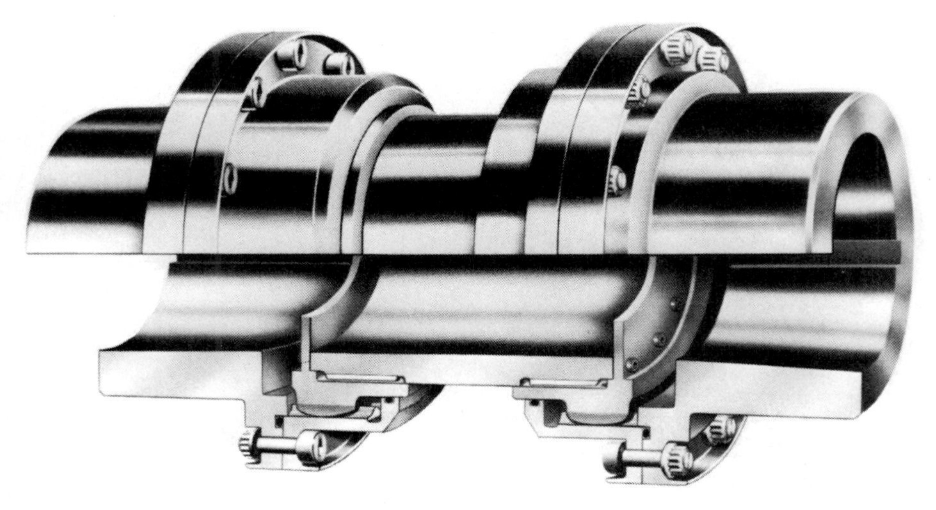

(C)

Figure 2.9 (Continued)

(courtesy of Zurn Industries, Inc., Mechanical Drives Division); (C) high-speed gear coupling (courtesy of Koppers Company, Inc., the Engineered Metal Products Group); (D) high-angle spindle gear coupling (courtesy of Zurn Industries, Inc., Mechanical Drives Division).

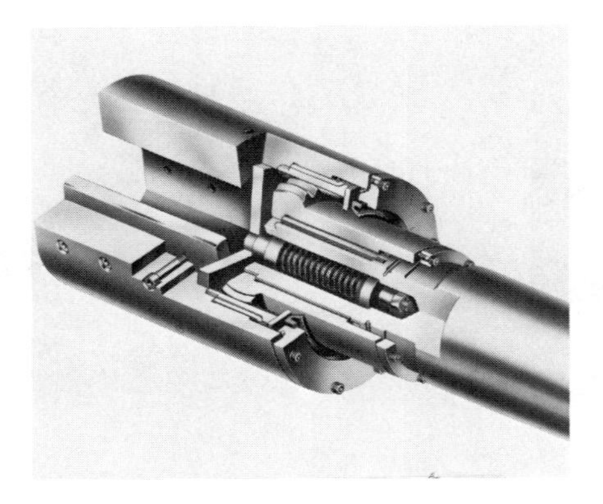

(D)

Figure 2.9 (Continued)

(A)

Figure 2.10 Types of chain couplings: (A) chain coupling (courtesy of Dodge Division of Reliance Electric); (B) nylon chain coupling (courtesy of Morse Industrial Corporation).

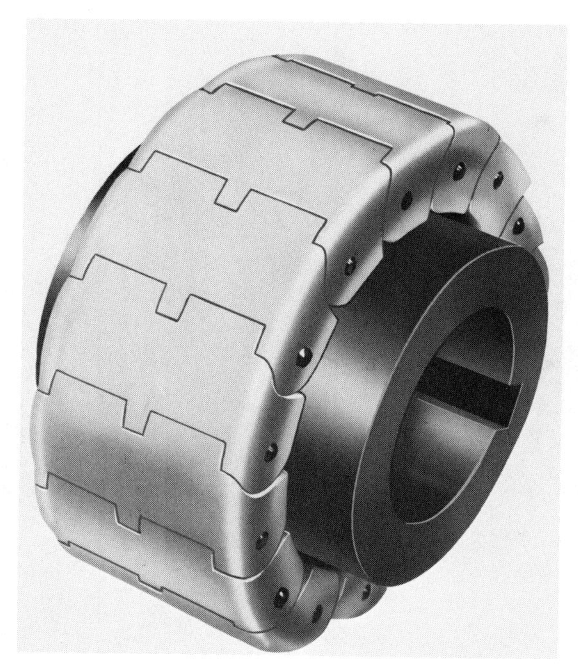

(B)

Figure 2.10 (Continued)

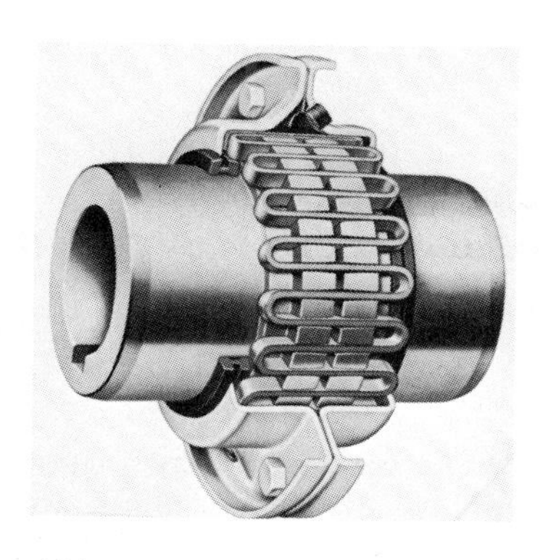

(A)

Figure 2.11 Types of grid couplings: (A) verticle split-cover grid coupling; (B) horizontal split-cover grid coupling (courtesy of Falk Corporation).

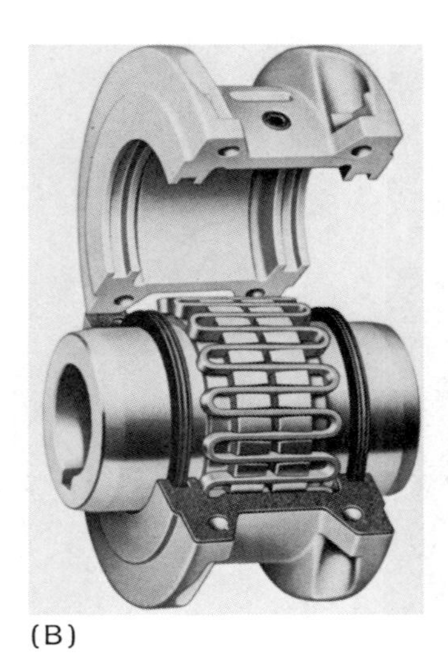

(B)

Figure 2.11 (Continued)

(A)

Figure 2.12 Types of universal joints: (A) universal joints for vehicles (courtesy of Spicer Universal Joint, Division of Dana Corporation); (B) universal joints for industrial equipment (courtesy of Voithe Transmit GmbH); (C) constant-velocity universal joint (courtesy of Parrish Power Products, Inc.); (D) universal joints for mechanisms (courtesy of Browning Manufacturing).

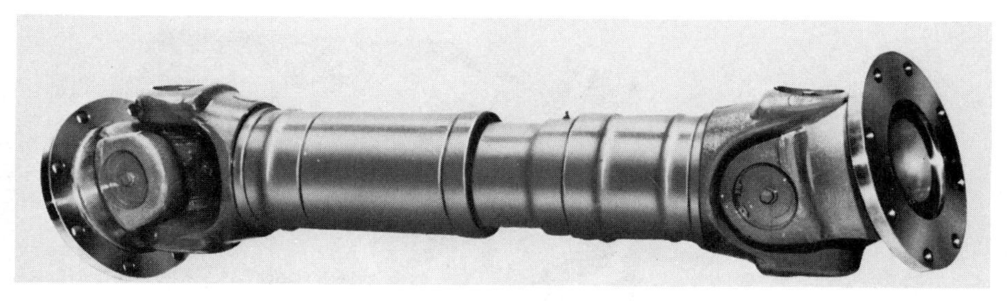

(B)

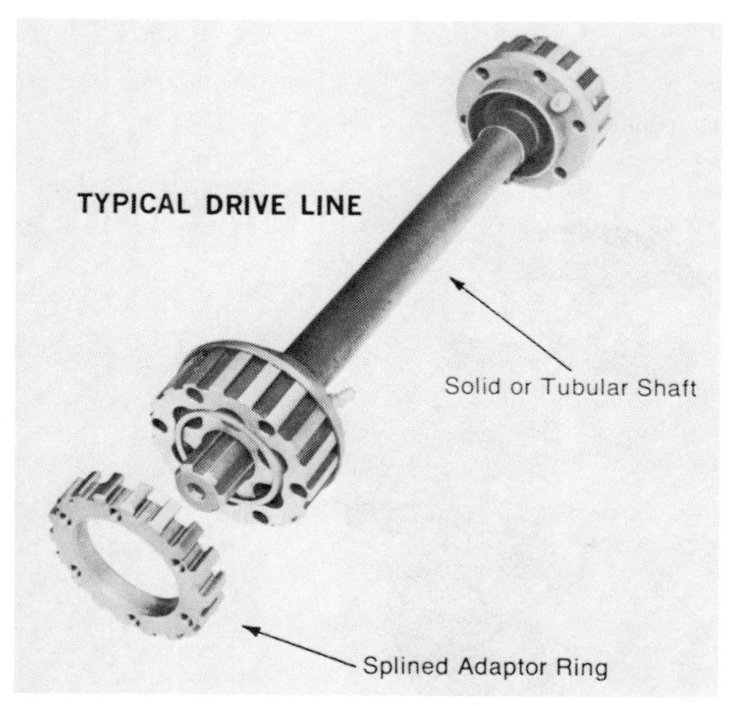

(C)

Figure 2.12 (Continued)

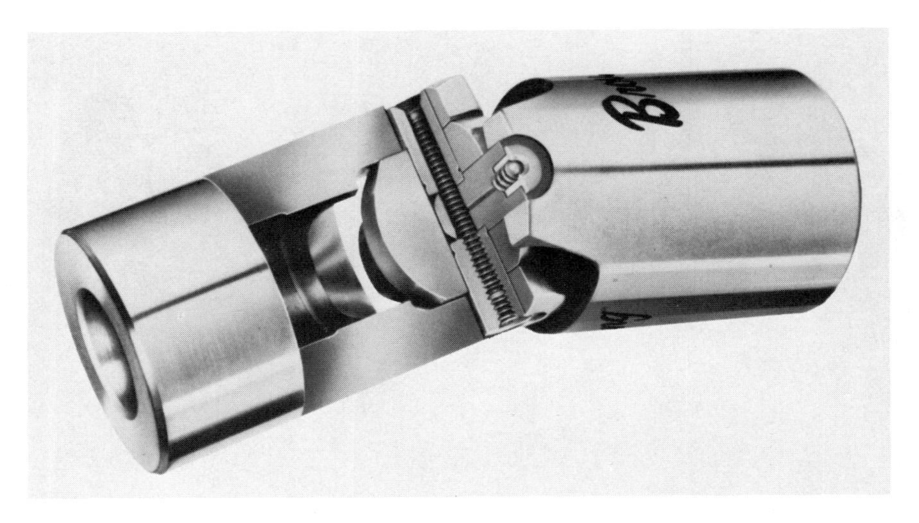

(D)

Figure 2.12 (Continued)

(A)

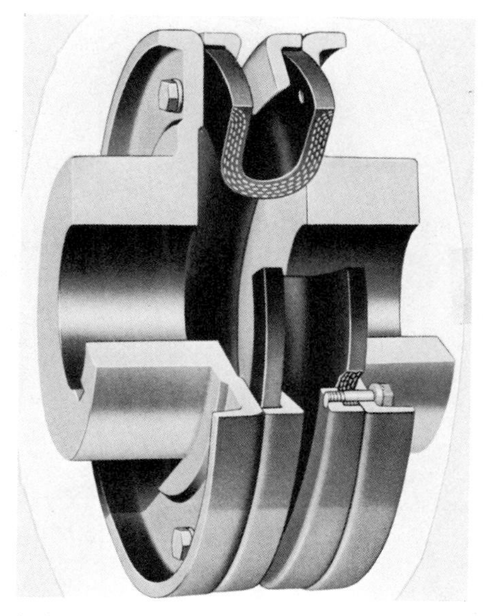

(B)

(C)

Figure 2.13 (Continued)

Figure 2.13 Types of Elastomeric shear couplings: (A) rubber tire coupling (courtesy of Dayco Corporation); (B) rubber tire coupling (courtesy of Falk Corporation); (C) bonded shear coupling (courtesy of Lord Corporation); (D) donut elastomeric shear coupling (courtesy of T. B. Wood's Sons Company).

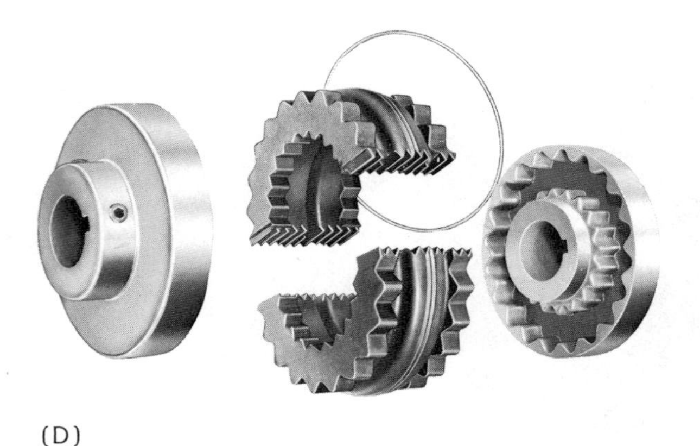

(D)

Figure 2.13 (Continued)

(A)

Figure 2.14 Types of elastomeric compression couplings: (A) jaw elastomeric compression coupling (courtesy of Lovejoy, Inc.); (B) block elastomeric compression coupling (courtesy of Holset Engineering Co. Ltd.); (C) pin and bushing elastomeric compression coupling (courtesy of Morse Industrial Corporation); (D) donut lug elastomeric compression coupling (courtesy of Koppers Company, Inc., Engineered Metal Products Group); (E) donut lug elastomeric compression coupling (courtesy of Lovejoy, Inc.);

(B)

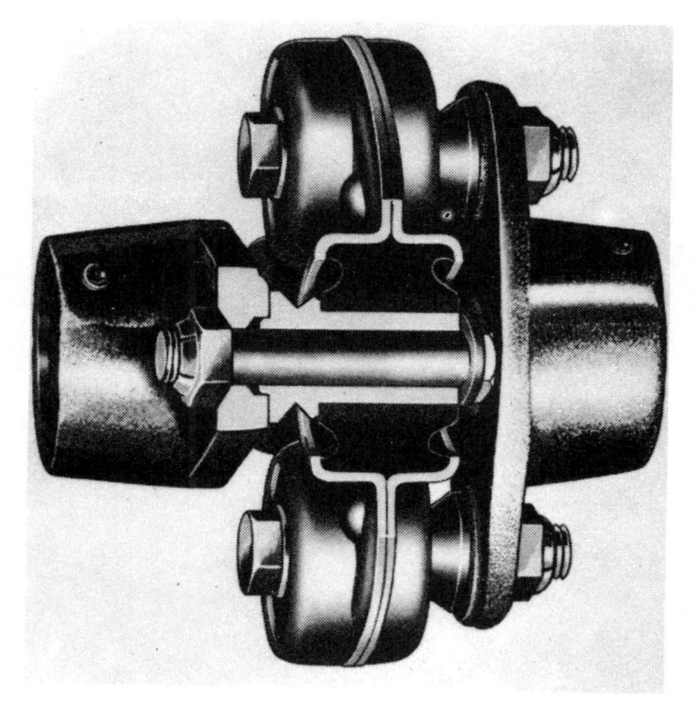

(C)

(F) donut ring elastomeric compression coupling (courtesy of Dodge Division of Reliance Electric).

(D)

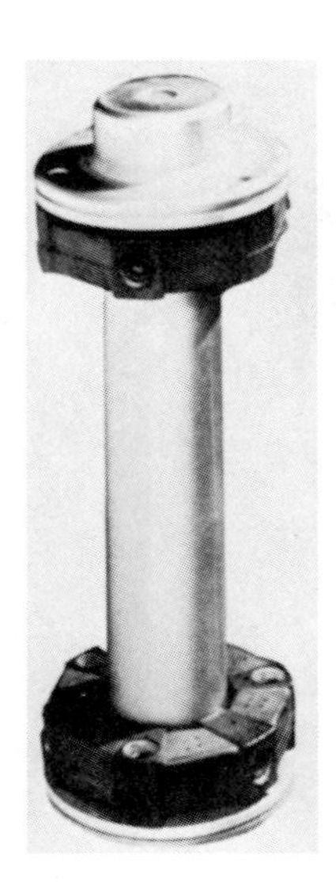

(E)

Figure 2.14 (Continued)

(E) (Continued)

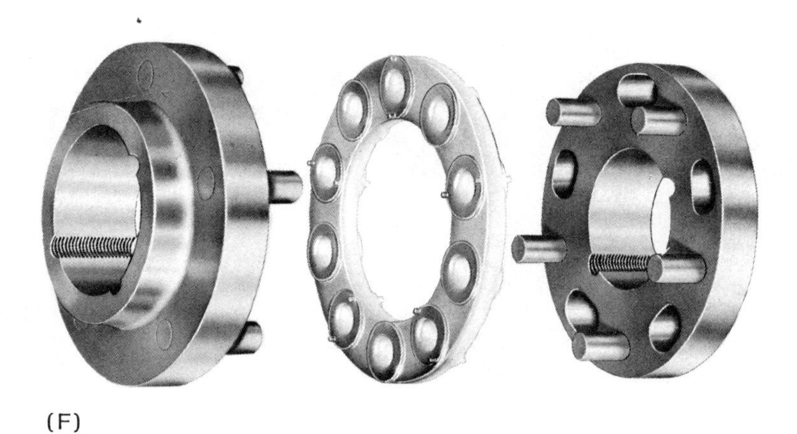

(F)

Figure 2.14 (Continued)

(A)

(B)

Figure 2.15 Types of disk couplings: (A) Disk coupling (circular) (courtesy of Coupling Division of Rexnord, Thomas Couplings); (B) square disk coupling (courtesy of Formsprag Division of Dana Corporation); (C) hexagonal disk coupling (courtesy of Flexibox International, Inc., Metastream Couplings); (D) flexible frame coupling (courtesy of Kamatics Corporation, Kaflex Couplings); (E) link disk coupling (courtesy of TGW Thyssen Getriebe).

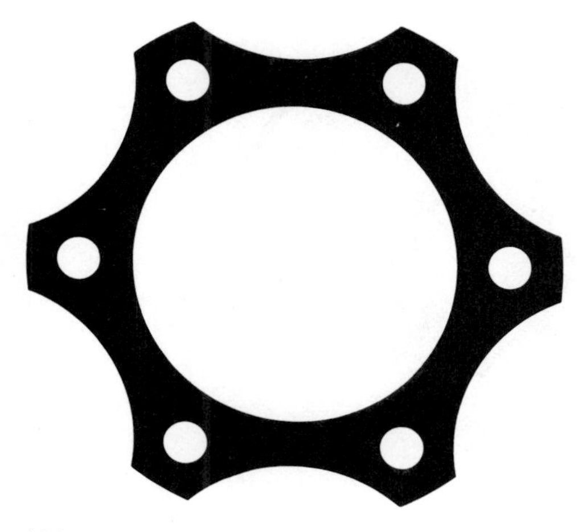

(C)

Figure 2.15 (Continued)

(D)

(E)

Figure 2.15 (Continued)

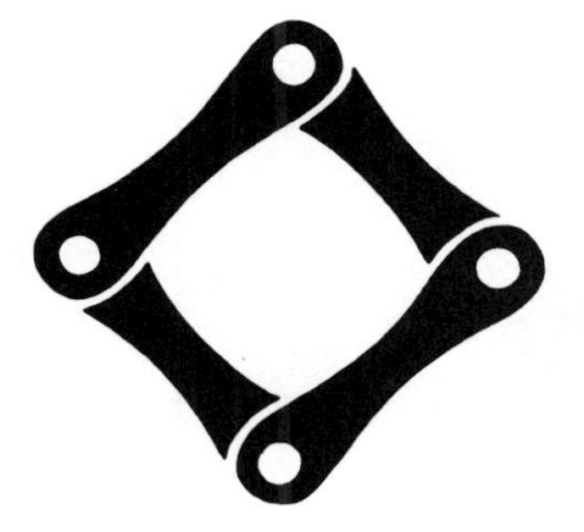

(E) (Continued)

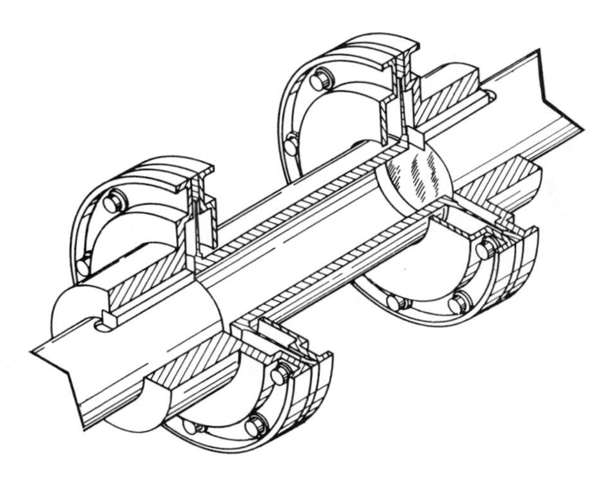

(A)

Figure 2.16 Types of diaphragm couplings: (A) tapered contoured diaphragm (courtesy of Bendix, Fluid Power Division); (B) multiple straight diaphragm coupling (courtesy of Flexibox International, Metastream Couplings); (C) multiple convoluted diaphragm coupling (courtesy of Zurn Industries, Inc., Mechanical Drives Division).

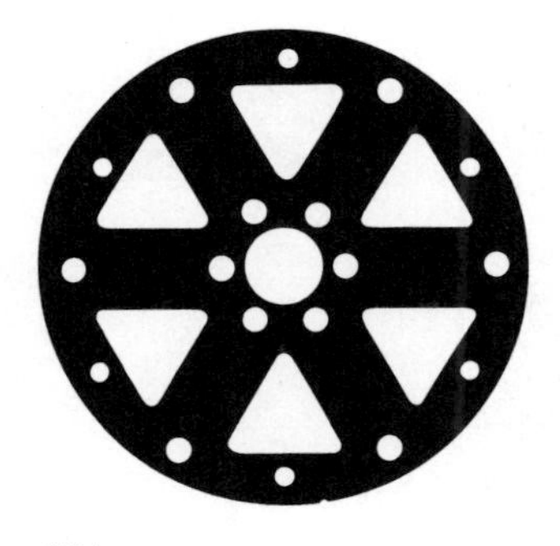

(B)

Figure 2.16 (Continued)

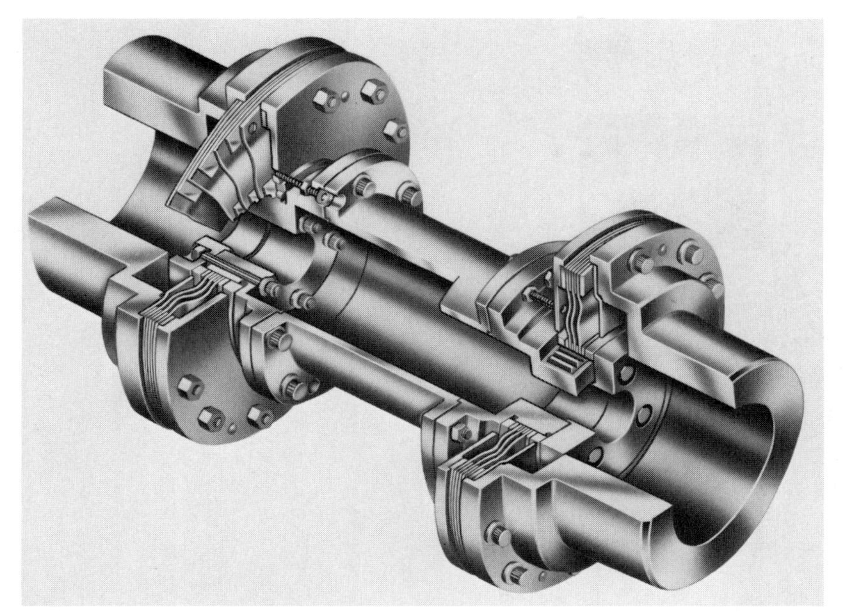

(C)

Figure 2.16 (Continued)

(A)

Figure 2.17 Miscellaneous types of couplings: (A) pin and bushing coupling (courtesy of Renolds, Inc., Engineered Products Division); (B) Schmidt coupling (courtesy of Schmidt Couplings, Inc.); (C) spring coupling (courtesy of Coupling Division of Rexnord, Thomas Couplings); (D) spring coupling (courtesy of Panamech Company); (E) slider block coupling (courtesy of Zurn Industries, Inc., Mechanical Drives Division).

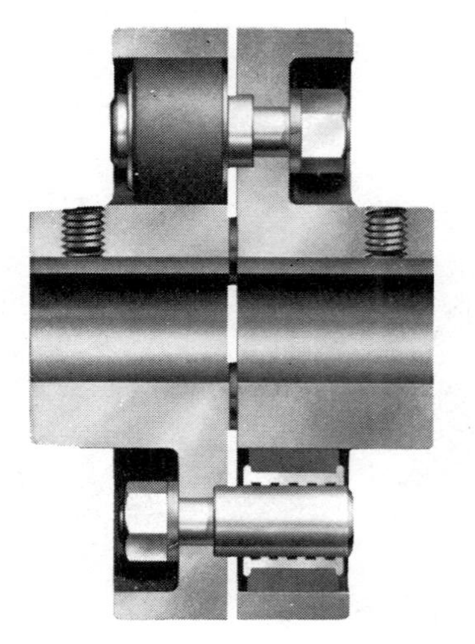

(A) (Continued)

(B)

Figure 2.17 (Continued)

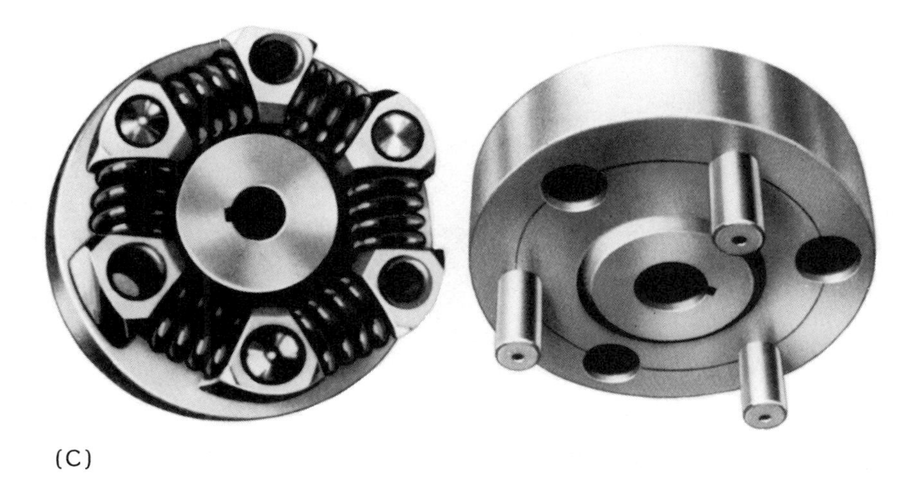

(C)

(D)

Figure 2.17 (Continued)

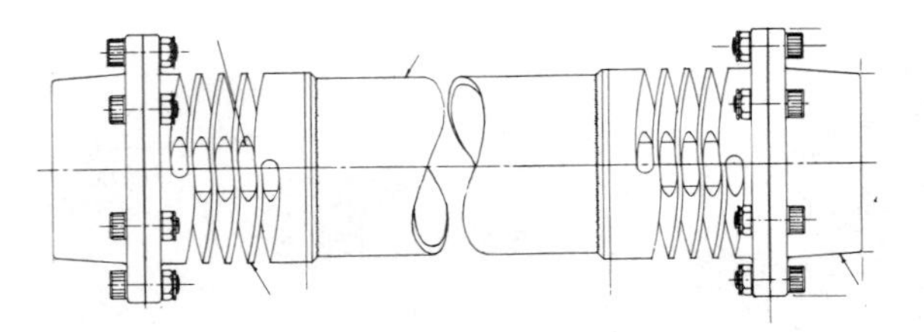

(D) (Continued)

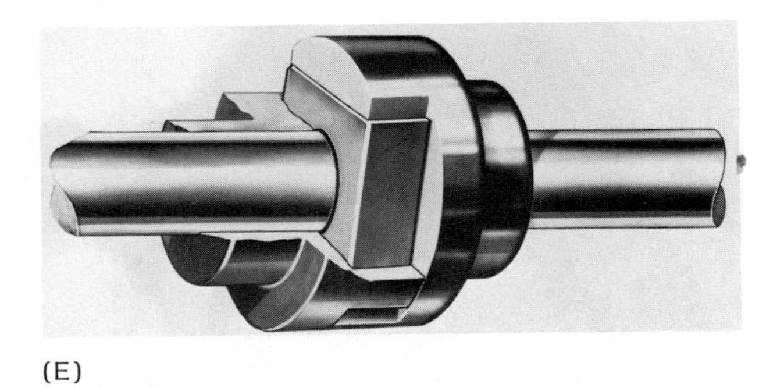

(E)

Figure 2.17 (Continued)

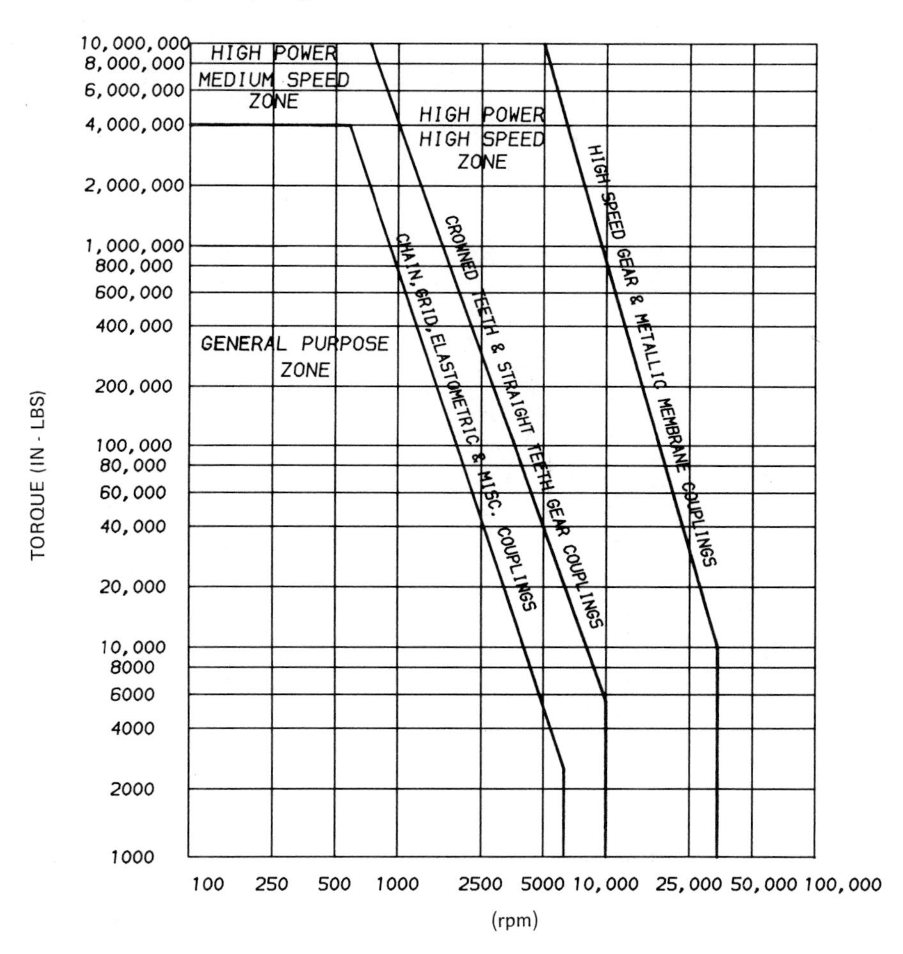

Figure 2.18 Torque versus speed (rpm) envelope of various types of couplings.

Functional capacities	Mechanically flexible						
	Straight gear	Crowned gear	High-performance gear	Spindle	Chain	Grid	Universal joint
Max. continuous torque in.-lb. $\times 10^6$	40	44	13	25	1.5	4	25
Max. speed (rpm)	12,000	14,500	40,000	2,000	6,500	4,000	8,000
Max. bore (in.)	36	36	12	26	10	20	20
Angular mis-alignment (degrees)	1/2	1 1/2	1/4	7	2	1/3	20
Parallel offset (in./in.)	0.008	0.026	0.004	0.12	0.035	0.006	0.34
Axial travel (in.)	1/8-1[a] 12[c] ∞[d]	1/8-1[a] 12[c] ∞[d]	1/8-1[a]	1/2-3[a] 12[c] ∞[d]	±1/4	±3/16	24[b]

Functional capacities	Elastomeric		Metallic membrane			
	Shear	Compression	Laminated disk	Multiple diaphragm	Convolute diaphragm	Tapered diaphragm
Max. continuous torque in.-lb. $\times 10^6$	0.5	15	4	4	6	6
Max. speed (rpm)	5,000	8,000	30,000	5,000	30,000	30,000
Max. bore (in.)	10	24	12	12	12	12
Angular misalignment (degrees)	3	3	1/4	1	1/2	1/2
Parallel offset (in./in.)	0.05	0.05	0.004	0.017	0.008	0.008
Axial travel (in.)	±5/16	±1/16	±3/8	±1/8	±1	±2

[a]Typical axial travels.
[b]With telescoping shaft.
[c]Limit without very special designs.
[d]Limited by manufacturing today to ≅ 36 in.

Figure 2.19 Coupling functional capacities.

Coupling characteristics	Mechanically flexible						
	Straight gear	Crowned gear	High-performance gear	Spindle	Chain	Grid	Universal joint
Lubrication	Yes	Yes	Yes	Yes	Yes	Yes	Yes
Backlash	Med. high	Med.	Low	High	High	Med.	None[a] Med.[b]
Overhung moment	Med.	Med.	Low	High	Med.	Med.	High
Unbalance	Med.	Med.	Low	High	Med. high	Med. high	High
Bending moment	High	High	High	High	High	Med.	High
Axial force	High	High	High	High	High	Med.	High
Torsional stiffness	High	High	High	High	High	Med.	High
Damping	Low	Low	Low	Low	Med.	Med. high	Low

Coupling characteristics	Elastomeric		Metallic membrane			
	Shear	Compression	Laminated disk	Multiple diaphragm	Convolute diaphragm	Tapered diaphragm
Lubrication	No	No	No	No	No	No
Backlash	None	Low	None	None	None	None
Overhung moment	High	High	Low	Med.	Med. low	Med. low
Unbalance	Med. high	Med. high	Low med.	Low med.	Low	Low
Bending	Low	Med. low	Med. low	Med. low	Med. low	Med.
Axial force	Low	Med. low	Med. low	Med. low	Med. low	Med.
Torsional stiffness	Low	Med. low	High	High	High	High
Damping	High	Med. high	Low	Low	Low	Low

[a]Without telescoping shaft.
[b]With telescoping shaft.

Figure 2.20 Coupling characteristics.

3
Selection and Design

I. SELECTION FACTORS

Flexible couplings are a vital part of a mechanical power transmission system. Unfortunately, many system designers treat flexible couplings as if they were a piece of hardware.

The amount of time spent in selecting a coupling and determining how it interacts with a system should be a function not only of the cost of the equipment but also how much downtime it will take to replace the coupling or repair a failure. In some cases the process may take only a little time and be based on past experience; however, on sophisticated systems this may take complex calculations, computer modeling, and possibly even testing. A system designer or coupling user just cannot put any flexible coupling into a system with the hope that it will work. It is the system designer or the user's responsibility to select a coupling that will be compatible with the system. Flexible coupling manufacturers are not usually system designers; rather, they size, design, and manufacture couplings to fulfill the requirements supplied to them. Some coupling manufacturers can offer some assistance based on past exposure and experience. They are the experts in the design of flexible couplings, and they use their know-how in design, materials, and manufacturing to supply a standard coupling or a custom-built one if requirements so dictate. A flexible coupling is usually the least expensive major component of a rotating system. It usually costs less than 1%, and probably never exceeds 10%, of the total system cost.

Flexible couplings are called upon to accommodate for the calculated loads and forces imposed on them by the equipment and their support structures. Some of these are motions due to thermal growth deflection of structures due to temperature, torque-load variations, and many other conditions. In operation, flexible couplings are sometimes called on to accommodate for the unexpected. Sometimes these may be the forgotten conditions or the conditions unleashed when the coupling interacts with the system, and this may precipitate coupling or system problems. It is the coupling selector's responsibility to review the interaction of the coupling with that system. When the coupling's interaction with the system is forgotten in the selection process, coupling life could be shortened during operation or a costly failure of coupling or equipment may occur.

A. Selection Steps

There are usually four steps that a coupling selector should take to assure proper selection of a flexible coupling for critical or vital equipment.

1. The coupling selector should review the initial requirements for a flexible coupling and select the type of coupling that best suits the system.
2. The coupling selector should supply the coupling manufacturer with all the pertinent information on the application so that the coupling may be properly sized, designed, and manufactured to fit the requirements.
3. The coupling selector should obtain the flexible coupling's characteristic information so the interactions of the coupling with the system can be checked to assure compatibility such that no determinal forces and moments are unleashed.
4. The coupling selector should review the interactions and if the system conditions do change, the coupling manufacturer should be contacted so that a review of the new conditions and their effect on the coupling selected can be made. This process should continue until the system and the coupling are compatible.

The complexity and depth of the selection process for a flexible coupling depend on how critical and how costly downtime would be to the ultimate user. The coupling selector has to determine to what depth the analysis must go. As an example, a detailed, complex selection process would be abnormal for a 15-hp motor-pump application, but would not be for a multimillion-dollar 20,000-hp motor, gear, and compressor train.

B. Types of Information Required

Coupling manufacturers will know only what the coupling selector tells them about an application. Missing information required to select and size couplings can lead to an improperly sized coupling and possible failure. A minimum of three items are needed to size a coupling: horsepower, speed, and interface information. This is sometimes the extent of the information supplied or available. For coupling manufacturers to do their best job, the coupling selector should supply as much information as is felt to be important about the equipment and the system. See Figure 3.1 for the types of information required. Figures 3.2 and 3.3 list specific types of information that should be supplied for the most common types of interface connections: the flange connection and the cylindrical bore.

C. Interaction of a Flexible Coupling with a System

A very important but simple fact often overlooked by the coupling selector is that couplings are connected to the system. Even if they are selected, sized, and designed properly, this does not assure trouble-free operation. Flexible couplings generate their own forces and can also amplify system forces. This may change the system's original characteristics or operating conditions. The forces and moments generated by a coupling can produce loads on equipment that can change misalignment, decrease the life of a bearing, or unleash peak loads that can damage or cause failure of the coupling or connected equipment.

Listed below are some of the coupling characteristics that may interact with the system. The coupling selector should obtain from the coupling manufacturer the values for these characteristics and analyze their affect on the system.

1. Torsional stiffness
2. Torsional damping
3. Amount of backlash
4. Weights
5. Coupling flywheel effect*
6. Center of gravity
7. Amount of unbalance
8. Axial force

---

*The flywheel effect of a coupling (WR^2) is the product of the coupling weight times the square of the radius of gyration. The radius of gyration is that radius at which the mass of the coupling can be considered to be concentrated.

9. Bending moment
10. Lateral stiffness
11. Coupling axial critical†
12. Coupling lateral critical†

D. Final Check

The coupling selector should use the coupling characteristics to analyze the system axially, laterally, thermally, and torsionally. Once the analysis has been completed, if the system's operating conditions change, the coupling selector should supply this information to the coupling manufacturer to assure that the coupling selection has not been sacrificed and that a new coupling size or type is not required. This process of supplying information between the coupling manufacturer and the coupling selector should continue until the system and coupling are compatible. This process is the only way to assure that a system will operate successfully.

†Coupling manufacturers will usually calculate coupling axial and lateral critical values with the assumption that the equipment is infinitely rigid. A coupling's axial and lateral critical values comprise a coupling's vibrational natural frequency in terms of rotational speed (rpm).

1. Horsepower
2. Operating speed
3. Interface connect information in Figs. 3.1, 3.2, etc.
4. Torques
5. Angular misalignment
6. Offset misalignment
7. Axial travel
8. Ambient temperature
9. Potential excitation or critical frequencies
   a. Torsional
   b. Axial
   c. Lateral
10. Space limitations (drawing of system showing coupling envelope)
11. Limitation on coupling generate forces
    a. Axial
    b. Moments
    c. Unbalance
12. Any other unusual condition or requirements or coupling characteristics - weight, torsional stiffness, etc.

    Note: Information supplied should include all operating or characteristic values of equipment for minimum, normal, steady state, momentary, maximum transient, and the frequency of their occurrence.

Figure 3.1 Types of information required to properly select, design, and manufacture a flexible coupling.

1. Size of bore including tolerance or size of shaft and amount of clearance or interference required.
2. Lengths
3. Taper shafts
   a. Amount of taper
   b. Position and size of o-ring grooves if required.
   c. Size, type and location of hydraulic fitting.
   d. Size and location of oil distribution grooves.
   e. Max. pressure available for mounting.
   f. Amount of hub draw-up required.
   g. Hub OD requirements.
   h. Torque capacity required (should also specify the coefficient of friction to be used).
4. Minimum strength of hub material or its hardness
5. If keyways in shaft
   a. How many
   b. Size and tolerance
   c. Radius required in keyway (minimum and maximum).
   d. Location tolerance of keyway respective to bore and other keyways.

Figure 3.2 Information required for cylindrical bores.

1. Diameter of bolt circle and true location.
2. Number and size of bolt holes
3. Size, grade and types of bolts required.
4. Thickness of web and flange.
5. Pilot dimensions.
6. Others.

Figure 3.3 Types of interface information required for bolted joints.

II. DESIGN EQUATIONS AND PARAMETERS

The intent of this section is to give the reader (coupling selector or user) some basic insights into coupling ratings and design. The equations and allowable values set forth in this section should be used only for comparison, not for design. They will help the reader compare "apples to apples"—specifically in this case, couplings to couplings.

A. Coupling Ratings

1. Torque Ratings. One of the most confusing subjects is that of coupling ratings. One reason for this is that *there is no common rating system*.

a. What Torque Ratings Mean. Some coupling manufacturers rate couplings at limits based on the *yield strength* of the component's material and then require the application of *service factors*. Some couplings have ratings based on the endurance strength of their component parts and require the use of small or no service factors at all. The third basis for some coupling ratings is *life*, and this is the most confusing because it is based on experience and empirical data. We will not attempt to cover life ratings but leave their derivation and explanation to the specific coupling manufacturer.

b. Which Coupling Rating Is Correct. All coupling ratings are correct—that is, as long as you use the recommended service factors and the selection procedure outlined by the coupling manufacturer.

c. How to Compare Couplings. First, the following should be compared:

1. The safety factor related to the endurance strength (for cyclic-reversing loading) or yield strength (for steady-state loading) for normal operating torque and conditions
2. The safety factor related to yield strength for infrequent peak operating torques and conditions

Such a comparison will probably permit the coupling selector to select the safest coupling for the system. To complete the selection analysis, the coupling selector should ask the coupling manufacturer the following:

1. The material properties of the major components for tensile and shear
   a. Ultimate strength
   b. Yield strength
   c. Endurance limit
2. The stresses in the major components
   a. At normal torque and loads
   b. At peak torque and loads
3. The stress concentration factors (this is important to know when cyclic and reversing torques and/or loading are present)
4. What they typically use or recommend as a safety factor

Most coupling manufacturers will supply this information to the coupling selector. Usually, coupling manufacturers will not supply design equations to calculate the stresses of the flexing element (gear teeth, disk, diaphragms, etc.) since these are usually proprietary.

2. Speed Ratings. The true maximum speed limit at which a coupling can operate cannot be divorced from the connected equipment. This makes it very difficult to rate a coupling because the same coupling can be used on different types of equipment. Let's forget the system for a moment and consider only the coupling.

a. Maximum Speed Based on Centrifugal Stresses. The simplest method of establishing a coupling maximum speed rating is to base it on centrifugal stress (S_t):

$$S_t = \frac{\rho V^2}{g} \tag{3.1}$$

where

ρ = density of material (lb/in.^{3})
V = velocity (in./sec)
g = acceleration (386 in./sec)

$$V = \frac{D_o \times \pi \times \text{rpm}}{60} \tag{3.2}$$

$$\text{rpm} = \frac{375\sqrt{\frac{S_t}{\rho}}}{D_o} \tag{3.3}$$

where D_o is the outside diameter (in.) and

Material	ρ (lb/in.^{3})
Steel	0.283
Aluminum	0.100

$$S_t = \frac{S_{yld}}{F.S.} \tag{3.4}$$

where

F.S. = factor of safety (can be from 1 to 2, typically 1.5)
S_{yld} = yield strength of material (psi)

b. Maximum Speed Based on Lateral Critical Speed. For long couplings the maximum coupling speed is usually based on the lateral critical speed of the coupling. Couplings are part of a drive-train system. If a system is very rigid and stiff, the maximum speeds might be greater than calculated, and if a system is soft and has a long shaft overhand, it may be much lower than the calculated value. Figure 3.4A is a schematic of a coupling as connected to a system. Figure 3.4B is a free-body diagram depicting the mass-spring system that models the coupling as connected to the system. The center section of a coupling can be represented as a simply supported beam.

$$N_c = 211.4\sqrt{\frac{1}{\Delta}} \tag{3.5}$$

$$\Delta = \frac{5WL^3}{384EI} \tag{3.6}$$

$$I = 0.049(D_o^4 - D_i^4) \tag{3.7}$$

$$K_c = \frac{W}{g}\left(\frac{\pi N_c}{30}\right)^2 \tag{3.8}$$

$$\frac{1}{K_e} = \frac{1}{2(K_c/2)} + \frac{1}{2K_L} \tag{3.9}$$

where

N_c = critical shaft whirl rotational frequency of the coupling shaft [cycles per minute (cpm)]
Δ = shaft end deflection (in.)
L = effective coupling length (in.)
I = moment of inertia of the coupling spacer (in.^{4})
W = center weight (supported) of the coupling (lb)
E = 30×10^6 for steel (psi)
D_o = outside diameter of shaft (in.)
D_i = inside diameter of shaft (in.)
g = 386 in./sec^2
K_c = stiffness of coupling spacer (lb/in.)
K_L = lateral stiffness of coupling (lb/in.)
K_e = effective spring rate (lb/in.) for the system shown in Figure 3.4B and calculated using equation (3.10)

These equations ignore the stiffness of the system which is usually how coupling manufacturers determine critical speed calculations. [*Note*: For some couplings (e.g., gear couplings), K_L can be ignored.] If we consider the stiffness of the connected shafts and bearings:

$$\frac{1}{K_e} = \frac{1}{2(K_c/2)} + \frac{1}{2K_L} + \frac{1}{2K_s} + \frac{1}{2K_B} \tag{3.10}$$

where

K_s = equipment shaft stiffness (lb/in.)
K_B = equipment bearing stiffness (lb/in.)

Finally, the maximum speed based on the critical speed becomes

$$N_c = \frac{1}{2\pi \sqrt{\dfrac{K_e g}{W}}} \tag{3.11}$$

$$N_s = \frac{N_c}{F.S.} \tag{3.12}$$

F.S. = factor of safety (ranges from 1.5–2.0, usually 1.5)

c. *Other considerations*. Many other considerations may affect a coupling's actual maximum speed capabilities, including:

1. Heat generated from windage, flexing, and/or sliding and rolling of mating parts
2. Limitations on lubricants used (e.g., separation of grease)
3. Forces generated from unbalance
4. System characteristics

3. Service Factors. Service factors have evolved from experience and based on past failures. That is, after a coupling failed, it was determined that by multiplying the normal operating torque by a factor and then sizing the coupling, the coupling would not fail. Since coupling manufacturers use different design criteria, many different service factor charts are in use for the same types of couplings. Therefore, it becomes important when sizing a coupling to follow and use the ratings and service factors recommended in each coupling manufacturer's catalog and not to intermix them with other manufacturers' procedures and factors. If a coupling manufacturer is told the load conditions—normal, peak, and their duration—a more detailed and probably more accurate sizing could be developed. Service factors become less significant when the load and duty cycle are known. It is only when a system is not analyzed in depth that service factors must be used. The more that is known about the operating conditions, the closer to unity the service factor can be.

4. Ratings in Summary. The important thing to remember is that the inverse of a service factor or safety factor is an ignorance factor. What this means is that when little is known about the operating spectrum, a large service factor should be applied. When the operating spectrum is known in detail, the service factor can be reduced. Similarly, if the properties of a coupling material are not exactly known or the method of calculating the stresses are not precise, a large safety factor should be used. However, when the material properties and the method of calculating the stresses are known precisely, a small safety factor can be used.

Sometimes the application of service and safety factors to applications where the operating spectrum, material properties, and stresses are known precisely can cause the use of unnecessary large and heavy couplings. Similarly, not knowing the operating spectrum, material properties, or stresses and not applying the appropriate service factor or safety factor can lead to undersized couplings. Ultimately, this can result in coupling failure or even equipment failure.

B. Coupling Components Strength Limits

1. Allowable Limits. Depending on type of loading, the basis for establishing allowable stress limits differ. Usually, loads can be classified into three groups;

1. Normal-steady state
2. Cyclic-reversing
3. Infrequent-peak

Therefore, for comparison purposes we can conservatively establish three approaches to allowable stress limits.

a. Normal-steady state loads:

$$S_t = \frac{S_{yld}}{F.S.} \tag{3.13}$$

$$\tau = \frac{0.6\ S_{yld}}{F.S.} \tag{3.14}$$

where

S_t = allowable tensile stress (psi)
τ = allowable shear stress (psi)
S_{yld} = yield strength (psi)
F.S. = factor of safety

Factors of safety are 1.25 to 2, usually 1.5

b. Cyclic-reversing, where loading produces primarily tensile or shear stresses:

$$S_{te} = \frac{S_{end}}{F.S. \times K} \tag{3.15}$$

$$\tau_e = \frac{0.6\ S_{end}}{F.S. \times K} \tag{3.16}$$

where

S_{te} = allowable endurance limit tensile (psi)
S_{end} = tensile endurance limit (psi) = $.5S_{ult}$
K = stress concentration factor (see Peterson, 1974)
τ_e = allowable endurance limit shear (psi)

Factors of safety are 1.5 to 2, usually 1.5. K usually ranges from 1.5 to 3.5.

Where combined tensile and shear stresses are present:

$$S_D \text{ or } S_M = \frac{S_t}{2} + \sqrt{\left(\frac{S_t}{2}\right)^2 + (\tau)^2} \qquad (3.17)$$

$$\frac{1}{F.S.} = \frac{S_D}{S_{end}} + \frac{S_M}{S_{ult}} \qquad (3.18)$$

where

S_D = combined dynamic stresses (psi)

S_M = combined steady-state stresses (psi)

S_{ult} = ultimate strength (psi)

Factors of safety are 1.5 to 2.

c. Infrequent-peak loads (F.S. ranges from 1.0–1.25, typically 1.1):

$$S_t = \frac{S_{yld}}{F.S.} \qquad (3.19)$$

$$\tau = \frac{0.6\ S_{yld}}{F.S.} \qquad (3.20)$$

2. Typical Material Properties of Material for Couplings. Figure 3.5 shows typical materials used for various components of couplings. The following are listed: material hardness, tensile ultimate, tensile yield, shear ultimate, and shear yield.

3. Compressive Stress Limits for Various Steels. Figure 3.6 shows compressive stress limits for various hardnesses of steels. These limits are strongly dependent on whether relative motion takes place, type of lubricant used, and whether lubrication is present. The limits given in this table are typical limits for nonsliding components (e.g., keys, splines). For sliding parts the allowable value would be approximately one-fourth to one-third of the given values. It should also be noted that in some applications with good lubricants and highly wear resistant materials (nitrided or carburized) for parts that roll and/or slide (e.g., high-angle low-speed gear couplings), the values in the table can be approached.

4. Allowable Stress Limits for Various Coupling Materials. Figure 3.7 shows typical allowable limits for various coupling materials using the following values for F.S. and K.

Steady state-normal conditions:
Tensile allowable F.S. = 1.5
Shear allowable F.S. = 2
Cyclic-reversing conditions:
Tensile: K = 1.5, F.S. = 1.5
Shear: K = 2, F.S. = 1.5
Infrequent-peak conditions:
Tensile: F.S. = 1.1
Shear: F.S. = 1.1

C. General Equations

1. Torque

$$\text{Torque (T)} = \text{horsepower (hp)} \times 63{,}000/\text{rpm} = \text{in.-lb} \quad (3.21)$$

$$\text{Torque (T)} = \text{killowatts (kW)} \times 84{,}420/\text{rpm} = \text{in.-lb} \quad (3.22)$$

$$\text{Torque (T)} = \text{newton-meters (N} \cdot \text{M)} \times 8.85 = \text{in.-lb} \quad (3.23)$$

2. Misalignment

a. Angular Misalignment. With the exception of some elastomeric couplings, a single-element flexible coupling can usually accommodate only angular misalignment (α) (see Figure 3.8A). A double-element flexible coupling can handle both angular and offset misalignment (see Figure 3.8B).

$$\text{Angular } \alpha = \alpha_1 + \alpha_2 \quad (3.24)$$

b. Offset Misalignment (see Figure 3.8C).

$$S = L \tan \alpha \quad (3.25)$$

where

S = offset distance
L = length between flex points

α (deg)	Offset in./in. shaft separation	α (deg)	Offset in./in. shaft separation
1/4	0.004	2	0.035
1/2	0.008	3	0.052
3/4	0.012	4	0.070
1	0.018	5	0.087
1 1/2	0.026	6	0.100

c. *Combined Offset and Angular Misalignment* (see Figure 3.8D).

$$\alpha_{allowable} = \alpha_{rated} - \tan^{-1}\left(\frac{S}{L}\right) \qquad (3.26)$$

3. Weight (W), Flywheel Effect (WR^2), and Torsional Stiffness (K).
   Solid disk:

Weight = lb

$$\text{Steel } W = 0.223LD_o^2 \qquad (3.27)$$

$$\text{Aluminum } W = 0.075LD_o^2 \qquad (3.28)$$

WR^2 = lb-in.^{2}

$$WR^2 = \frac{WD_o^2}{8} \qquad (3.29)$$

where

D_o = outside diameter (in.)
L = length (in.)

Torsional stiffness (K) = in.-lb/rad

$$\text{Steel } K = 1.13 \times 10^6(D_o^4) \qquad (3.30)$$

$$\text{Aluminum } K = 0.376 \times 10^6(D_o^4) \qquad (3.31)$$

Tubular (disk with a hole):

Weight = lb

$$\text{Steel } W = 0.223L(D_o^2 - D_i^2) \qquad (3.32)$$

$$\text{Aluminum } W = 0.075L(D_o^2 - D_i^2) \qquad (3.33)$$

WR^2 = lb-in.^{2}

$$WR^2 = \frac{W}{8}(D_o^2 + D_i^2) \qquad (3.34)$$

where

D_i = inside diameter (in.)

Torsional stiffness (K) = in.-lb/rad

$$\text{Steel } K = \frac{1.13 \times 10^6 (D_o^4 - D_i^4)}{L} \tag{3.35}$$

$$\text{Aluminum } K = \frac{0.375 \times 10^6 (D_o^4 - D_i^4)}{L} \tag{3.36}$$

4. Bending Moment. Most couplings exhibit a bending stiffness. If you flex them, they produce a reaction. Bending stiffness (K_B) is usually expressed as

$$K_B = \text{in.-lb/deg} \tag{3.37}$$

Therefore, the moment reaction (M) per flex element is

$$M = K_B \times \alpha = \text{in.-lb} \tag{3.38}$$

[*Note*: Some couplings (e.g., gear couplings) exhibit bending moments that are functions of many factors (torque, misalignment, load distribution, and coefficient of friction).]

5. Axial Force. Two types exist: force from the sliding of mating parts and force from flexing of material.

Sliding force (F) (e.g., gear and slider block couplings):

$$F = \frac{T\mu}{R} = \text{lb} \tag{3.39}$$

where

T = torque (in.-lb)
μ = coefficient of friction
R = radius at which sliding occurs (in.)

Flexing force (F) (e.g., rubber and metallic membrane couplings):

$$F = K_A \times S = \text{lb} \tag{3.40}$$

where

K_A = axial stiffness (lb/in. per mesh)

S = axial deflection (in.)

Note: For two meshes, $K_A' = K_A/2$.

D. Component Stress Equations

1. Shaft Stresses

a. Circular shaft

Solid (Figure 3.9A):

$$\tau_s = \frac{16T}{\pi D_o^3} \tag{3.41}$$

Tubular (Figure 3.9B):

$$\tau_s = \frac{16T}{\pi(D_o^4 - D_i^4)} \tag{3.42}$$

b. Noncircular tubing

Rectangular section (Figure 3.9C):

$$\tau_s = \frac{(15b + 9h)T}{5b^2h^2} \tag{3.43}$$

Square solid (Figure 3.9D):

$b = h$

$$\tau_s = \frac{24T}{5h^3} \tag{3.44}$$

Any shape:

$$\tau_s = \frac{T}{2At} \tag{3.45}$$

where

A = area enclosed by centerline
t = wall thickness

2. Spline Stresses (Figure 3.10).
Compressive stress:

$$S_c = \frac{2T}{n(P.D.)Lh} \tag{3.46}$$

Shear stress:

$$S_s = \frac{4T}{\pi(P.D.)^2 L} \tag{3.47}$$

Bending stress:

$$S_b = \frac{2T}{(P.D.)^2 L} \tag{3.48}$$

Bursting stress (hub):

$$S_t = \frac{T}{(P.D.)L}\left(\frac{\tan\theta}{t} + \frac{2}{P.D.}\right) \tag{3.49}$$

Shaft shear stress (shaft):

$$S_s = \frac{16TD_o}{\pi(D_o^4 - D_i^4)} \tag{3.50}$$

where

T = torque
n = number of teeth
P.D. = pitch diameter
L = effective face width (in.)
h = height of spline tooth
θ = pressure angle
t = hub wall thickness
D_o = diameter at root of shaft tooth
D_i = inside diameter (ID) of shaft (zero for solid shaft)

3. Key Stresses
Square or rectangular keys (Figure 3.11):

$$T = \frac{FDn}{2} \tag{3.51}$$

$$F = \frac{2T}{Dn} \tag{3.52}$$

$$\tau = \text{shear stress} = \frac{F}{wL} = \frac{2T}{WLDn} \tag{3.53}$$

$$S_c = \text{compressive stress} = \frac{2F}{hL} = \frac{4T}{hLDn} \tag{3.54}$$

where

L = length of key (in.)
n = number of keys
D = shaft diameter (in.)
w = key width (in.)
h = key height (in.)
T = torque (in.-lb)
F = force (in.-lb)

Allowable key stress limits (see Figure 3.12).

E. **Flange Connections** (Figure 3.13)

1. Friction Capacity

$$T = F\mu R_f \tag{3.55}$$

F = clamp load from bolts

$$F = \left(\frac{T_B}{\mu_B d}\right)n \tag{3.56}$$

where

T_B = tightening torque of bolt (in.-lb)
n = number of bolts
d = diameter of bolt (in.)
μ_B = coefficient of friction at bolt threads (0.2 dry threads)
μ = coefficient of friction at flange faces (0.1 to 0.2, usually 0.15)
R_f = friction radius (in.)

$$R_f = \frac{2}{3}\left(\frac{R_o^3 - R_i^3}{R_o^2 - R_i^2}\right) \tag{3.57}$$

where

R_o = outside radius of clamp (in.)
R_i = inside radius of clamp (in.)

2. Bolt Stresses for Flange Couplings (see Figure 3.13)

a. Shear Stress

Allowables:

$$\tau_A \text{ at normal operation} = \frac{0.30_{ult}}{1.5} = 0.20S_{ult} \tag{3.58}$$

$$\tau_A \text{ at peak operation} = 0.55S_{yld} \tag{3.59}$$

where

S_{ult} = ultimate strength of material (psi)
S_{yld} = yield strength of material (psi)

Equation:

$$\tau_s = \frac{8T}{(D.B.C.)(\%n)\pi d} \tag{3.60}$$

where

T = operating torque (in.-lb)
D.B.C. = diameter bolt circle (in.)
d - bolt body diameter
n = number of bolts
%n = percentage of bolts loaded: fitted bolts 0.000 to 0.006 loose (100%); clearance bolts 0.006 to 0.032 loose (75%).

For very loose bolts with over 0.032 clearance or very long bolts of approximately 10d, the analysis above may not be applicable. For very loose and long bolts, the bending stresses produced in a bolt should be considered.

b. Tensile Stress

Allowables:

$$S_t = 0.8S_{yld} \tag{3.61}$$

where S_t = stress from tightening bolt

Equation:

$$S_t = \frac{F_n}{A_s} \tag{3.62}$$

where

A_S = bolt thread root stress area (in.)
F_n = clamp load produced by bolt (lb)

$$S_t = 0.25 \text{ to } 0.75 \text{ of } S_{yld} \quad (3.63)$$

usually 0.5.

c. Tightening Torque. From this we can determine the tightening torque:

$$F_n = 0.5S_{yld} \times A_S = \text{lb} \quad (3.64)$$

$$T_B = F_n \times \mu \times d = \text{in.-lb} \quad (3.65)$$

$$T_B = 0.2\, F_n \times d = \text{in.-lb} \quad (3.66)$$

where

T_B = tightening torque of bolt (in.-lb)
μ = coefficient of friction at threads

For different lubrication condition of threads, use Figure 3.14 to determine the value T_B.

$$T_B' = T_B \times C \quad (3.67)$$

where

C = lubrication condition (Figure 3.14)

Figure 3.15 shows properties of the four most common grades of bolts used with couplings.

d. Combined Bolt Stress for Flange Connections.

$$S_A(\text{allowable}) = 0.80S_{yld} \quad (3.68)$$

$$S_T = \frac{S_t}{2} + \sqrt{\left(\frac{S_t}{2}\right)^2 + (\tau_s)^2} \quad (3.69)$$

3. Flange Stresses (Figure 3.16). Most flexible couplings have flanges (rigids, spacers, etc.) that are loaded, usually with

moments and also in shear. Generally, the stresses at point A can be approximated by the following equations when $t \simeq w$.

Bending stress (S_B):

$$S_B = \frac{MC}{I} \times K \tag{3.70}$$

$$I = 0.049 \, (D_o^4 - D_i^4) \tag{3.71}$$

$$C = \frac{D_o + D_i}{2}$$

where

K = stress concentration (usually 3 or less; see Peterson, 1974)
M = applied moments (any external moment or the moment generated by the flexing element of the coupling (in.-lb) (see Section II.C.4)

Shear stress:

$$\tau = \frac{16TD_oK}{\pi(D_o^4 - D_i^4)} \tag{3.72}$$

Combined stress:

$$S_M \text{ (steady-state stress)} = \frac{S_t}{2} + \sqrt{\left(\frac{S_t}{2}\right)^2 + (\tau)^2} \tag{3.73}$$

where S_t is any steady state tensile stress [e.g., centrifugal stress $S_t = \rho V^2/g$ or any other axial loading on the flange (external or internal)].

$$\frac{1}{\text{F.S.}} = \frac{S_M}{S_{ult}} + \frac{S_B}{S_{end}} \tag{3.74}$$

where

F.S. = factor of safety
S_{ult} = ultimate strength of material (psi)

F. Hub Stresses and Capacities

1. Hub Bursting Stress for Hubs with Keyways (Figure 3.17). The following approach is applicable when the interference is less than 0.0005 in./in. of shaft diameter and X is greater than $(D_o - D_i)/8$.

Stress:

$$S_t = \frac{T}{AYn} \tag{3.75}$$

where

n = number of keys
Y = radius of applied force (in.)
A = area = XL[L = length of hub (in.)]

$$X = \sqrt{\left(\frac{D_o}{2}\right)^2 - \left(\frac{W}{2}\right)^2} - \frac{h}{2} - \sqrt{\left(\frac{D_i}{2}\right)^2 - \left(\frac{w}{2}\right)^2} \tag{3.76}$$

$$Y = \frac{(2D_i) + h}{4} \tag{3.77}$$

where

D_o = outside diameter
D_i = inside diameter
h = height of key
w = width of key

2. Hub Stress and Capacity for Keyless Hubs (Shrink Drives) (see Figure 3.18). Shrink fit or interference fits are generally based on Lamé's equation for thick-walled cylinder under internal pressure.

Stresses:

$$S_t = P\left(\frac{D_o^2 + D_i^2}{D_o^2 - D_i^2}\right) \tag{3.78}$$

$$P = \frac{Ei}{D_i^2}\left[1 - \left(\frac{D_i}{D_o}\right)^2\right] \tag{3.79}$$

where

D_o = outside diameter (psi)
D_i = inside diameter (psi)
P = pressure (psi)
E = modulus of elasticity for steel, 30×10^6 (psi)
i = diametral interference

Use maximum interference for stress calculations. Use minimum interference for torque capacity.

Torque capacity:

$$T = \frac{\pi D_i^2 L \mu P}{2} \tag{3.80}$$

where L is the length of the hub or sections of it (in.) and μ = 0.1 to 0.2, typically 0.15.

3. Torque Capacity. For hubs with steps or flanges (Figure 3.19): Section 1 requires maximum amount of pressure for mounting:

$$P_1 = \frac{Ei}{D_3^2}\left[1 - \left(\frac{D_3}{D_o}\right)^2\right] \tag{3.81}$$

$$S_{t_1} = P_1\left(\frac{D_o^2 + D_3^2}{D_o^2 - D_3^2}\right) \tag{3.82}$$

$$T_1 = \frac{\pi D_3^2 L_1 \mu P_1}{2} \tag{3.83}$$

Section 2:

$$P_2 = \frac{Ei}{D_3^2}\left[1 - \left(\frac{D_3}{D_2}\right)^2\right] \tag{3.84}$$

$$S_{t_2} = P_2\left(\frac{D_2^2 + D_3^2}{D_2^2 - D_3^2}\right) \tag{3.85}$$

$$T_2 = \frac{\pi D_3^2 L_2 \mu P_2}{2} \tag{3.86}$$

Section 3:

$$P_3 = \frac{Ei}{D_3^2}\left[1 - \left(\frac{D_3}{D_1}\right)^2\right] \tag{3.87}$$

$$S_{t_3} = P_3\left(\frac{D_1^2 + D_3^2}{D_1^2 - D_3^2}\right) \tag{3.88}$$

$$T_3 = \frac{\pi D_3^2 L_3 \mu P_3}{2} \tag{3.89}$$

Total torque capacity:

$$T_1 = T_1 + T_2 + T_3 = \text{in.-lb} \tag{3.90}$$

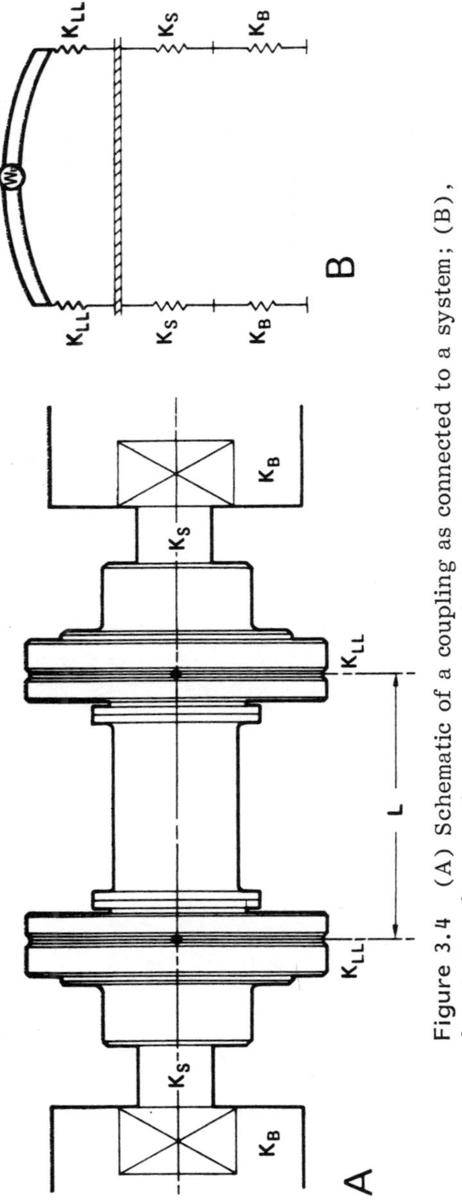

Figure 3.4 (A) Schematic of a coupling as connected to a system; (B), free body of a coupling as connected to a system.

	Hardness bhn	Tensile ultimate S_{ult} psi	Tensile yield S_{yld} psi	Shear ultimate τ_{ult} psi	Shear yield τ_{yld} psi
Aluminum die	50	25,000	12,000	18,500	7,200
casting	75	50,000	25,000	35,000	15,000
Aluminum bar—					
bar forging	100	60,000	35,000	45,000	21,000
heat treated	150	80,000	70,000	60,000	42,000
Carbon steels	110	50,000	30,000	37,500	18,000
	160	85,000	50,000	64,000	30,000
	250	110,000	70,000	82,000	42,000
Alloy steels	300	135,000	110,000	100,000	66,000
	330	150,000	120,000	110,000	72,000
	360	160,000	135,000	120,000	81,000
Surface hardened	450	200,000	170,000		
alloy steels	550	250,000	210,000		
	600	275,000	235,000		

Figure 3.5 Properties of typical coupling components. $\tau_{ult} \simeq .75S_{ult}$; $\tau_{yld} \simeq .6S_{yld}$.

bhn	Normal psi	Maximum psi
110	32,500	65,000
160	45,000	90,000
250	65,000	130,000
300	72,500	145,000
330	80,000	160,000
360	90,000	180,000
450	125,000	250,000
550	137,500	275,000
600	150,000	300,000

Figure 3.6 Compressive stress limits of steel.

	bhn	Steady-state allowables psi		Cyclic-reversing allowables		Peak allowables psi	
		S_t	τ	S_t	τ	S_t	τ
Aluminum	50	8,000	3,600	5,500	2,500	11,000	6,500
Die casting	75	16,500	7,500	11,000	5,000	22,500	13,500
Aluminum							
Bar-forging	100	23,500	10,500	13,500	6,000	32,000	19,000
Heat treated	150	46,500	21,000	18,000	8,000	63,500	38,000
Carbon steels	110	20,000	9,000	11,000	5,000	27,000	16,500
	160	33,500	15,000	19,000	8,500	45,000	27,500
	250	46,500	21,000	24,500	11,000	63,500	38,000
Alloy steels	300	73,500	33,000	30,000	13,500	100,000	60,000
	330	80,000	36,000	33,500	15,000	110,000	65,500
	360	90,000	40,500	35,500	16,000	125,000	73,500
Surface hardened	450	113,500		45,500		155,500	
Alloy steels	550	140,000		55,500		191,000	
	600	156,500		61,000		213,500	

Figure 3.7 Typical allowable limits for various materials for coupling components.

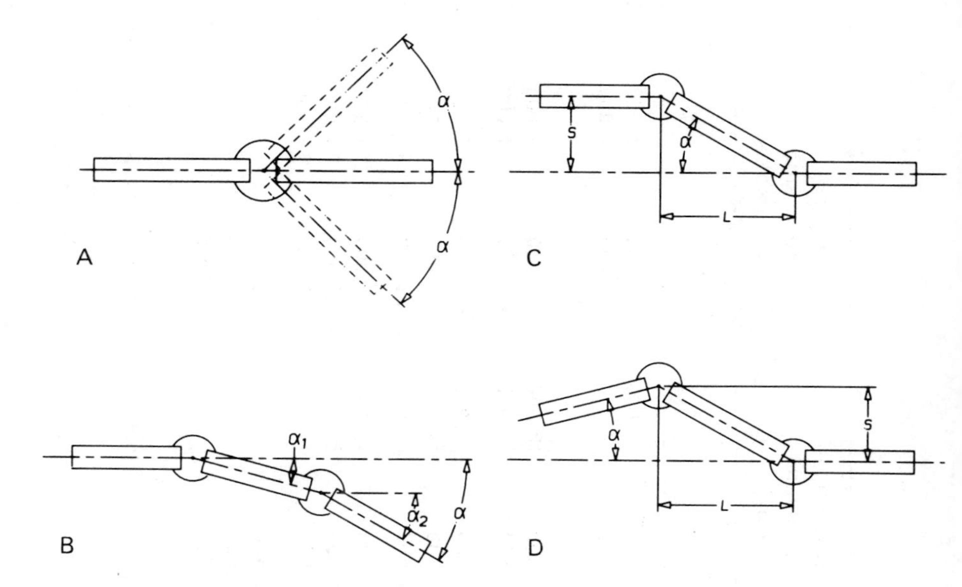

Figure 3.8 Types of shaft misalignments: (A) angular misalignment for single element; (B) angular misalignment for double element; (C) offset misalignment; (D) offset and angular misalignment combined.

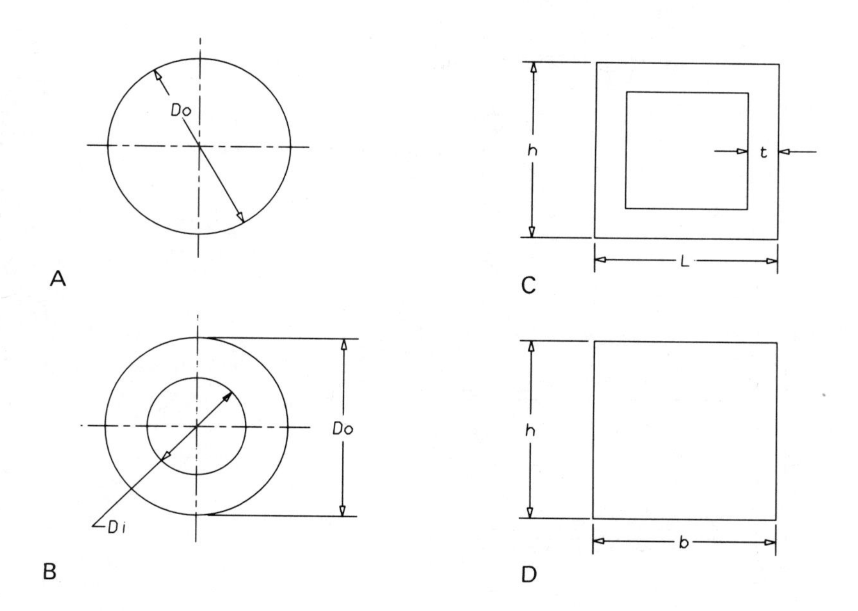

Figure 3.9 Types of shaft shapes: (A) circular shaft, solid; (B) circular shaft, tubular; (C) rectangular shaft, tubular; (D) rectangular shaft, solid.

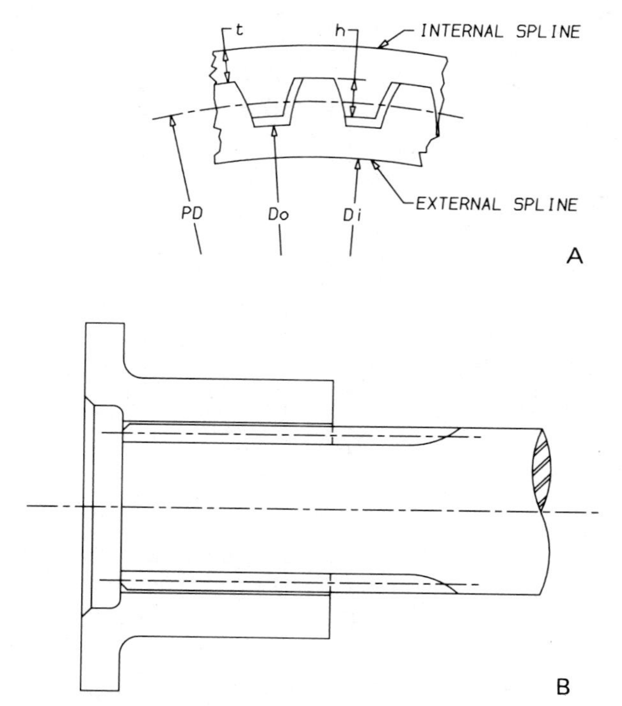

Figure 3.10 Spline dimensions: (A) spline teeth dimensions; (B) splined hub dimensions.

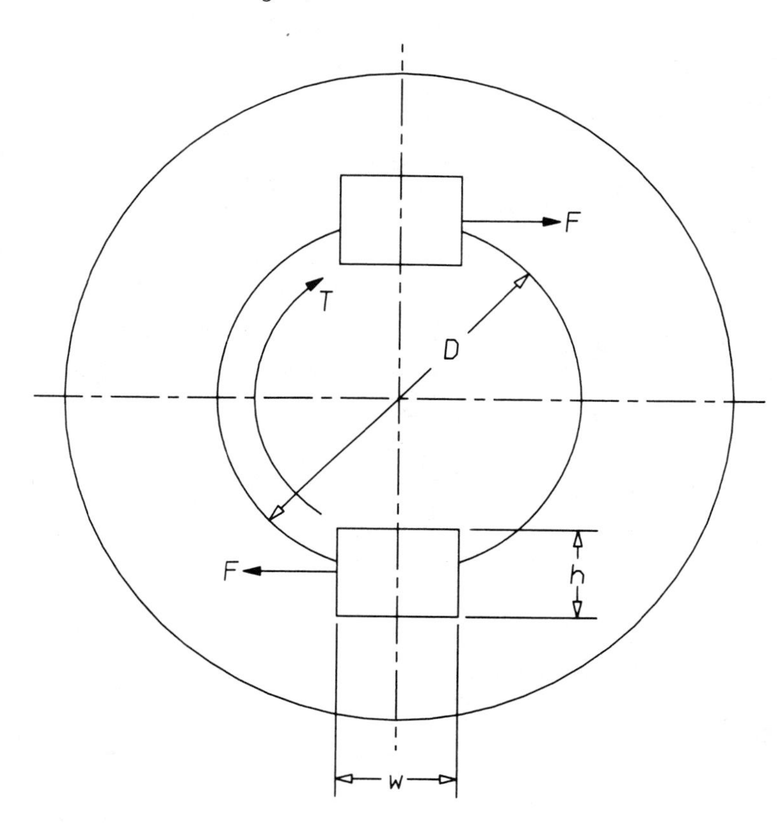

Figure 3.11 Square or rectangular key loading and dimensions.

	Shear τ normal psi	Compressive S_c normal psi	Shear τ peak psi	Compressive S_c peak psi
Commercial key stocks	12,500	32,500	23,000	65,000
1045 (170 bhn)	15,000	45,000	27,500	90,000
4140-4130 HT (270 bhn)	30,000	70,000	55,000	140,000

Figure 3.12 Key limits.

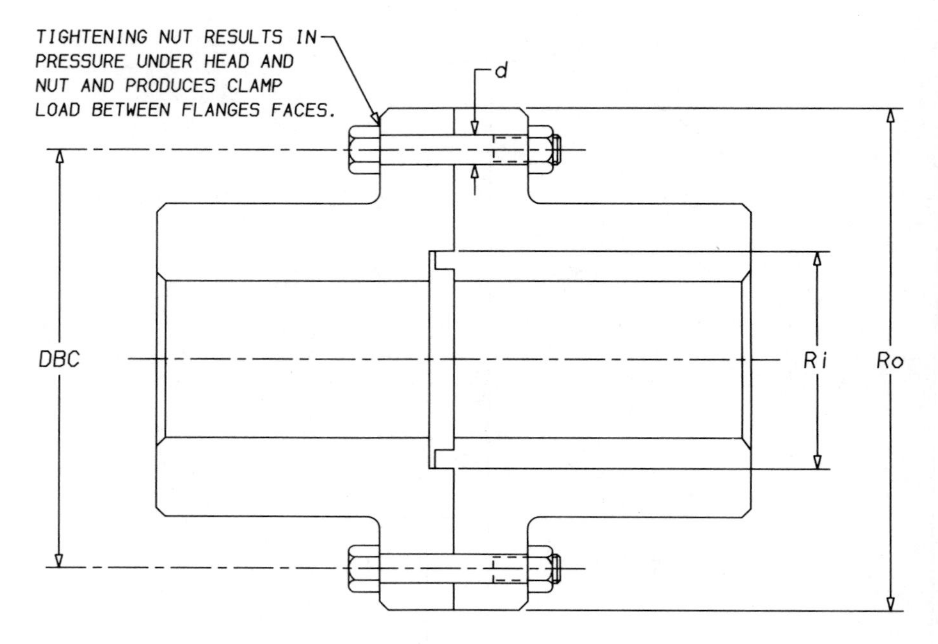

Figure 3.13 Dimensions for bolted flange connection.

Multiply torque (T_B) by C	Condition
1 $\mu = .2$	Dry threads—unplated or black oxided bolts
.75% $\mu = .15$	Lube threads—unplated or black oxided bolts
.75% $\mu = .15$	Dry threads—cadmium plated bolts
.56 $\mu = .11$	Lube threads—cadmium plated bolts

Figure 3.14 Frictional factors for various bolt tread conditions.

S.A.E. GRADE	HEADSTYLE	BOLT SIZE DIA. IN.	PROOF LOAD	TENSILE STRENGTH	HARDNESS ROCKWELL	MATERIAL - HEAT TREATMENT
2		Up to 1/2 incl. Over 1/2 to 3/4 Over 3/4 to 1-1/2	55,000 52,000 28,000	69,000 64,000 65,000	100 B Max. 100 B Max. 95 B Max.	Low or medium carbon steel; C 0.28 max., and S 0.05 max. For bolts over 6 in. long, or over 3/4 in. dia., carbon may be as high as 0.55.
5		Up to 3/4 incl. Over 3/4 to 1 Over 1 to 1-1/2	85,000 78,000 74,000	120,000 115,000 105,000	23-32 C 27-32 C 19-30 C	Medium carbon steel; C 0.28 to 0.55 P 0.04 max., and S 0.05 max., Quenched and tempered at a minimum temperature of 800°F.
8		Up to 1-1/2 incl.	120,000	150,000	32-38 C	Medium carbon fine grain alloy steel (b); C 0.28 to 0.55. p 0.04 max. and S 0.05 max., providing sufficient hardenability to have a minimum oil quenched hardness of 47 RC at the center of the threaded section one diameter from the end of the bolt. Oil quenched and tempered at a minimum temperature of 800°F.
12-point Aircraft Style		Up to 1-1/4	144,000	160,000	34-40 C	Alloy steel usually 4140, 4340, and others.

Figure 3.15 Properties of common grades of coupling bolts.

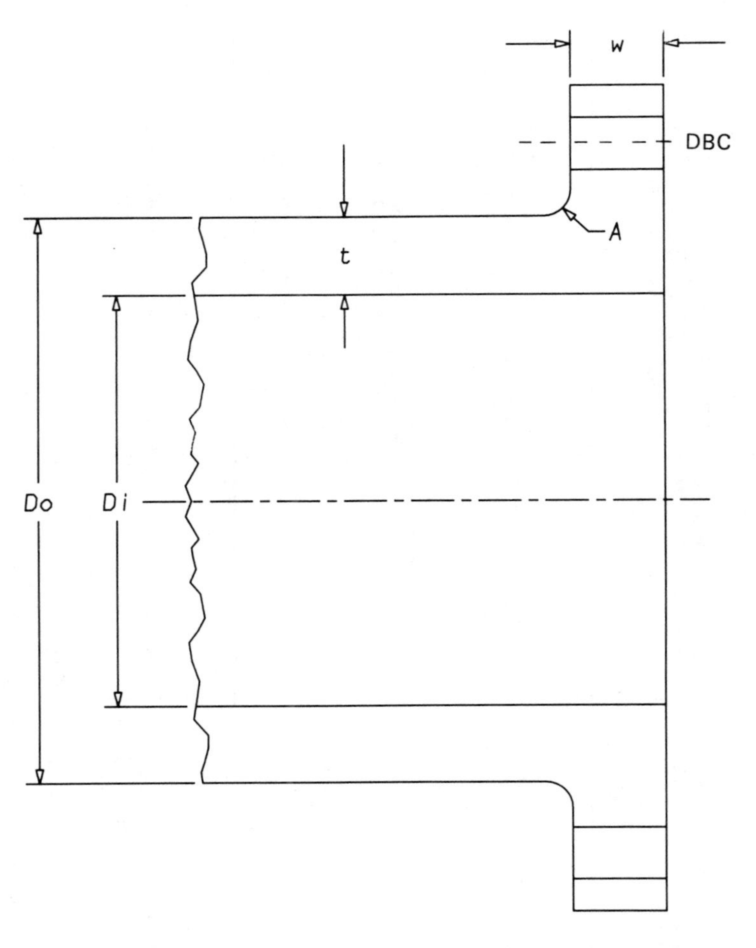

Figure 3.16 Flange dimensions for calculating flange stresses.

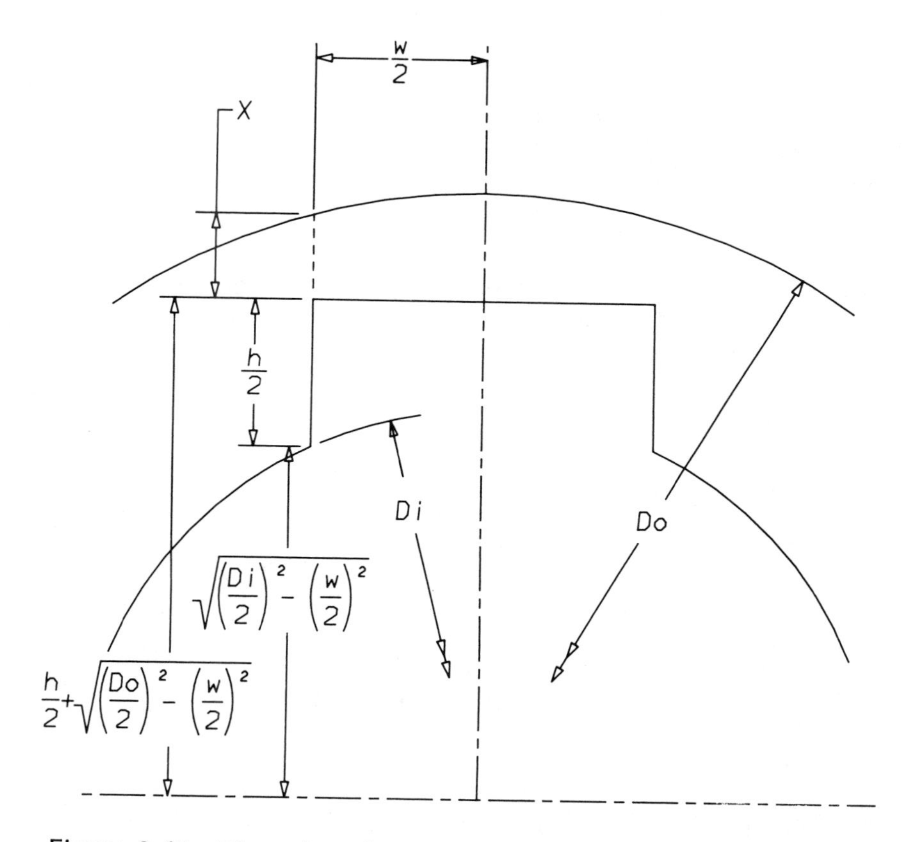

Figure 3.17 Dimensions for calculating hub bursting stress.

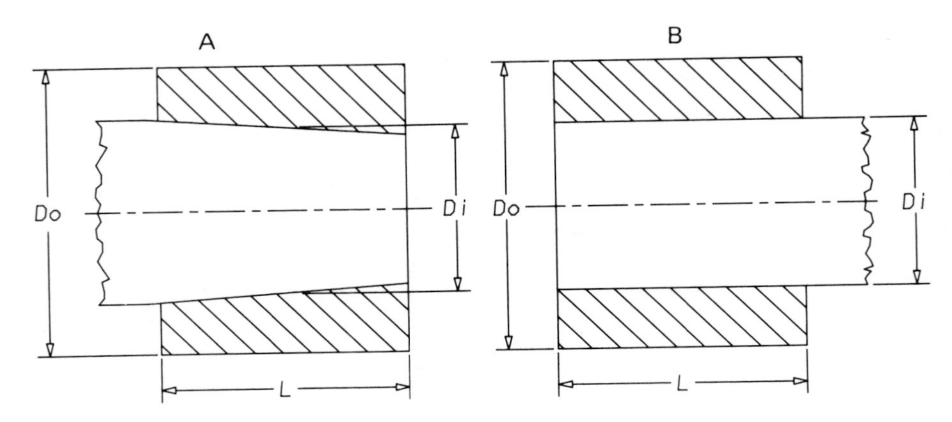

Figure 3.18 Dimensions for keyless hubs: (A) taper shaft; (B) straight shaft.

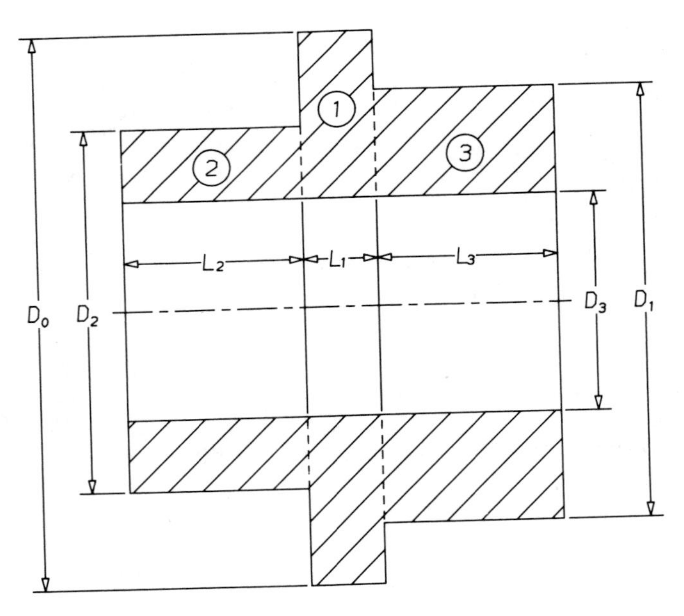

Figure 3.19 Dimensions for determining torque capacity of stepped or flanged hubs.

III. BALANCING OF COUPLINGS

Like coupling ratings, when and to what level to balance a coupling appear to be almost as mystifying. In reality, however, they are rather simple once a basic understanding is achieved of what contributes to unbalance and how it affects a system. Once this understanding is obtained, the arbitrary balance limits can be put into perspective for a system's needs. The values given in this section are only guidelines. Many of the terms, definitions, and information in this section are from American Gear Manufacturers Association (AGMA) 515.

A. Basics of Balancing

1. Units for Balancing or Unbalance. The unbalance in a piece of rotating equipment is usually expressed in terms of unbalance weight (ounces) and its distance from the rotating centerline (inches). Thus we get unbalance (U) in oz-in. terms.

Example 1

Unbalance weight is 2 oz; the distance from the rotating centerline is 2 in. (see Figure 3.20). Unbalance (U) equals 4 oz-in.

For couplings, it is more convenient to express unbalance in terms of mass displacement. This usually helps bring balance limits into perspective with tolerances, fits, and the "real world." Mass displacement is usually specified in microinches, mils, or inches [1000 microinches (μin.) = 1 mil = 0.001 in.].

Example 2

If a 100-lb part (1600 oz) is displaced 0.001 in. (1 mil or 1000 μin.) from the centerline (Figure 3.21), a 16 oz-in. unbalance (1600 oz × 0.001 in.) would occur.

2. Balancing. Balancing is a procedure by which the mass distribution of a rotating part is checked and, if required, adjusted (usually by metal removal, but on low-speed couplings, metal may be added) so that the unbalance force or equipment vibration is reduced.

3. Types of Unbalance

a. Single-Plane Unbalance (Sometimes Called Static Balance). This unbalance condition exists when the center of gravity of a rotating part does not lie on the axis of rotation (Figure 3.22). This part will not be in static equilibrum when placed at random positions about its axis. Single-plane balancing can be done without rotation (much like static balancing of an automobile wheel), but most couplings are balanced on centrifugal balancing machines, which produce a much better balance. Single-plane balancing is usually limited to relatively short parts, usually with a length less than $1.5 \times D_o$ (depending on part configuration), where D_o is the outside diameter of the part being balanced.

b. *Two-Plane (Moment, Coupling, or Dynamic) Unbalance.* This type of unbalance is present when the unbalance existing in two planes is out of phase, as shown in Figure 3.23, but not necessarily 180° out of phase. The principal intertial axis closest to the axis of rotation is displaced from the axis of rotation, and these two axes are askew with respect to each other. If not restrained by bearings, a rotating part with this type of unbalance will tend to rotate about its principal inertial axis closest to its geometric axis. If the moments are equal and opposite, they are referred to as a coupling unbalance. Figure 3.23 illustrates this case of two-plane unbalance.

4. Rigid Rotor. A rotor is considered rigid when it can be corrected in any two planes and after the part is balanced, its unbalance does not significantly exceed the unbalance tolerances limit at any other speed up to the maximum operating speed and when running conditions closely approximate those of the final system. A flexible coupling is an assembly of several components having diametral clearances and eccentricities between pilot surfaces of those components. Therefore, it is not appropriate to apply standards and requirements that were written to apply to rigid rotors in lieu of something more appropriate.

5. Axis of Rotation. The axis of rotation is a line about which a part rotates as determined by journals, fit, or other locating surfaces.

6. Principal Inertial Axis Displacement. The displacement of the principal inertial axis is the movement of the inertial axis that is closest to the axis of rotation with respect to the axis of rotation. In some special cases these two axes may be parallel. In most cases, they are not parallel and are therefore at different distances from each other in the two usual balancing planes.

7. Amount of Unbalance. The amount of unbalance is the measure of unbalance in a part (in a specific plane) without relation to its angular position.

8. Residual Unbalance. Residual unbalance is the amount of unbalance left after a part has been balanced. It is equal to or usually less than the balance limit tolerance for a part. (*Note*: A check to determine whether a part is balanced by removing it from the balancing machine and then replacing it will not necessarily produce the same residual unbalance as that originally measured. This is due to the potential differences in mounting and/or indicating surface runouts.)

9. Potential Unbalance. The potential unbalance is the maximum amount of unbalance that might exist in a coupling assembly after balancing (if corrected) when it is dissassembled and then reassembled.

10. Balance Limit Tolerance. The balance limit tolerance specifies a maximum value below which the state of unbalance of a coupling is considered acceptable. There are two types of balance limit tolerances:

1. Balance limit tolerance for residual unbalance
2. Balance limit tolerance for potential unbalance

11. Mandrel. A shaft on which the coupling components or assembly is mounted for balancing purposes is called a mandrel (see Figure 3.24).

12. Bushing Adapter. A bushing adapter (see Figure 3.24) or adapter assembly is used to mount the coupling components or the coupling assembly on the mandrel.

13. Mandrel Assembly. A mandrel assembly (see Figure 3.24) consists of one or more bushings used to support the part mounted on a mandrel.

14. Mounting Surface. A mounting surface is the surface of a mandrel, bushing, or mandrel assembly on which another part of the balancing tooling, a coupling component, or the coupling assembly is mounted. This surface determines the rotational axis of the part being balanced.

15. Mounting Fixtures. Mounting fixtures are the tooling that adapt to the balancing machine and provide a surface on which a component or coupling assembly is mounted.

16. Indicating or Aligning Surface. The indicating surface is the axis about which a part is aligned for the purpose of balancing. The aligning surface is the axis from which a part is located for the purpose of balancing. (*Note*: Some coupling manufacturers do not component-balance. Not all couplings can be component- or assembly-balanced without mandrels or mounting fixtures.)

17. Pilot Surface. A pilot surface is that supporting surface of a coupling component or assembly upon which another coupling component is mounted or located. Examples are shown in Figure 3.25.

B. Why a Coupling Is Balanced

One important reason for balancing a coupling is that the forces created by unbalance could be detrimental to the equipment, bearings, and support structures. The amount of force generated by unbalance is

$$F = 1.77 \times \left(\frac{\text{rpm}}{1000}\right)^2 \times \text{oz-in.} = \text{lb} \tag{3.91}$$

$$F = \frac{\text{microinches}}{1{,}000{,}000} \times \text{weight} \times 16 \times \left(\frac{\text{rpm}}{1000}\right)^2 = \text{lb} \tag{3.92}$$

As you can see from these equations, the amount of force generated by unbalance is proportional to the square of the speed.

Example 3

Assume 2 oz-in. of unbalance at 2000 rpm and at 4000 rpm:
F at 2000 rpm - 14.1 lb
F at 4000 rpm = 56.6 lb
Therefore, if speed doubles, the same amount of unbalance produces four times the force.

Another important reason for balancing is vibration. Sometimes this unwanted vibration produces poor product quality. For example, spindle coupling vibration can produce chatter and rough finish on the rolled strip or sheet being produced, making the product unacceptable. The next question is: If balance is all that important, why not forget all these complications and just balance all couplings? However, unless you have an unlimited budget, please read on.

There are several reasons why manufacturers do not balance all couplings. To balance a coupling in a balancing machine as an assembly, a shaft or a fixture is needed. This means that the coupling must be assembled on this special shaft or fixture and then balanced. (*Note*: Some couplings, including disk and diaphragm couplings, can be rigidized and balanced without a shaft or fixture.) This is expensive; also, the shaft or fixture does not exactly resemble the equipment shaft, so inherent errors are introduced into the balancing. Sometimes this is greater than the original "as manufactured" potential unbalance. Some coupling manufacturers do not assemble the entire coupling on a shaft or fixture, but component-balance parts by aligning the indicating diameters of these individual parts—which again is costly and introduces some inherent error from this eccentricity of the aligning diameters.

We must always remember that a coupling is an assembly of components that are taken apart and put back together for assembly of equipment, maintenance of the coupling and the equipment, and so on. With the assembly and disassembly of coupling components the relative position of mating parts can change and therefore the coupling's state of unbalance will change. Both component and assembly balance can usually produce a coupling with equal potential unbalance, sometimes not what the coupling selector asked for but generally sufficient.

On very sensitive equipment the inherent errors introduced from the balancing method and assembly/disassembly unrepeatability may be far greater than the balance tolerance limits placed on the coupling and in some cases the actual requirements. When the real balance needs exceed practicality (potential unbalance capabilities), the coupling must be balanced on the actual equipment, which is very expensive and time consuming.

In summary, the basic reason why all couplings are not balanced is because a balanced coupling costs more, not only because of balance time and equipment but in most cases to assure some repeatability of the coupling potential unbalance, the tolerances and fits must be tightened, which can greatly increase the cost of the coupling. For example, to manufacture gear couplings with half the standard tolerances would cost 150 to 200% more.

C. What Contributes to Unbalance in a Coupling

1. Contributors to Potential Unbalance of Uncorrected Couplings

a. *Inherent unbalance*: If the coupling assembly or components are not balanced, an estimate of inherent unbalance caused by manufacturing tolerances may be based on one of the following:
   (1) Statistical analysis of balance data accumulated for couplings manufactured to the same tolerances, or
   (2) Calculations of the maximum possible unbalance that could theoretically be produced by the tolerances placed on the parts.

b. *Coupling pilot surface eccentricity*: This is any eccentricity of a pilot surface that permits relative radial displacement of the mass axis of mating coupling parts or subassemblies.

c. *Coupling pilot surface clearance*: This is the clearance that permits relative radial displacement of the mass axis of the coupling components or subassemblies. [*Note*: Some couplings must have clearances in order to attain their flexibility (e.g., gear couplings).]

d. *Hardware unbalance*: Hardware unbalance is the unbalance caused by all coupling hardware, including fasteners, bolts, nuts, lockwashers, lube plugs, seal rings, gaskets, keys, snap rings, keeper plates, thrust plates, and retainer nuts.

e. *Others*: Many other factors may contribute to coupling unbalance. The factors mentioned are those of primary importance.

2. Contributors to Potential Unbalance of Balanced Couplings

a. *Balance tolerance limits*: These limits are the largest amount of unbalance (residual) for which no further correction need be made.

b. *Balance fixtures of mandrel assembly unbalance*: This is the combined unbalance caused by all components used to balance a coupling, including mandrel, flanges, adapters, bushings, locking devices, keys, setscrews, nuts, and bolts.

c. *Balance machine error*: The major sources of balancing machine error are overall machine sensitivity and error of the driver itself. This unbalance is usually very minimal and can generally be ignored.

d. *Mandrel assembly, mounting surface eccentricity*: This is the eccentricity, with respect to the axis of rotation, of the surface of a mandrel assembly upon which the coupling assembly or component is mounted.

e. *Component or assembly indicating surface eccentricity*: This is the eccentricity with respect to the axis of rotation of a surface used to indicate or align a part on a balancing machine. On some parts this surface may be machined to the axis of rotation rather than indicated on balancing machines.

f. *Coupling pilot surface eccentricity*: This is an eccentricity of a pilot surface that permits relative radial displacement of the mass axis of another coupling part or subassembly, upon assembling subsequent to the balancing operation. This eccentricity is produced by manufacturing before balancing or by alternation of pilot surfaces after balancing. For example, most gear couplings that are assembled balanced are balanced with an interference fit between gear major diameters and after balancing are remachined to provide clearance. This may produce eccentricity.

g. *Coupling pilot surface clearance*: This is the clearance that permits relative radial displacement of the mass axes of the coupling component or subassemblies on disassembly/reassembly. The radial shift affecting potential unbalance is equal to half this clearance. (*Note*: If this clearance exists in a coupling when it is balanced as an assembly, the potential radial displacement affecting potential unbalance is equal to the full amount of the diametral clearance that exists in the assembly at the time of balance.)

h. *Hardware unbalance*: This is the unbalance caused by all coupling hardware, including fasteners, bolts, nuts, lockwashers, lube plugs, seal rings, gaskets, keys, snap rings, keeper plates, thrust plates, and retainer nuts.

i. *Others*: Many other factors may contribute to coupling unbalance.

D. How to Bring a Coupling into Balance

Couplings can be brought into balance by four basic methods:

1. Tighter manufacturing tolerances
2. Component balancing
3. Assembly balancing
4. Field balancing on the equipment

1. Tighter Manufacturing Tolerances. The majority of unbalance in most couplings comes from the tolerances and the clearance fits that are placed on components so that they can be competitively produced yet be interchangeable. Most couplings are not balanced.

The amount of unbalance can be greatly improved by tightening fits and tolerances. For example, by cutting the manufacturing tolerances and fits approximately in half, only half the amount of potential unbalance remains. If a coupling is balanced without changing the tolerance, the unbalance is improved by only 5 to 20%. Remember that the tighter the tolerances and the fits, the more couplings are going to cost. A real-life practical limit is reached at some point. As tolerances are tightened, the price of the coupling not only increases, but interchangeability is lost, delivery is extended, and in some instances the choice of couplings and potential vendors is limited. Figure 3.26 shows how tolerances affect cost.

2. Component Balancing. Component balance is usually best for couplings that have inherent clearances between mating parts or require clearances in balancing fixtures. Component balance offers interchangeability of parts usually without affecting the level of potential unbalance. In most cases component-balanced couplings can approach the potential unbalance limits of assembly-balanced couplings. In some cases it can produce potential unbalance levels lower; this is particularly true where large, heavy mandrels or fixtures must be used to assembly-balance a coupling.

3. Assembly Balancing of Couplings. Assembly balancing may provide the best coupling balance. This is true when no clearance exists between parts (e.g., disk or diaphragm couplings). The balancing fixture and mandrels are lightweight and are manufactured to extremely tight tolerances: 0.0003 to 0.0005 *total indicator runout* (TIR). This is 300 to 500 μin., or the coupling is somehow locked or rigidized without the use of a fixture or a mandrel. For relatively large assemblies the mandrels and fixtures may introduce more error than if the coupling were not balanced at all. It is possible to balance into the coupling more unbalance than the original unbalanced coupling had. Assembly balanced couplings are matchmarked and components should not be interchanged or replaced without rebalancing.

4. Field Balancing. On very high speed and/or lightweight equipment, it may not be possible to provide a balanced coupling to meet the requirements due to the inherent errors introduced when balancing a coupling on a balancing machine. When this occurs, the coupling manufacturer can provide a means whereby weight can easily be added to coupling so that coupling can be balanced by trial and error on the equipment itself. Couplings can be provided

with a balancing ring with radial setscrews and tapped holes, so bolts with washers can be added or subtracted. There are also other means that can be used.

E. When to Balance and to What Level

The amount of coupling unbalance that can be satisfactorily tolerated by any system is dictated by the characteristics of the specific connected machines and can best be determined by detailed analysis or experience. Systems that are insensitive to coupling unbalance might operate satisfactorily with values of coupling balance class lower than those shown. Conversely, systems or equipment that are usually sensitive to coupling unbalance might require a higher class than suggested. Factors that must be considered in determining the system's sensitivity to coupling unbalance include:

1. *Shaft end deflection*: Machines having long and/or flexible shaft extensions are relatively sensitive to coupling unbalance.
2. *Bearing loads relative to coupling weight*: Machines having lightly loaded bearings or bearing loads primarily by the overhung weight of the coupling are relatively sensitive to coupling unbalance. Machines having overhung rotors or weights are often sensitive to coupling unbalance.
3. *Bearings, bearing supports, and foundation rigidity*: Machines or systems with flexible foundations or supports are relatively sensitive to the coupling unbalance.
4. *Machine separation*: Systems having a long distance between machines often exhibit coupling unbalance problems.
5. *Others*: Other factors may influence coupling unbalance sensitivity.

Figure 3.27 is taken from American Gear Manufacturers Association (AGMA) 515. In general, selection bands can be put into the following speed classifications:

Low speed: A and B
Intermediate speed: C, D, and E
High speed: F and G

Figure 3.28 is also taken from AGMA 515. Superimposed on this graph are the three most common speed classifications. The graph has also been extended to 2000 lb.

1. Coupling Balance Limits. The balance limit placed on a coupling should be its *potential* unbalance and not its residual unbalance

limit. The residual unbalance limit usually has little to do with the *true* coupling unbalance (potential). It can be seen in many cases that by cutting the residual unbalance limit in half, the coupling potential unbalance may change by only 5%. The best method of determining the potential unbalance of a coupling is by the square root of the sum of the squares of the maximum possible unbalance values. These unbalances are mostly from the eccentricities between the coupling parts but include any other factors that produce unbalance. The coupling balance limit is defined by AGMA 515 as a range of unbalance expressed in microinches. The potential unbalance limit classes for couplings are given in Figure 3.29. They are given in maximum root-mean-square (rms) microinches of displacement of the inertia axis of rotation at the balance plane. Limits are given as per displacement plane. The residual unbalance limit of a part or an assembly balanced coupling is shown in Figure 3.30. Tolerances tighter than these usually do very little to improve the overall potential unbalance of a coupling assembly.

F. Other Coupling Balancing Criteria

What is meant by "arbitrary" balancing criteria? The limits (potential and residual) described in the preceding section give the most realistic values for unbalance limits. There are other criteria presently being used, generally referred to as *arbitrary limits*. They are used in several coupling specifications. The most common is to express unbalance (U) as

$$\text{oz-in. (unbalance)} = \frac{K \times W}{N} = U \text{ per balanced plane} \qquad (3.93)$$

where

K = 40 to 120 for potential unbalance limits; 4 to 12 for residual unbalance limits
W = weight of the part per balance plane (lb)
N = operating speed of coupling (rpm)

The two most common values for these arbitrary limits are given in American Petroleum Institute (API) Standard 671:

Residual limit:

$$\text{oz=in.} = \frac{4W}{N} \qquad (3.94)$$

Potential limit:

$$\text{oz-in.} = \frac{40W}{N} \tag{3.95}$$

In general, these limits are not too bad, but in some cases if users are not careful they may get a very expensively balanced coupling when they do not need one or pay extra and not gain anything. In other cases they will be specifying limits beyond the real world of practicality and will be faced with arguments, delays, and so on, while everyone involved regroups and tries to blame everyone else for not specifying correctly and/or not balancing correctly.

1. Arbitrary Potential Unbalance Limits. Applying the arbitrary limits to the three coupling speed classes results in the following values:

Speed class	Unbalance limit (oz-in.)
Low speed	120 W/N
Intermediate speed	80 W/N
High speed	40 W/N

Stated as before, it is best to put unbalance in terms of microinches. If this is done, another arbitrary criterion results, but now if a low limit tolerance is applied (in microinches), we can assure that the balance limit has some relationship to the real world of manufacturing the coupling with the requisite tolerances and fits.

Speed class	Unbalance limit[a] (μin.)
Low speed	7,500,000/N or 4000
Intermediate speed	5,000,000/N or 2000
High speed	2,500,000/N or 500

[a]The highest values become limits.

2. Arbitrary Residual Unbalance Limits. The residual unbalance limits are approximately one-tenth of the potential unbalance limit.

Speed class	Unbalance limit (oz-in.)	Unbalance limit[a] (μin.)
Low speed	12 W/N	750,000/N or 400
Intermediate speed	8 W/N	500,000/N or 200
High speed	4 W/N	250,000/N or 50

[a]The highest values become limits.

G. Types of Coupling Balance

Most coupling manufacturers can and will supply couplings that are brought into balance. Some coupling manufacturers have what they call their "standard balancing procedure and practice" and if you request other than their standard practice they will charge extra. For example, some coupling manufacturers prefer to supply component-balanced couplings, whereas others prefer assembly balanced couplings. What do these balance types mean?

1. *As manufactured*: Most couplings are supplied as manufactured with no balancing.
2. *Controlled tolerances and fits*: Usually provides the most significant improvement in the potential unbalance of coupling. This can also substantially increase the price of a coupling if carried too far.
3. *Component balance* (see Figures 3.31 and 3.32): Can usually produce potential unbalanced values equal to assembly balanced couplings. Offers the advantage of being able to replace components (as related to balance—some couplings are not interchangeable for other reasons) without rebalancing.
4. *Assembly balancing with mandrels or fixtures* (see Figures 3.33 and 3.34): Usually offers the best balance, but is usually expensive because the mandrels and/or fixtures must be made extremely accurately. The coupling is basically rigidized with the mandrel or fixtures and then balanced. On assembly-balanced couplings, parts cannot be replaced without rebalancing the coupling.
5. *Component balancing with selective assembly*: Sometimes offers the best possible balance attainable without field balancing on the equipment. Parts are component-balanced and then runout (TIR) is checked. The highs of the TIR readings between controlling diameters for mating parts are marked. At final coupling assembly, the high spots are assembled 180° out

of phase. This tends to negate eccentricities and reduces the potential unbalance of the coupling. The parts are still interchangeable as long as replacement parts are inspected and marked for their high TIRS.

6. *Assembly balancing without a mandrel* (see Figure 3.35): This is usually limited to disk, diaphragm, and some types of gear couplings. The coupling is locked rigid with various locking devices, which are usually incorporated into the coupling design. The coupling is rolled on rolling surfaces that are aligned or machined to the coupling bores or alignment pilots. This type of balance can usually provide a better balance than with a mandrel. This is because there is no weight added to the assembly when it is balanced. On very large and long couplings a mandrel assembly can weight almost one-half that of a coupling. This can introduce very significant balancing errors.
7. *Field or trim balancing on the equipment*: Offers the best balance, but it is usually the most costly, because of the trial and error and the time involved. The coupling cannot be disassembled/reassembled without rebalancing.

H. Example of Coupling Balance Selection

The best way to help you understand how to determine the coupling balance requirement is to go through an example case and some calculations.

1. *Conditions*: You are a pump manufacturer and have two units that consist of a motor, gear, and pump. You are to supply the couplings between the gears and the pumps.

2. *Operating data*: Unit 1 operates at 6000 rpm. Unit 2 operates at 9000 rpm.

3. *Coupling data*: You have done your homework and have selected a gear coupling. You have received quotes from brand Y and brand Z. Brand Y's and Z's coupling both weight 100 lb. The brand Z manufacturer said in their quote that the coupling conforms to an AGMA class 9 balance level. The brand Y manufacturer did not give you any information on the balance quality of their coupling so you go back and ask. They then tell you that brand Z also conforms to AGMA class 9.

4. *What level of balance do you need for these couplings?* Using Figure 3.28, you can find which band these applications fall into. Unit 1 falls into band D. Unit 2 falls into band E.

Next, you go to Figure 3.27 to find which balance class is required. This requires that you decide how sensitive the equipment is. You decide, based on your past experience with similar equipment, that it is a relatively average system. Therefore, you can now determine the coupling balance class required. Unit 1 requires a class 9 or, from Figure 3.29, 2000 μin. Unit 2 requires a balance class 10 or, from Figure 3.29, 1000 μin.

5. *What type of coupling do you need?* Unit 1 can use the coupling as originally quoted. Unit 2 needs a class 10.

So you decide that you had better buy a balanced coupling. You go back to the manufacturers of brands Y and Z and tell them to quote on a balanced coupling and tell you what balance class it would be. The new quotes state that these couplings would cost 25 to 50% more but that the balance class is still a class 9.

6. *You now need calculations*: The brand Y calculations are shown in Examples 4 and 5. The brand Z calculations are shown in Examples 4 and 6. You review the calculations and find:

a. As manufactured, both brand Y and brand Z couplings have a potential unbalance of approximately 1765 μin.
b. An assembly-balanced coupling has a potential unbalance of 1456 μin (brand Y).
c. A component balanced coupling has a potential unbalance of 1405 μin (brand Z).
d. Therefore, it becomes evident that just balancing these couplings will not meet class 10; in fact, it improves the potential unbalance by only 17 to 20%.

7. *Now what?* Let's get the coupling manufacturers to balance to tighter tolerances. You do this on the pump rotors. Let's plug this into the calculations: Tighter balancing tolerances for an assembly-balanced coupling gives 1452; tighter balancing tolerances for a component-balanced coupling gives 1402. This cannot be. You have improved balance by only 0.15 to 0.2%.

8. *This is becoming tedious*. So you sit and scratch your head and review all those calculations again. You see that the biggest influence on the coupling potential unbalance are the eccentricities and the pilot clearances inherent in these couplings.

Let's see what happens if you cut these tolerances in half: Coupling not balanced, 962 μin.; coupling assembly-balanced 943 μin.; coupling component-balanced 859 μin.

9. *Get a quote for a tighter-toleranced couplings*. You ask for quotes on couplings that meet AGMA class 10 (these would be couplings built with tighter tolerances). The quote comes back with couplings costing 250 to 300% more. However, what choice do you have?

You review the calculations (see Figure 3.36). From these it is clear that you really do not need to balance this new type of coupling, but you are going to have trouble explaining this. So you buy a component-balanced coupling made to tighter tolerances and hope nobody asks you why a balanced coupling cost so much.

It is hoped that this example helped you better understand coupling balance requirements and limits. The balance limits (potential or residual) for coupling should be specified in microinches of mass displacement. This helps put balance limits into a real-world perspective. Specifying arbitrary limits without analyzing the system's real needs may increase the coupling cost unnecessarily or preclude the use of some perfectly acceptable couplings.

It is important to remember that a coupling is an assembly of parts that must be taken apart for assembly, maintenance of equipment, and self-maintenance. Its balance level is most affected by its ability to be assembled and disassembled repeatedly without changing the various mass eccentricities of its parts.

Example 4: Uncorrected Gear Coupling

Coupling geometry and assumed data (see Example Figure 1 (p. 133))

1. Weights:
   a. Coupling assembly = 100 lb
   b. Coupling component weights
      Hubs: 2 at 23 lb = 46 lb
      Sleeves: 2 at 26 lb = 52 lb
      Hardware: 1 set = 2 lb
2. Coupling pilot surface eccentricity:
   Max. eccentricity of hub major diameter = 0.004 in. TIR
3. Coupling pilot surface clearance:
   Hub major diameter to sleeve root clearance (radial) = 0.0015 in.
4. Inherent unbalance of coupling components: Potential inherent unbalance of the hub and sleeve components is calculated for the assumed manufacturing tolerances shown on the sketches of the components. The fasteners are assumed to have a uniform weight, but are installed in holes which allow them to be eccentric by one-half the eccentricity shown for datum -K-. This includes 0.005 in. inaccuracy in the bolt hole position.

Calculated potential unbalance of coupling hub (see Example Figures 2 and 3 (p. 134)):

Datum	Maximum runout to -B- (TIR)	Maximum eccentricity to -B-	Weight of section (lb)	Maximum potential unbalance	
				oz-in.	oz-in.^{2}
A	0.004	0.002	12	0.384	0.147
B	—	—	—	—	—
C	0.008	0.004	4	0.256	0.065
D	0.004	0.002	4	0.128	0.0164
E	0.004	0.004	3	0.192	0.0368
Totals				0.96	0.265

Rms value of potential unbalance of hub = $\sqrt{0.265}$ = 0.5147 oz-in.

Datum	Maximum runout to -J- (TIR)	Maximum eccentricity to -J-	Weight of section (lb)	Maximum potential unbalance	
				oz-in.	oz-in.^{2}
F	0.004	0.002	8	0.256	0.065
G	0.006	0.003	5	0.240	0.057
H	0.006	0.003	3	0.144	0.021
J	—	—	—	—	—
K	0.014	0.007	1[a]	0.112	0.0126
L	0.008	0.004	10	0.64	0.4096
Totals				1.402	0.565

Rms value of potential unbalance of sleeve and hardware = $\sqrt{0.565}$ = 0.752 oz-in.

[a]one lb is one-half the weight of the hardware.

Calculation	Unbalance per correction plane	
	oz-in.	oz-in.^{2}
Inherent unbalance of uncorrected coupling		
Hub	0.5147	0.265
Sleeve and hardware	0.752	0.565
Coupling pilot surface eccentricity: Runout of hub major diameter (26 + 1)(16)(0.004/2)	0.864	0.746
Coupling pilot surface clearance: tooth pilot clearance (radial) after mounting hub (26 + 1)(16)(0.0015)	0.648	0.419
Hardware unbalance	—	—
Others	—	—
Totals	2.778	1.995

Rms value of potential unbalance = $\sqrt{1.995}$ = 1.412 oz-in.

Rms displacement of principal inertia axis = 1.412/50 × 16 = 1765 μin.

Example 5: Gear Coupling Balanced as an Assembly

Coupling geometry and assumed data (see Example Figure 4 (p. 135)):

1. Weights:
   a. Total coupling weight = 100 lb
   b. Coupling component weights
      Flex hubs: 2 at 23 lb = 46 lb
      Sleeves: 2 at 26 lb = 52 lb
      Hardware: 1 set = 2 lb
   c. Balancing tooling weights
      Mandrel = 30 lb
      Bushings: 2 at 2 lb = 4 lb
2. Balance correction tolerance = 100 μin. per plane
3. Balance machine error:
   a. Sensitivity = 10 μin. per plane
   b. Driver error assumed to be negligible

4. Balancing mandrel assembly unbalance = 10 μin per plane
5. Dial indicator graduation = 0.0001 in.
6. Mandrel assembly mounting surface eccentricity:
   Mandrel runout at mounting surface for bushing = 0.0005 in. TIR
   Bushing bore to bushing mounting surface for hub runout = 0.0004 in. TIR
   Mandrel to bushing clearance = 0.0005 in.
   Bushing mounting surface to hub bore clearance = 0.0005 in.
7. Coupling pilot surface eccentricity: It is assumed that parts are match marked and therefore the balancing operation compensates for these runouts (hub bore to tooth tip and sleeve root to sleeve pilot for each sleeve). After balancing the tips of hubs are machined to hub bore to 0.004 TIR.
8. Coupling pilots surface clearance:
   Tooth tip pilot clearance = 0.003 after balancing, 0.000 before balancing
9. Hardware unbalance:
   Fasteners (deviation per set) = 0.2 g
   Bolt circle diameter = 9.75 in.

Calculation	Unbalance per correction plane	
	oz-in.	$oz\text{-}in.^2$
Balance correction tolerance		
$\frac{100}{2} \times 16 \times 0.0001$	0.08	0.0064
Balance machine error		
Sensitivity		
$\frac{134}{2} \times 16 \times 0.00001$	0.01072	0.0001149
Driver error assumed to be negligible		
Balancing mandrel assembly unbalance		
Mandrels	not applicable	not applicable
Bushings		
Mandrel assembly		
$\frac{30 + 2 + 2}{2} \times 16 \times 0.00001$	0.00272	0.0000074

Calculation	Unbalance per correction plane	
	oz-in.	oz-in.^{2}
Mandrel assembly mounting surface eccentricity		
Mandrel runout		
$\frac{100}{2} + 2 \times 16 \times \frac{0.0001 + 0.0005}{2}$	0.2496	0.0623
Mandrel to bushing clearance		
$52 \times 16 \times \frac{0.0005}{2}$	0.208	0.043
Bushing bore to mounting surface runout		
$\frac{100}{2} \times 16 \times \frac{0.0001 + 0.0004}{2}$	0.200	0.0400
Bushing to hub bore clearance		
$\frac{100}{2} \times 16 \times \frac{0.0005}{2}$	0.200	0.040
Coupling pilot surface eccentricity (see assumed data, item 7)	—	—
Coupling pilot surface clearance: tooth tip clearance		
$\frac{26 + 26 + 2}{2} \times 16 \times \frac{0.003}{2}$	0.648	0.419
Eccentricity of hub after remachining		
$(26 + 1) \times 16 \times \frac{0.004}{2}$	0.862	0.746
Hardware unbalance		
$\frac{0.2 \text{ g } 4.875}{28.35} \times \frac{\sqrt{2}}{2}$	0.0244	0.00059
Others		
Totals	2.485	1.357

Rms value of potential unbalance = $\sqrt{1.357}$ = 1.165 oz-in.

Rms displacement of principal inertial axis = 1.165/50 × 16 = 1456 μin.

Example 6: Gear Coupling Balanced as Individual Components
Coupling geometry and assumed data (see Example Figure 4):

1. Weights:
   a. Total coupling weight = 100 lb
   b. Coupling component weights
      Flex hubs: 2 at 23 lb = 46 lb
      Sleeves: 2 at 26 lb = 52 lb
      Hardware: 1 set = 2 lb
2. Balance correction tolerance = 100 μin. per plane
3. Balance machine error:
   a. Sensitivity = 10 μin. per plane
   b. Driver error assumed to be negligible
4. Dial indicator graduation = 0.0001 in.
5. Coupling pilot surface eccentricity: Maximum eccentricity of hub major diameter = 0.004 in. TIR
6. Coupling pilot surface clearance: Tooth tip pilot clearance = 0.003 in.
7. Hardware unbalance:
   Fasteners (deviation per set) = 0.2 g
   Bolt circle diameter = 9.75 in.

Calculation	Unbalance per correction plane oz-in.	Unbalance per correction plane oz-in.^{2}
Balance correction tolerance		
$\frac{100}{2} \times 16 \times 0.0001$	0.08	0.0064
Balance machine error		
Sensitivity		
$\frac{100}{2} \times 16 \times 0.00001$	0.008	0.00006
Indicating surface eccentricity		
Hub bore		
$23 \times 16 \times \frac{0.0001 + 0.001}{2}$	0.2024	0.0409
Sleeve major diameter		
$26 \times 16 \times \frac{0.0001 + 0.001}{2}$	0.2288	0.0523

Calculation	Unbalance per correction plane	
	oz-in.	$oz\text{-}in.^2$
Coupling pilot surface eccentricity:		
Runout of hub major diameter		
$(26 + 1) \times 16 \times \frac{0.004}{2}$	0.864	0.746
Coupling pilot surface clearance:		
Tooth pilot clearance		
$(26 + 1) \times 16 \times \frac{0.003}{2}$	0.648	0.419
Hardware unbalance		
$\frac{0.2\ g \times 4.875}{28.35} \times \frac{\sqrt{2}}{2}$	0.0244	0.00059
Others		
Totals	2.055	1.265

Rms value of potential unbalance = $\sqrt{1.265}$ = 1.124 oz-in.

Rms displacement of principal inertia axis = 1.124/50 × 16 = 1405 μin.

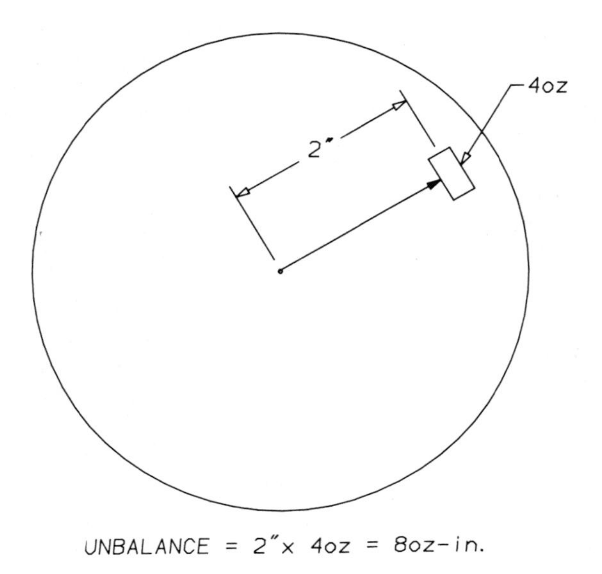

Figure 3.20 Unbalance units.

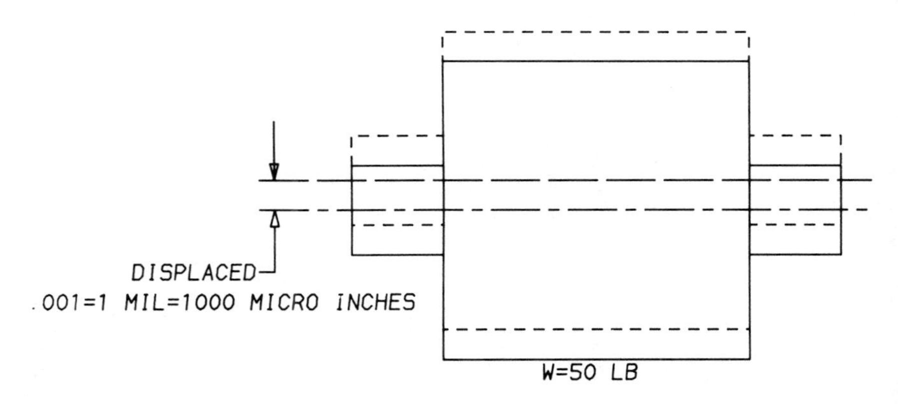

Figure 3.21 Displaced mass.

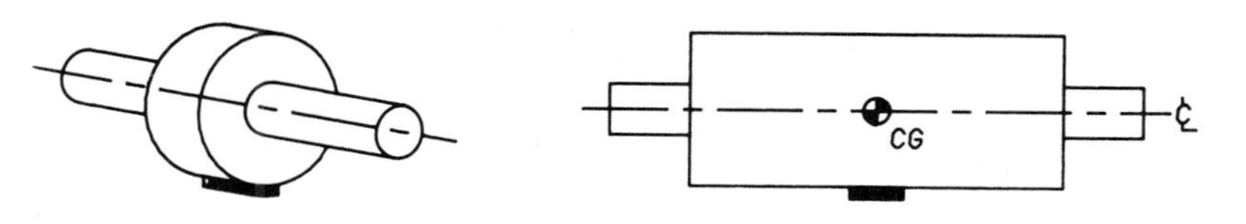

Figure 3.22 Single-plane unbalance.

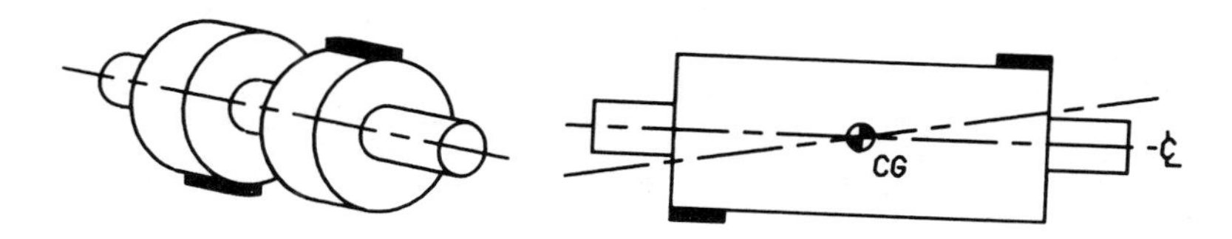

Figure 3.23 Two-plane unbalance.

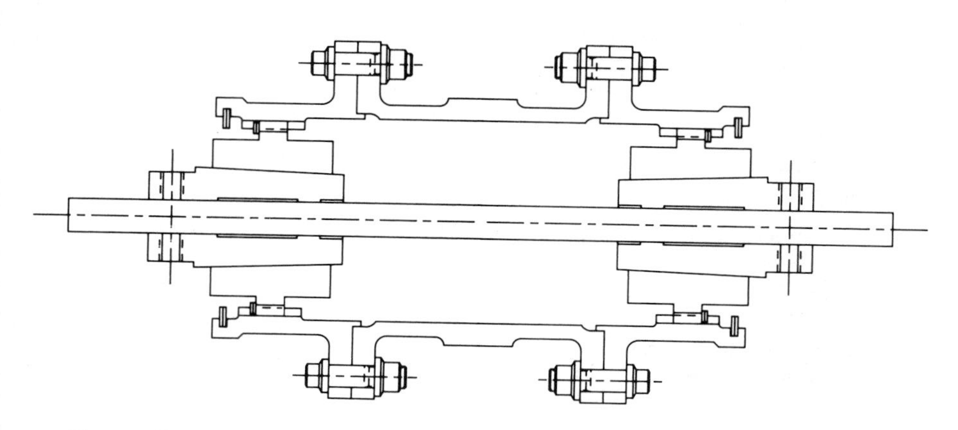

Figure 3.24 Mandrel assembly for balancing a gear coupling.

Components	Usual pilot surfaces
Rigid hub	Bore, rabbet diameter, bolt circle
Flex hub (gear type)	Bore, hub body OD, tooth tip diameter
Flanged sleeve (gear type)	Tooth root diameter, end ring I.D., rabbet diameter, bolt circle
Flanged adapter	Rabbet diameter or bolt circle
Flanged stub end adapter	Stub end (shaft) diameter, rabbet diameter, bolt circle
Spool spacer (gear type)	Tooth tip diameter
Flanged spacer	Rabbet diameter or bolt circle
Ring spacer	Rabbet diameter or bolt circle

Typical examples of coupling pilot surface

Figure 3.25 Types of coupling pilot surfaces.

	General tolerances[a]	Important tolerances[b]	DBC true location	Approximate cost
Low speed	± 1/64	± .005	± 1/64	1
Intermediate speed	± .005	± .002	± .005	1.5–2
High speed	± .002	± .001	± .005	3–4

[a]General tolerance applied to non-critical diameters and lengths (example: coupling OD).
[b]Important tolerances apply to critical diameters and lengths (example: bores, pilots).

Figure 3.26 Coupling tolerances versus cost.

Selection Band	Typical AGMA Coupling Balance Class; System Sensitivity to Coupling Unbalance		
	Low	Average	High
A	5	6	7
B	6	7	8
C	7	8	9
D	8	9	10
E	9	10	11
F	10	11	12
G	11	12	—

Figure 3.27 Coupling balance classes versus selection bands.

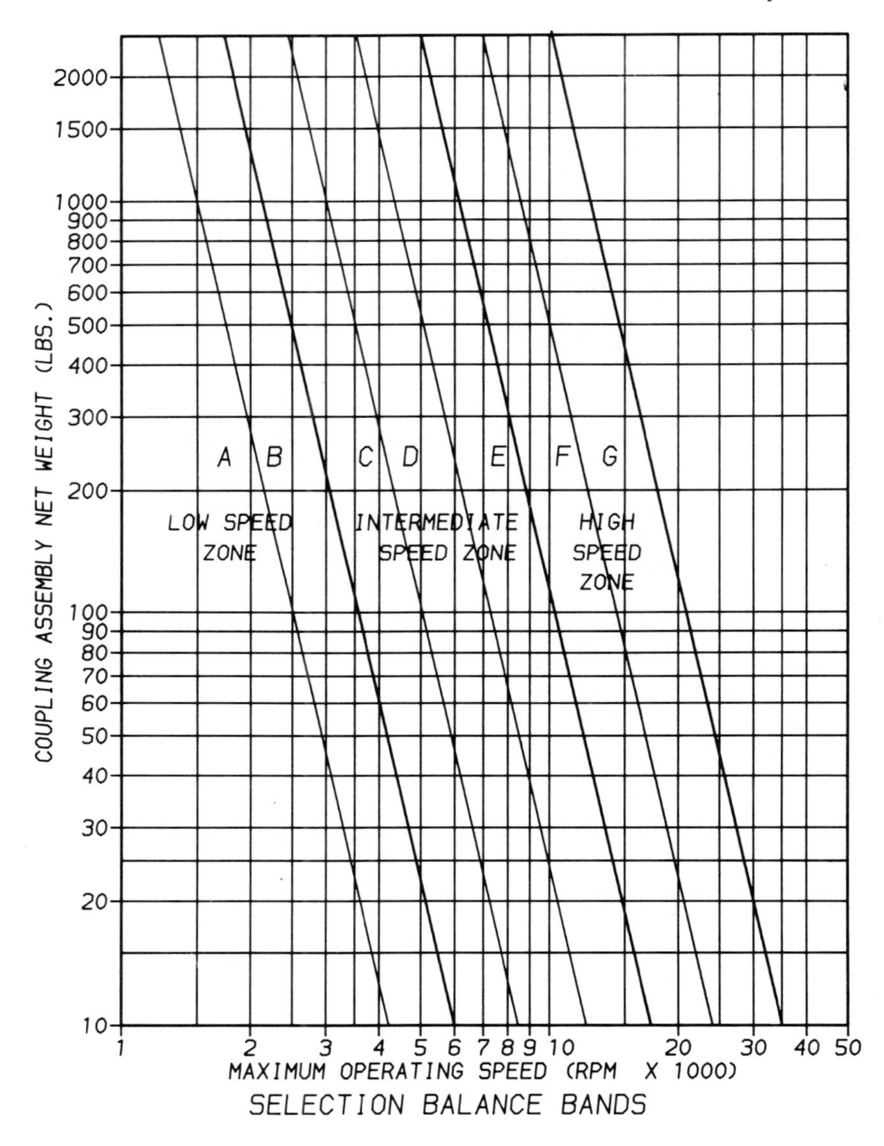

Figure 3.28 Balance selection weight versus speed.

AGMA coupling balance class	Maximum displacement of principal inertia axis at balancing planes (rms microinches)
4	Over 32,000
5	32,000
6	16,000
7	8,000
8	4,000
9	2,000
10	1,000
11	500
12	250

Figure 3.29 AGMA coupling balance classes.

Speed class	Residual unbalance limits (microinches)
Low speed couplings	500[a]
Intermediate speed couplings	200
High speed couplings	50

[a]Low speed couplings are usually not balanced.

Figure 3.30 Practical residual unbalance limits.

Figure 3.31 Component balancing of a geared spacer (courtesy of Zurn Industries, Inc., Mechanical Drives Division).

Figure 3.32 Component balancing of a hub and spacer for a gear coupling (courtesy of Zurn Industries, Inc., Mechanical Drives Division).

Figure 3.33 Assembly balancing of a gear coupling (courtesy of Sier-Bath Gear Co., Inc.).

Figure 3.34 Assembly balancing of a universal joint (courtesy of Spicer Universal Joint Division of Dana Corporation).

Figure 3.35 Assembly balancing of a diaghragm coupling (without mandrel) (courtesy of Zurn Industries, Inc., Mechanical Drives Division).

	Potential unbalance	Improvement
Unbalanced standard tolerances	1765	
Assembly balanced	1456	17.5%
Component balanced	1405	20.4%
Assembly balanced tighter tolerances on balance	1452	17.7%
Component balanced tighter tolerances on balance	1402	20.5%
Unbalanced tighter tolerances	962	45.5%
Assembly balanced tighter tolerances	943	46.6%
Component balanced tighter tolerance	859	51.3%

Figure 3.36 Summary of coupling balance selection example.

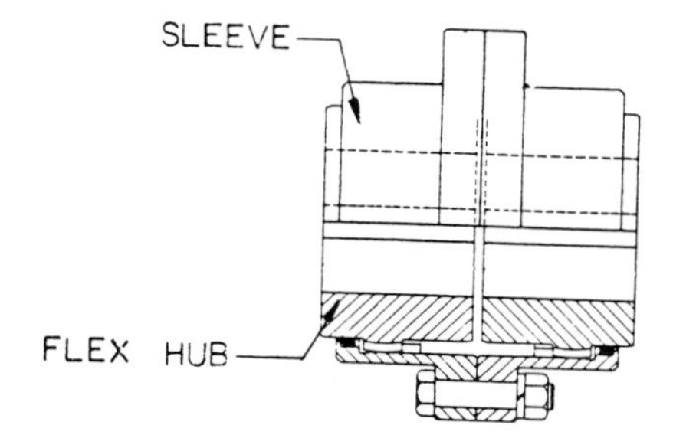

Example Figure 1

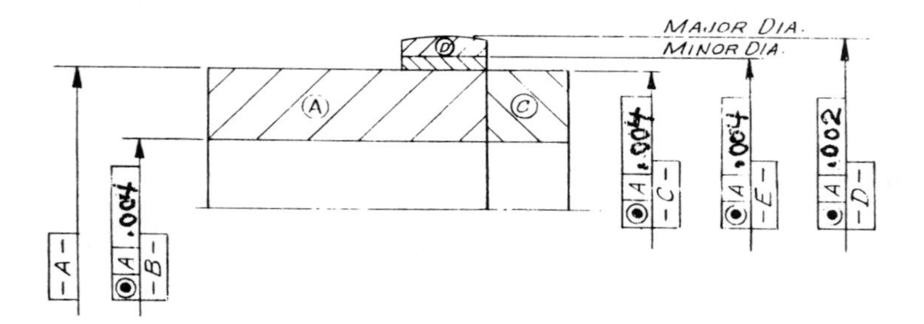

Example Figure 2

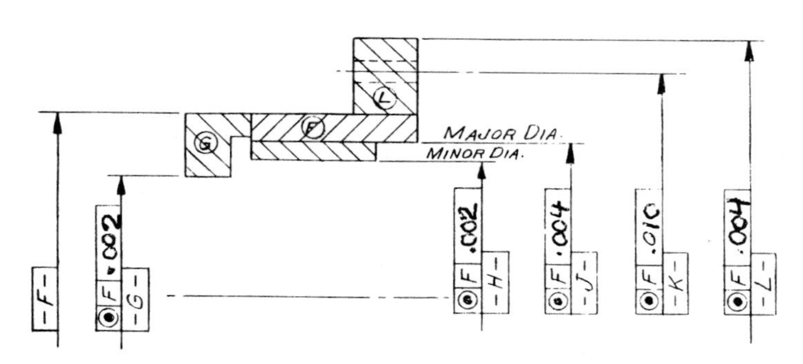

Example Figure 3

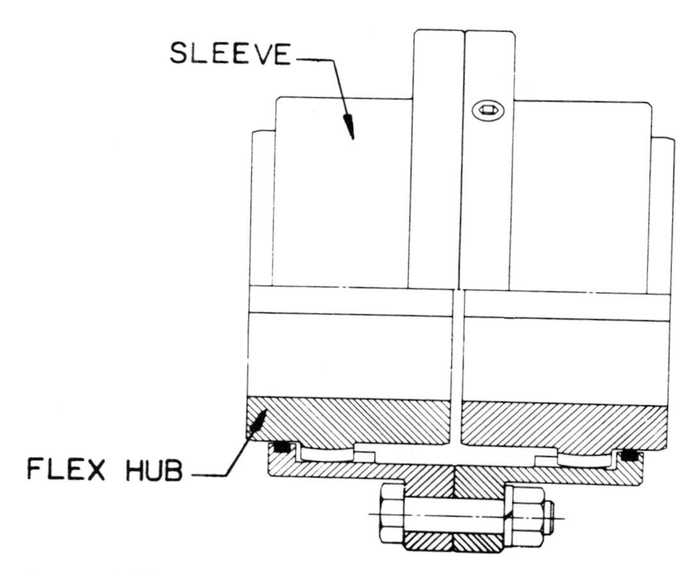

Example Figure 4

IV. INTERFACE CONNECTIONS

One area that is often overlooked or ignored because everyone feels that it is someone else's responsibility is the interface connection. What is an interface? It is the torque-transmission connection between the equipment and the coupling. As stated in Section 3.I, the responsibility for the interface connection must lie with the coupling selector (equipment manufacturer or user).

There are two basic types of interface connections, the cylindrical bore and the flange connection. Figure 3.37 lists some of the most common interfaces used. Shown is the relative torque capacity, ease of assembly/disassembly, reliability, machining requirements, usage, and relative cost of each. The most common type of interface connection is the cylindrical bore with key(s) and an interference fit between shaft and the coupling hub. The trend today in high-speed equipment indicates a preference for tapered hydraulic hubs. This section covers the interface connections listed in Figure 3.37, together with their usage, application, and standards for each (where they exist).

A. Shaft Connections

1. Clearance-Fitted Bores. This type of interface connection is very commonly used with small-horsepower couplings, usually under 150 hp. Some very small couplings (miniatures) may use just a setscrew to hold the hub on the shaft (both torsionally and axially). On larger units, however, the clearance fit is used in conjunction with a key and two setscrews.

The most common standard available for clearance fits for couplings is AGMA. This specification has two classes of clearance-fitted bores (see Figure 3.38):

Class I: This type of fit is more expensive and requires tighter control in shaft and coupling bore manufacture.

Class II: This type of fit is usually cheaper and provides almost twice the clearance of class I.

2. Interference-Fitted Bores. This type of interface connection is the most commonly used, particularly for medium-sized couplings and is generally used from 150 to 100,000 hp. It is typically used with one or two keys. If the interference is high enough, no key may be required (see Section IV.A.9). Interference-fitted bores minimize fretting corrosion of the hub bore surface that sometimes occurs with clearance-fitted bores.

The most common standard available for interference fits (with keys) for couplings is AGMA from which Figure 3.38 is extracted. For larger bore sizes than in AGMA, a nominal interference of 0.0005 in./in. of shaft diameter is commonly used for medium-carbon steel hubs with keys. For the alloy hubs (4140, 4340, etc.) usually used on high-speed couplings, the interference typically range from 0.00075 to 0.001 in./in. of shaft diameter.

3. Setscrews (see Figure 3.37A and B). Usually used to retain clearance-fitted bores. Commonly used with two setscrews, one over the key (if used) and the other 90° or 120° from it. Sometimes the shaft is flattened in the area of the setscrew to help increase the holding power. Setscrews can dig into a shaft and cause assembly and disassembly problems. To help minimize gouging of the shaft, brass setscrews or setscrews with brass tips are sometimes used.

The setscrew interface connection is relatively easy to install and remove as long as care is exercised to prevent galling of the shaft from the setscrews. It is important that the setscrews be secured with Nylok inserts or a locking compound so that the screw does not work loose. The lock keeps the coupling from spinning on the shaft and ruining it, and also prevents the setscrew, and possibly the coupling, from becoming a flying projectile.

4. Split Hubs (see Figure 3.37D). This type of interface connection is made with a clearance fit between the shaft and the hub bore. The hub is clamped and deformed by a bolt or bolts and provides an interference fit when installed to transmit torque and axially fix the hub on the equipment shaft. The split hub design makes it relatively easy to assemble and disassemble a coupling. You just loosen the bolt(s) and you have clearance to slide the coupling hub off the shaft.

The interface is usually used with relatively small couplings that transmit less than 150 hp. The hubs are usually complex and expensive to make. A hub's torque capacity is usually reduced, making it prone to failure. Due to its configuration, the hub usually becomes the weak link in the chain—in this case, the drive train.

5. Keys (see Figure 3.37C). Most couplings are supplied with hubs that are bored and keyed. Coupling manufacturers do not normally supply the keys used at the equipment interface. They are usually supplied and fitted to the equipment and the coupling by the equipment manufacturer. Low-speed and intermediate-speed

couplings are usually supplied with one keyway. High-speed and high-torque couplings are usually supplied with two keyways.

The better the fit between the key and the keyways, the more torque this interface connection can handle. Providing for two keyways, which may fit sloppily, does not necessarily increase the torque capacity of the interface connection. Improper fit of keys to keyways can lead to key roll and eventually splitting of the coupling hub (see Figure 3.39). The combination of tight-fitting keys and an interference-fitted bore provides for an interface connection with high reliability and high torque capabilities.

The most common standards used for key sizes, key fit, keyway dimensions, and tolerances are AGMA's. The following material has been adapted from AGMA. These standards define the recommended sizes and tolerances of keys and keyways used with industrial-type flexible couplings. The keyway tolerances and recommended sizes contained in these standards are intended for single-key applications only, but may be used for multikey connections where application conditions permit.

a. Types of Keys and Keyways. The tolerances in these standards apply to keyways for the following types of keys:

Square parallel keys
Rectangular (flat) keys
Plain taper keys

b. Shaft Diameter Range. These standards cover the keyway tolerances and recommended key sizes and tolerances for shafts ranging in diameter from 5/16 through 7 in.

c. Class of Key Fit. These standards recognize three classes of key fit:

1. Commercial
2. Precision
3. Fitted

Commercial. Commercial keys, used for most applications, have a clearance fit with the sides of the keyway and use commercial bar stock keys having widths ranging from exact size to minus tolerances. Keyway width and depth tolerances, recommended key tolerances, and the resulting fit dimensions are shown in Figure 3.40.

Precision. Precision keys have a transitional fit with the sides of the keyway and limited clearance over the top of the key. They require use of keystock having widths ranging from exact size to plus tolerances. For keyway width and depth tolerances, see Figure 3.40.

Fitted. Fitted keys require the use of an oversized key whose width is fitted to suit the keyway at assembly. The height may also be fitted where required by operating conditions. Keyway width and depth tolerances are shown in Figure 3.40.

d. Keyway Dimensions and Tolerances.

Dimensions. Keyway dimensions for square, rectangular, taper, and Woodruff keys are shown in Figure 3.41 and defined below:

$$C = 0.5D - \sqrt{(0.5D)^2 - (0.5W_2)^2}$$

where

C = chord height at keyway
D = diameter of bore
W_2 = width of keyway

Keyway width: for square, rectangular, and taper keys and for all classes of fit

$$W_1 = W_2$$

where W_1 is the width of the key.

Keyway depth:

1. For square, rectangular, and Woodruff keys and for all classes of fit:

$$h = 0.5H$$

$$A = D + 0.5H - C$$

where

A = dimension from top of keyway to opposite side of bore
H = depth of key
h = depth of keyway

2. For taper keys and for all classes of fit (measured at large end):

$$h = 0.5H - 0.020 \text{ in.}$$

$$A = D + 0.5H - C - 0.020 \text{ in.}$$

Tolerances

Keyway offset: Maximum offset (N) for keyways is determined as follows (see Figure 3.42A):

$$N = \frac{6 + W_2}{1000}$$

Keyway lead (see Figure 3.42B): Permissible lead (J) for keyways must not exceed the values tabulated in Figure 3.43.

Keyway parallelism (see Figure 3.42C): Parallelism is restricted by the keyway width and depth tolerances shown in Figure 3.40.

Finish of keyways: In general, good machine shop practice is indicated for the finish on keyways. It is recommended that keyways and sides and bottoms be finished to a maximum of 125 μin. or better.

Fillet radii in keyways: Couplings may or may not be furnished with fillet keyways. Where fillets are furnished, they will be in accordance with Figure 3.44 unless otherwise specified. Keys must be chamfered or rounded to clear the fillet radii.

6. Flattened Bores

a. Flattened Cylindrical Bores (See Figure 3.37E). This type of interface connection is usually used where equipment or couplings are removed or disassembled quite often. It is very common on steel mill drives, where couplings connect to the rolls and the rolls are constantly being changed. The use of the flattened bore for roll necks is shown in Figure 3.37E. Since the bore-to-shaft fit is usually quite loose, it is very common to provide lube fittings in the hub through which lubricants can be introduced, thus helping to minimize wear. Usually, replaceable wear plates are provided with this type of interface connection.

b. Square Bores. This type of interface connection is commonly used on small steel mill drives and on universal joints for agricultural equipment. If required, the hubs are axially fixed by pins through the hub and the shaft.

7. Bushings (Figure 3.37F). Intermediate bushings which are tapered, come in two basic configurations: with and without a flange. They come in many standard sizes of outside diameter with variable bore diameters. They are usually cheaper than coupling

hubs and thus can be stocked in a wide range of sizes for use on different pieces of equipment. This type of interface connection is easy to assemble. It has a clearance fit at assembly, but when properly assembled provides for an interference fit between hub and shaft. The use of tapered bushings tends to increase the initial cost of the coupling and sometimes increases the size of the coupling required. Standard sizes are available for bore sizes from approximately 1/2 to 10 in. Three of the most common types available are:

1. The Dodge taper lock
2. The Eaton QD bushing
3. The Browning XT bushing

8. Splines (Figure 3.37G). A spline is an interface connection consisting of integral keys and keyways equally spaced around a bore. In general, the teeth have a pressure angle of less than 45°. If the pressure angle is over 45°, it is usually called a serration rather than a spline.

a. Involute Splines. An involute spline is a spline with teeth having an involute profile. This form is the most common and has a relatively high torque capacity due to its configuration. Involute splines are economically manufactured by hobbing (external teeth) and shaping (external and internal teeth).

The various tooth proportions and tolerances are covered by a standard on involute splines and inspection [American National Standards Institute (ANSI) B92.1a]. The two most commonly used fits used for involute splines are:

1. The major diameter fit
2. The side fit

In the major diameter fit, the mating diameter fit between the mating parts is tightly fitted at the major diameter of the gear teeth. The sides of the teeth have some clearance. This type of fit requires fitting parts or tight-tolerance parts, which tends to increase cost.

The side-fitted spline is most commonly used. The teeth contact only at the sides; the side clearance is minimal. The major and minor diameters have relatively large clearances. In high-torque and/or high-speed applications the side-fitted spline is used with locating pilots usually at the front and back of the hub. The pilots help prevent shifting, rocking, and fretting of the spline.

9. Keyless Fitted Hubs. The trend in high-speed and/or high-torque applications is toward the keyless taper-fitted hub connection. The first type of keyless hub connection provided high torque capacity but presented problems when the hub had to be removed. It usually had to be cut off, destroying the hub and sometimes the equipment shaft itself. Several hydraulic methods for the removal of shrunk-on parts were introduced by SKF and Siemens (Figure 3.37H) in the late 1940s and early 1950s. They provided a means of applying oil under pressure between the hub and the shaft so that the hub could be pressurized and then pulled off. With the Siemens method, the pressure against the chamfered shaft pushed the hub off the larger portion of the shaft.

a. Tapered Shaft (Figure 3.37I). Taper hydraulic shafts are becoming very popular. This is because the hubs cannot only be removed hydraulically but can also be mounted. This is an important advantage, particularly in the refinery and petrochemical industries, where open flames and other means of heating hubs to expand them may present a safety problem.

This type of interface connection requires more attention than the straight bore because it is easier to machine cylindrical bores and shafts than tapered shafts and bores. It is very important that the mating shafts and hubs surfaces match. If the tapers do not match, the hubs can slip and/or produce high residual stresses that may cause the shaft to fail.

It is important that hubs on tapered shafts be properly advanced or pulled up on the shaft to provide the proper torque capacity of the interface. To determine the advance (pull-up) required to obtain a given interference, the following equation is used:

$$\text{Advance (S)} = 12\,\frac{i}{t}$$

where

i = diametral interference
t = taper (in./ft)

If the taper is given in degrees,

$$t = \frac{\text{degrees}}{57.28} \times 12 = \text{in./ft}$$

The dimensional interference can be found by multiplying the interference rate (in./in. of shaft diameter) by the shaft diameter (D).

Example 7

Given: 0.0015 in./in. of shaft diameter; shaft diameter = 10 in. Thus i = 0.015 in. (diametral interference)

Figure 3.45 gives the advance required for various tapers at different interferences.

As stated earlier, the contact between the bore of the hubs and the shaft is critical. The fit between the hub and the shaft should be checked by bluing. On hydraulic hubs, 70 to 80% contact is usually acceptable. If it is less than this, the part(s) may have to be remachined or lapped with special tools. The tapered shaft or hub should not be made without a plug and ring set so that parts can be checked with these master gages when they are being made. If this occurs, the likelihood of mismatched parts is minimized. When a mismatch is found, some people attempt to lap the hub to the shaft. This can be dangerous, as it can produce steps in the shafts and really makes the mismatched condition worse. Other people use the master gages for lapping the hub or shaft and end up destroying them so that the next time this set of gages is used, very odd looking contact patterns emerge.

When a mismatch occurs, the first thing to do is to check the plug gage with the ring gage for contact. If they do not fit properly, get a new set, because you have a problem. If the gages check out, check the hub and shaft with them. Once you find which part is wrong, have it fixed. Remachine (grind) it if possible, being careful to calculate a new advance and that any change does not cause problems with setup dimensions of equipment or the coupling. If machining the parts is not possible, make a lapping plug or ring from the master gages. If none of the above work, make a ring gage to fit the shaft and then make a special plug gage so that a new coupling hub can be made to fit this special shaft. There are many design variations and preferences with regard to hydraulic hubs.

b. *Sealing* (Figure 3.46). Are O-rings required or not? O-rings are required if the hub is to be hydraulically mounted and removed. If hubs are heat mounted and removed hydraulically, O-rings are not necessary. The hydraulic pressure causes nonuniform expansion of a hub, particularly at the ends of the hub, and this acts to prevent pressure lose and leakage so that the hub can be removed.

There are two common O-ring configurations (see Figure 3.46A and B); selection is a matter of preference. Most hydraulic hubs that use O-rings use backup rings (usually nylon or Teflon). The

O-rings must be inboard of the hub, facing each other. The backup rings must be installed outboard on the ends of the hubs (Figure 3.46C).

c. *Injection Holes*. Oil must be injected between the hub and the shaft. The oil can be introduced either through the hub or the shaft (see Figure 3.47).

d. *Stress Relieving* (Figure 3.48). The stress distribution generated by the contact pressure is higher at the ends of the hubs and is considerably higher at the inboard end of hub at point A in the figure. The easiest way to reduce these stresses is with a relief groove in the hub.

e. *Hubs* (Figure 3.49). Steps or flanges in hubs tend to present difficulties with hydraulic mounting or dismounting of hubs unless sufficient pressure is supplied to expand these areas. Flanges over shafts are difficult to avoid, particularly when reduced moment couplings are required. Large flanges should overhand the equipment shaft (see Figure 3.49B) so that the mounting pressure is reduced. If the face of flange is directly over the end of the shaft, a stress relief groove at the hub face may be required to help reduce stresses.

f. *Retaining Mechanisms* (Figure 3.50). Once a hub is installed, some type of axial retention should be used for safety. An improperly installed hub could slip down the shaft and even off the shaft, causing serious damage to the equipment.

g. *Hydraulic Mounting and Dismounting Equipment*. Figure 3.51 shows a typical mechanism of which there are several variations. Most provide a means of injecting oil and a means of pushing the hub up on the shaft. Since in most cases the hydraulic pressures are between 20,000 and 30,000 psi, special high-pressure tubing and fittings are required.

B. Flanges

This type of interface connection (see Figure 3.37) is very common on large-horsepower equipment (large steam turbines) and in applications to help reduce the overhung weight. It would be nice if there were only one flange connection, but there are so many different types that a standard would restrict the use of many of them.

Since there are so many different flange connections, it becomes important that exact specifications be supplied. In most cases, the equipment manufacturer's flange is fixed and the coupling must be made to bolt up to it, although in some cases, the equipment flanges are made to accommodate a coupling. Whichever is the case, the proper information has to be supplied; otherwise, the user will end up with a coupling that will not bolt up to the equipment.

1. Shrouded Versus Exposed Bolts. In the past, shrouded bolt couplings were used to protect people, but today all couplings have some type of guard. The shrouded coupling can in some instances reduce windage and noise, depending on the design of the coupling guard. The shrouded flange coupling is usually heavier than the exposed flange coupling. Exposed bolt couplings are usually cheaper. The noise level can be higher, but if the coupling guard is designed properly, the difference between the two types is usually insignificant compared to the rest of the system. In fact, on applications where the coupling guard is fitted tightly around the coupling, an exposed bolt coupling may produce less windage and horsepower loss than those produced by a shrouded coupling. This is true simply because the flange length of the shrouded bolt coupling is greater than that of the exposed bolt coupling.

2. Face Keys (Figure 3.37K) and Face Splines (Figure 3.37L). This type of interface connection is used on high-torque applications where bolts constantly loosen and break. When face keys are incorporated into a flange, care must be exercised so that the flanges are not weakened from the keyway. Usually, the keyway depths are approximately one-half of the flange thickness. Also important are keyway radii and key-to-keyway fit. Face splines provide even higher torque capacity than face keys and are relatively expensive.

Both face keys and face splines are used with heavy-duty drives such as universal joints, and are also used where a high-capacity connection must be disassembled with ease. The addition of face keys or splines allows for a reduction in the number of bolts used with the flange connection.

C. Summary

Interface connections are important and should be considered carefully by the coupling selector. The coupling manufacturer should be supplied with specifications detailing all aspects of the interface. This will prevent delays, misunderstandings, and possibly failure of the coupling and/or the equipment.

	TYPE	TORQUE CAPACITY	EASE OF ASSY. & DISASSY.	RELIABILITY	MACHINING REQUIREMENTS	USAGE	RELATIVE COST	DRAWING
A	CYLINDRICAL BORES WITH CLEARANCE & SETSCREWS.	POOR	EXCELLENT	?	LITTLE	COMMON (SMALL CPLGS.)	LOW	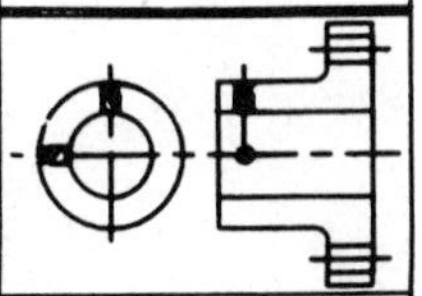
B	CYLINDRICAL BORES WITH CLEARANCE, KEYWAY, & SET-SCREWS.	GOOD	GOOD	GOOD	SOME	VERY COMMON (SMALL CPLGS.)	MODERATE	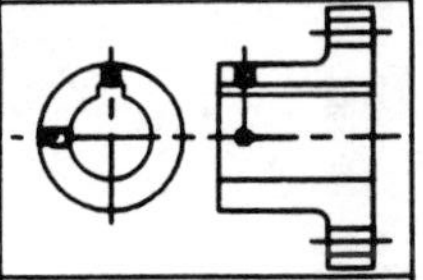
C	CYLINDRICAL BORES WITH INTERFERENCE & KEYWAYS.	VERY GOOD	FAIR	VERY GOOD	MODERATE	MOST COMMON (ALL CPLGS.)	MODERATE	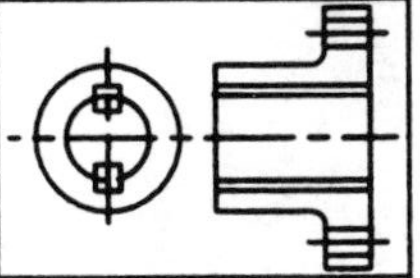
D	SPLIT HUBS	FAIR	EXCELLENT	?	HIGH	LITTLE to COMMON. (VERY SMALL CPLGS.)	MODERATE to HIGH	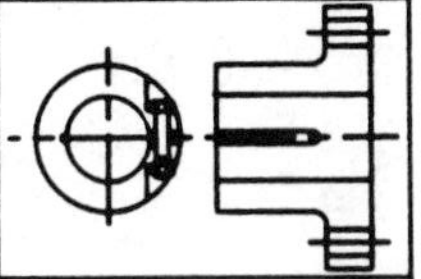
E	FLATTED BORES	VERY GOOD	VERY GOOD	GOOD	MODERATE to HIGH	LITTLE (LARGE CPLGS.)	HIGH	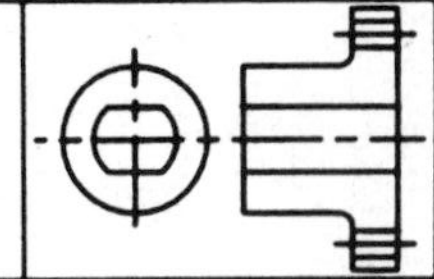
F	BUSHINGS	FAIR	VERY GOOD	GOOD	MODERATE	COMMON (LOW HP)	MODERATE to HIGH	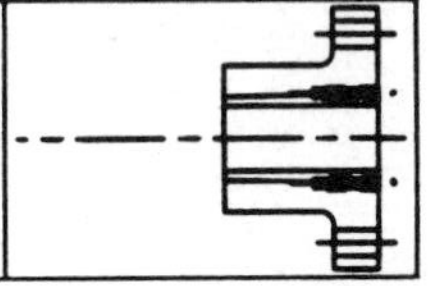

G	SPLINES	EXCELLENT	GOOD	VERY GOOD	HIGH	LITTLE to COMMON	HIGH	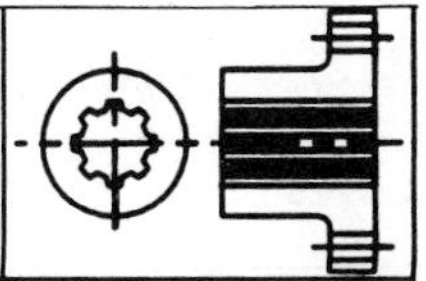
H	KEYLESS STRAIGHT BORES.	VERY GOOD	FAIR	VERY GOOD	MODERATE	LITTLE	MODERATE	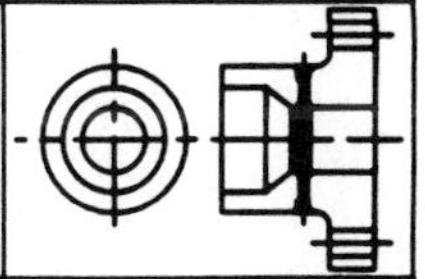
I	KEYLESS TAPERED BORES.	VERY GOOD	FAIR to GOOD	VERY GOOD	HIGH	BECOMING COMMON	HIGH	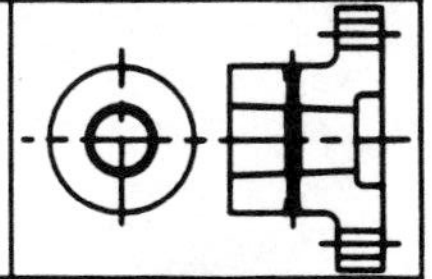
J	BOLTED FLANGE	GOOD	GOOD	GOOD	MODERATE	VERY COMMON	MODERATE	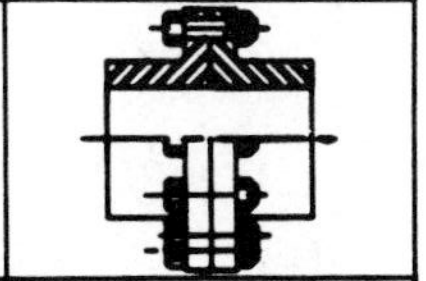
K	FACE KEYS	VERY GOOD	VERY GOOD	VERY GOOD	MODERATE	LITTLE	MODERATE to HIGH	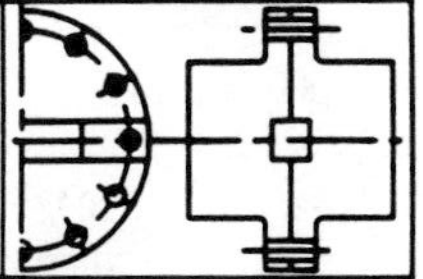
L	FACE SPLINES	EXCELLENT	VERY GOOD	EXCELLENT	HIGH	LITTLE	HIGH	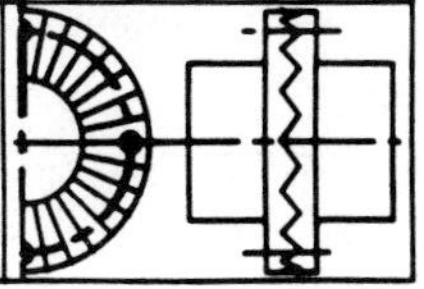

Figure 3.37 Comparison of coupling attachment methods.

Nom. Dia. Shaft & Bore	Shaft Tol. (minus)	Clearance Fit				Interference Fit				Keyway Sizes			
		Class 1		Class 2		Bore Tol. Range (minus)		Fit Tol. Range (minus)		Square		Reduced	
		Bore Tol. (plus)	Fit Tol. (plus)	Bore Tol. (plus)	Fit Tol. (plus)					W^a	H^b	W^a	H^b
.500	.0005	.001	.0015	.002	.0025	.0005	.001	.000	.001	1/8	1/16	1/8	3/64
.625	.0005	.001	.0015	.002	.0025	.0005	.001	.000	.001	3/16	3/32	3/16	1/16
.750	.0005	.001	.0015	.002	.0025	.0005	.001	.000	.001	3/16	3/32	3/16	1/16
.875	.0005	.001	.0015	.002	.0025	.0005	.001	.000	.001	1/4	1/8	1/4	3/32
.9375	.0005	.001	.0015	.002	.0025	.0005	.001	.000	.001	1/4	1/8	1/4	3/32
1.000	.0005	.001	.0015	.002	.0025	.0005	.001	.000	.001	1/4	1/8	1/4	3/32
1.125	.0005	.001	.0015	.002	.0025	.0005	.001	.000	.001	1/4	1/8	1/4	3/32
1.1875	.0005	.001	.0015	.002	.0025	.0005	.001	.000	.001	1/4	1/8	1/4	3/32
1.250	.0005	.001	.0015	.002	.0025	.0005	.001	.000	.001	1/4	1/8	1/4	3/32
1.375	.0005	.001	.0015	.002	.0025	.0005	.001	.000	.001	5/16	5/32	5/16	1/8
1.4375	.0005	.001	.0015	.002	.0025	.0005	.001	.000	.001	3/8	3/16	3/8	1/8
1.500	.0005	.001	.0015	.002	.0025	.0005	.001	.000	.001	3/8	3/16	3/8	1/8
1.625	.001	.001	.002	.002	.003	.001	.002	.000	.002	3/8	3/16	3/8	1/8
1.750	.001	.001	.002	.002	.003	.001	.002	.000	.002	3/8	3/16	3/8	1/8
1.875	.001	.001	.002	.002	.003	.001	.002	.000	.002	1/2	1/4	1/2	3/16
1.9375	.001	.001	.002	.002	.003	.001	.002	.000	.002	1/2	1/4	1/2	3/16
2.000	.001	.001	.002	.002	.003	.001	.002	.000	.002	1/2	1/4	1/2	3/16
2.125	.001	.0015	.0025	.002	.003	.001	.002	.000	.002	1/2	1/4	1/2	3/16

2.250	.001	.0015	.0025	.002	.003	.001	.002	.000	.002	1/2	1/4	1/2	3/16
2.375	.001	.0015	.0025	.002	.003	.001	.002	.000	.002	5/8	5/16	5/8	7/32
2.4375	.001	.0015	.0025	.002	.003	.001	.002	.000	.002	5/8	5/16	5/8	7/32
2.500	.001	.0015	.0025	.002	.003	.001	.002	.000	.002	5/8	5/16	5/8	7/32
2.625	.001	.0015	.0025	.002	.003	.001	.002	.000	.002	5/8	5/16	5/8	7/32
2.750	.001	.0015	.0025	.002	.003	.001	.002	.000	.002	5/8	5/16	5/8	7/32
2.875	.001	.0015	.0025	.002	.003	.001	.002	.000	.002	3/4	3/8	3/4	1/4
2.9375	.001	.0015	.0025	.002	.003	.001	.002	.000	.002	3/4	3/8	3/4	1/4
3.000	.001	.0015	.0025	.002	.003	.001	.002	.000	.002	3/4	3/8	3/4	1/4
3.250	.001	.0015	.0025	.003	.004	.0015	.003	.0005	.003	3/4	3/8	3/4	3/4
3.500	.001	.0015	.0025	.003	.004	.0015	.003	.0005	.003	7/8	7/16	7/8	5/16
3.625	.001	.0015	.0025	.003	.004	.0015	.003	.0005	.003	7/8	7/16	7/8	5/16
3.750	.001	.0015	.0025	.003	.004	.0015	.003	.0005	.003	7/8	7/16	7/8	5/16
4.000	.001	.0015	.0025	.004	.005	.0015	.003	.0005	.003	1	1/2	1	3/8
4.250	.001	.0015	.0025	.004	.005	.002	.0035	.001	.0035	1	1/2	1	3/8
4.500	.001	.0015	.0025	.004	.005	.002	.0035	.001	.0035	1 1/4	1/2	1 1/4	3/8
4.750	.001	.0015	.0025	.004	.005	.002	.0035	.001	.0035	1 1/4	5/8	1 1/4	7/16
5.000	.001	.0015	.0025	.004	.005	.002	.0035	.001	.0035	1 1/4	5/8	1 1/4	7/16
5.250	.001	.0015	.0025	.004	.005	.0025	.004	.0015	.004	1 1/4	5/8	1 1/4	7/16
5.500	.001	.0015	.0025	.004	.005	.0025	.004	.0015	.004	1 1/4	5/8	1 1/4	7/16
5.750	.001	.0015	.0025	.004	.005	.0025	.004	.0015	.004	1 1/4	5/8	1 1/4	7/16
6.000	.001	.0015	.0025	.004	.005	.0025	.004	.0015	.004	1 1/2	3/4	1 1/2	1/2

Figure 3.38 Hub-to-shaft fits and nominal keyway sizes.

Nom. Dia. Shaft & Bore	Shaft Tol. (minus)	Clearance Fit				Interference Fit				Keyway Sizes			
		Class 1		Class 2						Square		Reduced	
		Bore Tol. (plus)	Fit Tol. (plus)	Bore Tol. (plus)	Fit Tol. (plus)	Bore Tol. Range (minus)		Fit Tol. Range (minus)		W[a]	H[b]	W[a]	H[b]
6.250	.001	.0015	.0025	.004	.005	.0025	.004	.0015	.004	1 1/2	3/4	1 1/2	1/2
6.500	.001	.0015	.0025	.004	.005	.0025	.004	.0015	.004	1 1/2	3/4	1 1/2	1/2
6.750	.001	.0015	.0025	.004	.005	.0025	.004	.0015	.004	1 3/4	7/8	1 3/4	3/4
7.000	.001	.0015	.0025	.004	.005	.0025	.004	.0015	.004	1 3/4	7/8	1 3/4	3/4

[a]W = Hub Keyway Width.
[b]H = Hub Keyway Depth.

Figure 3.38 (continued)

(A)

(B)

Figure 3.39 Hub damage due to improperly fitted keys: (A), rolled over key; (B), (C), split hub.

(C)

Figure 3.39 (continued)

Nominal key width (Wk)			Commercial class fit			Precision class fit		
Over	To (incl.)	Keyway width tolerance	Recommended key tolerance width and height	Resulting side fit[a]	Recommended keyway depth tolerance	Recommended key tolerance width and height	Resulting side fit[a]	Recommended keyway depth tolerance
	5/16	+0.002 −0.000	+0.000 −0.002	CL 0.004 0.000	+0.016 −0.000	+0.001 −0.000	CL 0.002 Tight 0.001	+0.005 −0.000
5/16	1/2	+0.0025 −0.000	+0.000 −0.002	CL 0.0045 0.000	+0.016 −0.000	+0.001 −0.000	CL 0.0025 Tight 0.001	+0.005 −0.000
1/2	3/4	+0.003 −0.000	+0.000 −0.002	CL 0.005 0.000	+0.016 −0.000	+0.001 −0.000	CL 0.003 Tight 0.001	+0.005 −0.000
3/4	1	+0.003 −0.000	+0.000 −0.003	CL 0.006 0.000	+0.016 −0.000	+0.001 −0.000	CL 0.003 Tight 0.001	+0.005 −0.000
1	1 1/2	+0.0035 −0.000	+0.000 −0.003	CL 0.0065 0.000	+0.016 −0.000	+0.001 −0.000	CL 0.0035 Tight 0.001	+0.005 −0.000
1 1/2	2	+0.004 −0.000	+0.000 −0.005	CL 0.009 0.000	+0.016 −0.000	+0.001 −0.000	CL 0.004 Tight 0.001	+0.005 −0.000

Note: Sides of keyways to be parallel within the width tolerance—bottoms of keyways are to be parallel with centerline of bore (or bottom of bore opposite keyway) within the depth tolerance.
[a]CL, Clearance.

Figure 3.40 Keyway width and depth tolerances and width fit.

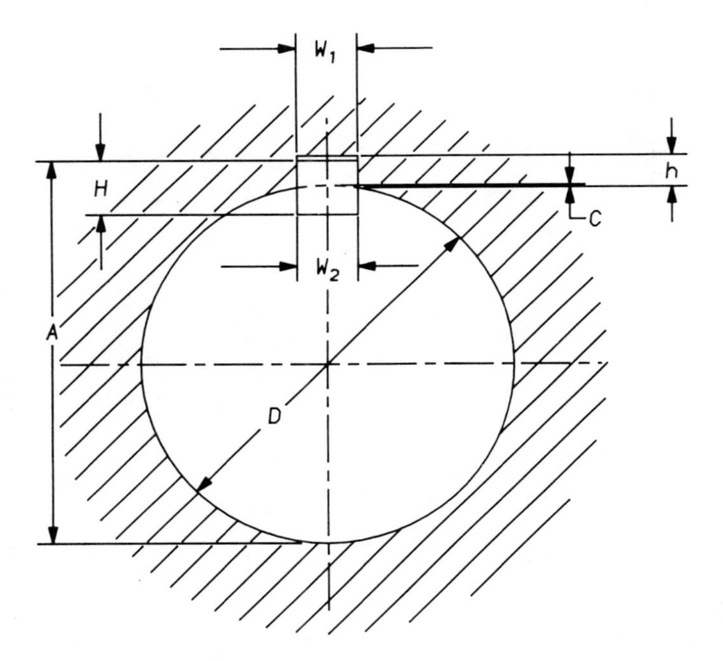

Figure 3.41 Key and keyway dimensions.

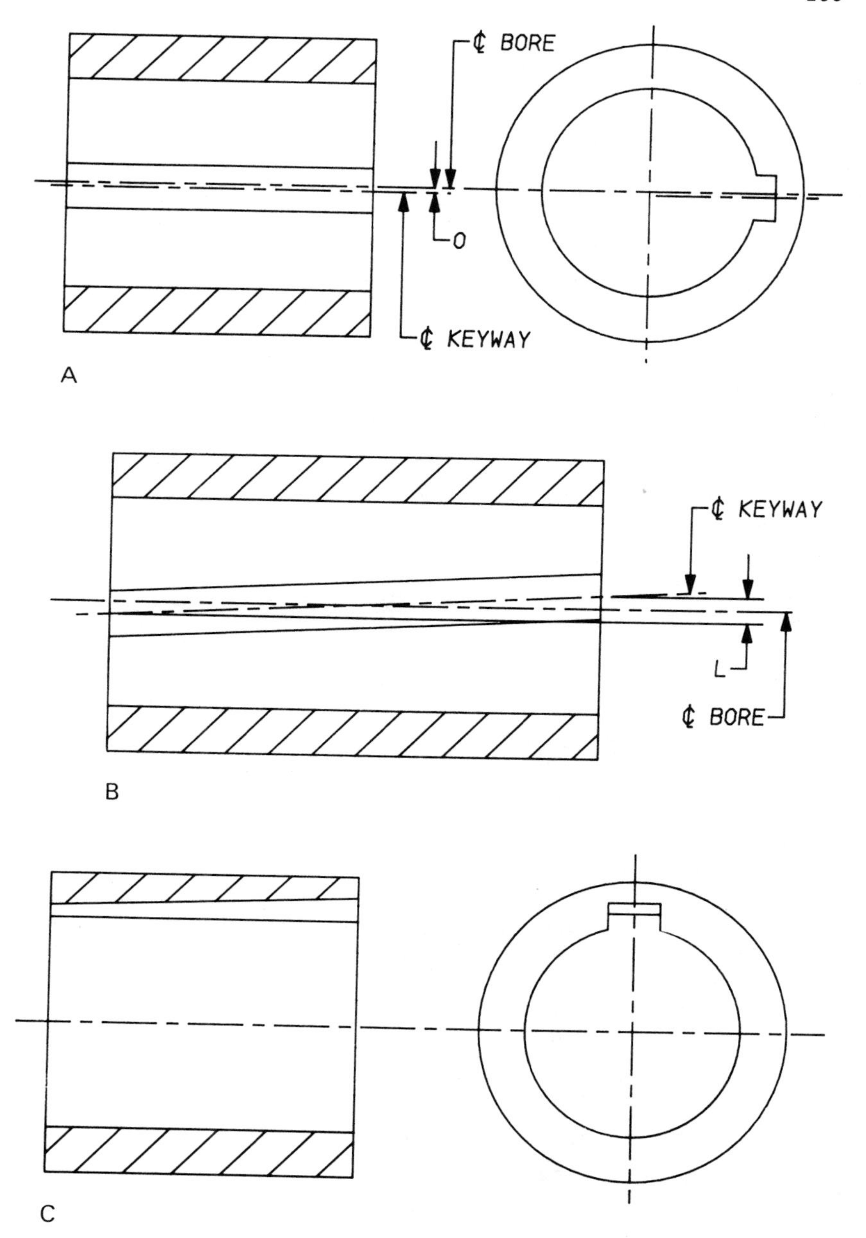

Figure 3.42 Keyway errors: (A) keyway offset; (B) key lead; (C) keyway parallelism.

Keyway length, (inches)		Maximum lead	
Over	To (incl.)	Commercial and fitted class	Precision class
–	2	0.002 in.	0.001 in.
2	10	0.001 in./in. of length	0.0005 in./in. of length
10	–	0.010 in.	0.005 in.

Figure 3.43 Permissible keyway lead error values.

Keyway depth		Keyway fillet radius	Suggested key chamfer (@45°)
Over	To (incl.)		
–	1/8	1/64	1/32
1/8	1/4	1/32	3/64
1/4	1/2	1/16	5/64
1/2	7/8	1/8	5/32
7/8	1 1/4	3/16	7/32

Figure 3.44 Values for keyway fillet radius.

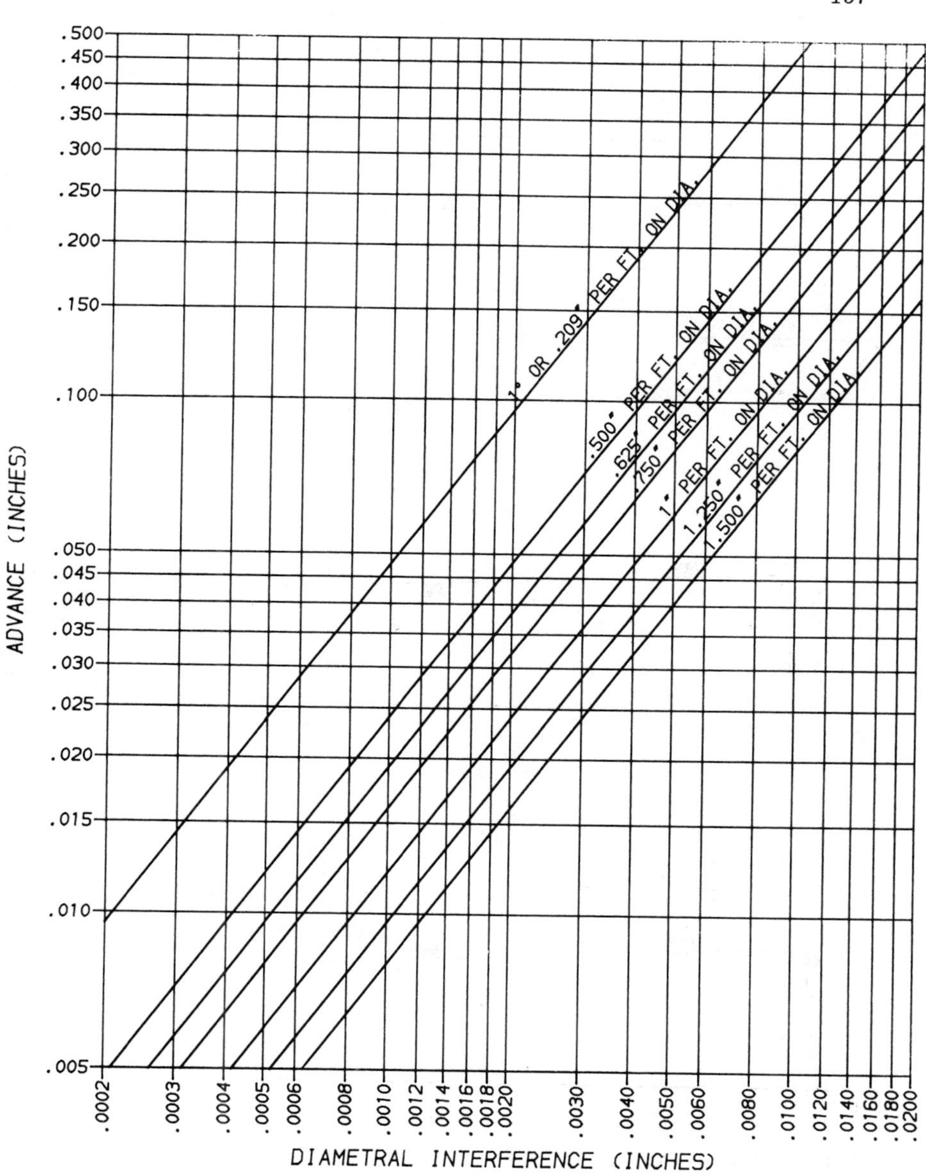

Figure 3.45 Amount of hub advance for various tapers.

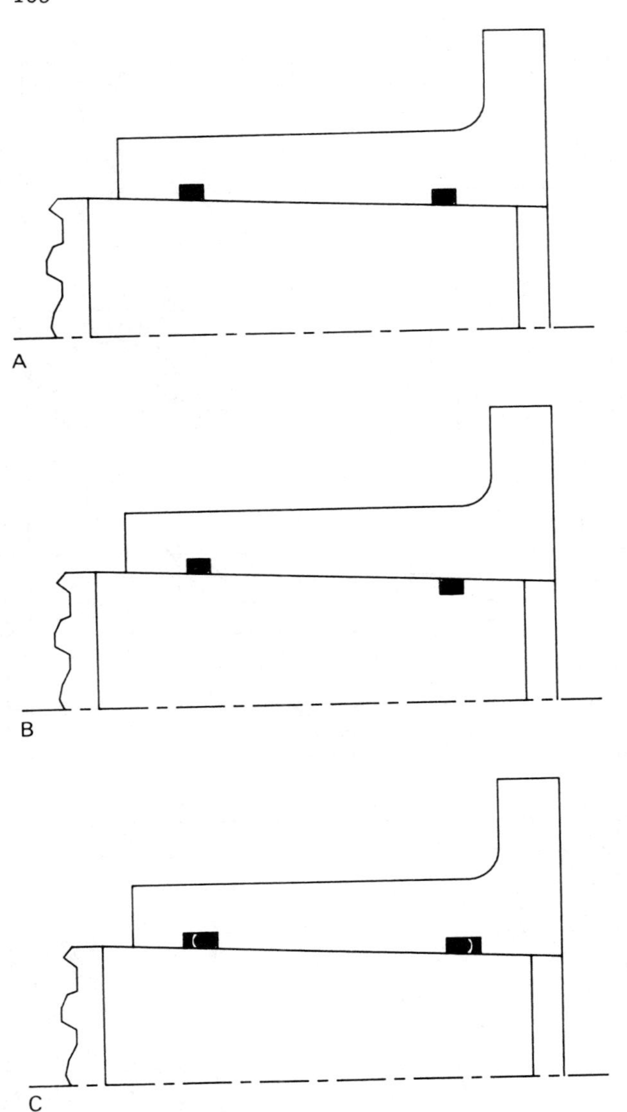

Figure 3.46 Various O-ring arrangements.

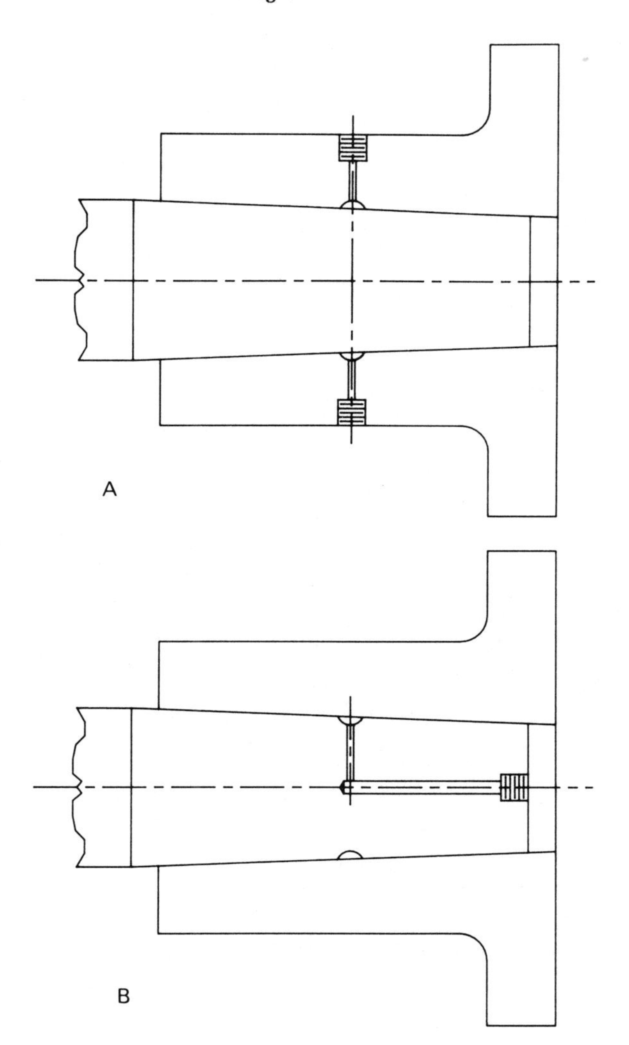

Figure 3.47 Types of oil injection methods: (A) through the hub; (B) through the shaft.

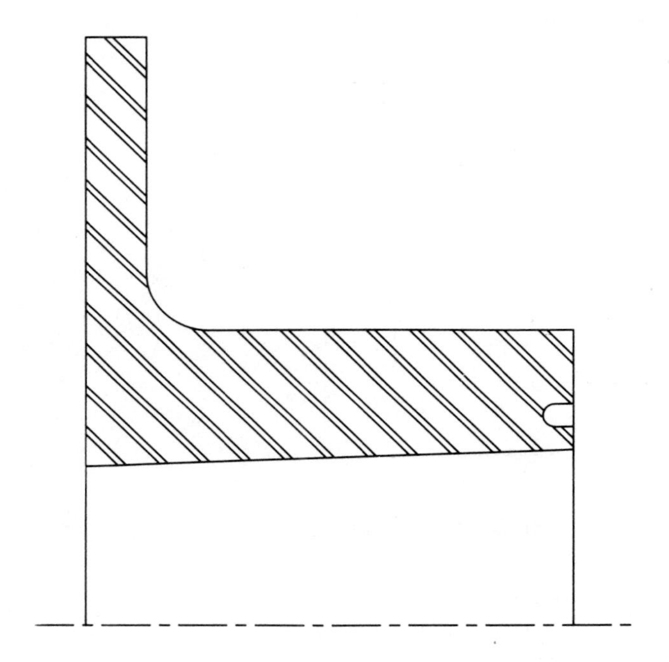

Figure 3.48 Stress-relief grooves for taper-bored hubs.

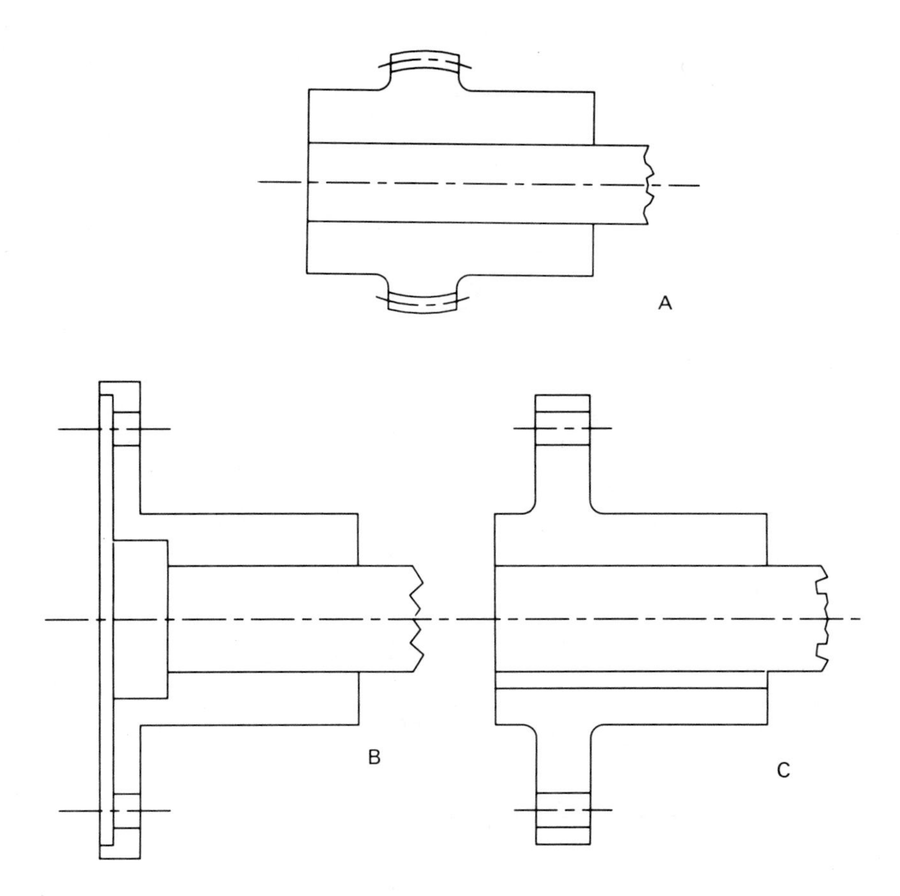

Figure 3.49 Various types of hub designs: (A) flex hub; (B) rigid hub flange overhanging shaft end; (C) rigid hub flange over shaft.

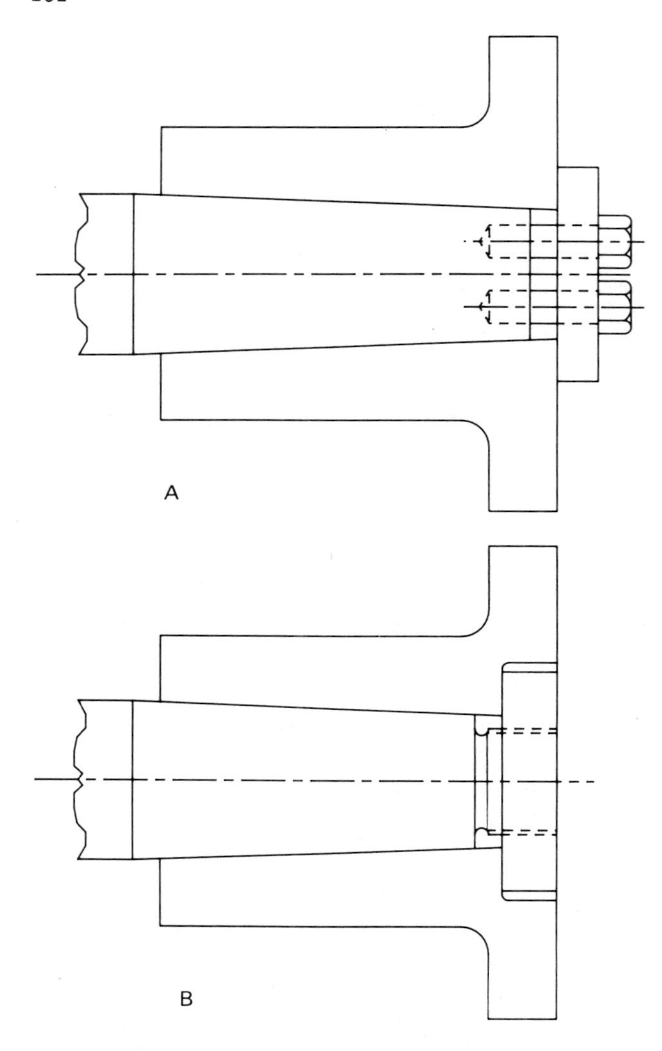

Figure 3.50 Types of hub retention: (A) bolted plate; (B) thread nut.

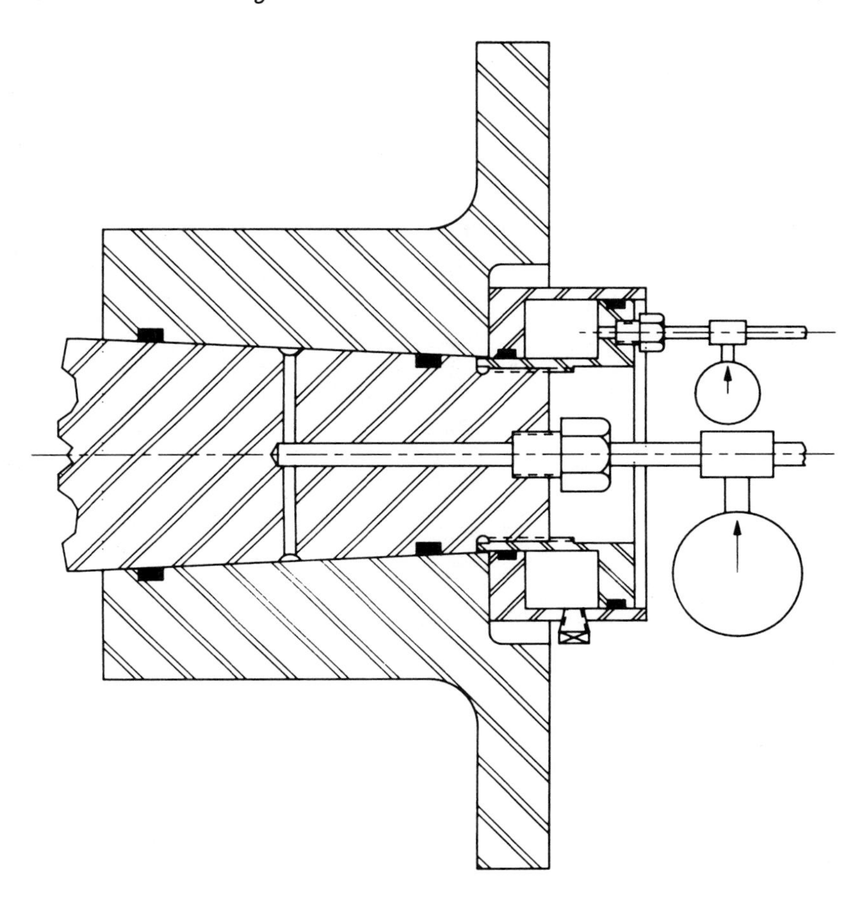

Figure 3.51 Hydraulic hub mounting equipment.

4

Installation and Maintenance

I. INTRODUCTION

In this chapter we discuss the installation and maintenance of flexible couplings in general terms and also the various coupling types available on the market. Most coupling manufacturers provide information and recommendations for installation, maintenance, and lubrication of their couplings. If a copy is not included with a coupling, the coupling manufacturer should be contacted before installation is begun. A good maintenance department has a file where this information is kept for reference.

So why should you read this chapter? Because it presents general procedures as well as some of the reasons for these practices and provides a good starting place for the person who has not installed a coupling before.

II. INSTALLATION OF COUPLINGS

A. Preparation

Upon receiving a coupling, proceed as follows:

1. Obtain a drawing (if one exists) of the coupling and a copy of the coupling manufacturer's and/or equipment builder's installation and maintenance manual.
2. Review and read the information. If any questions arise, contact the coupling manufacturer or equipment builder.

3. Inspect the components to assure that all parts that were ordered have been shipped.
4. If the coupling is going to be stored for a period of time, make sure that the parts are properly protected. Steel parts are usually coated with an oil or wax coating. Rubber parts are usually wrapped or packaged so that light and air exposure are minimized (light and air tend to harden rubber and some elastomers). If the period in storage is to be more than one year, check with the coupling manufacturer for special instructions. Particular attention should be directed to seals, rubber elements, and any prelubricated parts when long-term storage is required.
5. Prior to assembling the coupling, clean and disassemble the components. *Note*: Some subassemblies should not be taken apart. Check specific coupling instructions as to what to disassemble. *This is very important.*
6. Check for burrs and nicks on the mating surfaces. If they exist, remove (stone, file, sand).
7. Measure and inspect:
   a. Bore dimensions
   b. Keyways
   c. Coupling lengths
   d. Diameter of bolt circle (DBC), bolts, and holes of the mating flange
   e. Any other dimensions to assure that the coupling will mate properly to the equipment
8. Obtain hardware, keys, tools, and anything else not supplied but needed to complete the assembly.

B. Hub Installation

Normally, hubs are mounted before the equipment is aligned. In some instances, the hubs may be mounted by the equipment manufacturer. Described below are steps in the installation procedure for various fittings.

1. Straight Shafts

a. Clearance Fits. This type of installation is relatively simple.

1. Install the key(s) in the shaft keyway(s).
2. Make sure that any part that will not slide over the coupling hub is placed back on the shaft, such as seals, carriers, and covers, and on gear couplings, the sleeves.

3. Push the hub onto the shaft until the face of the hub is flush with the ends of the shaft. (*Note*: Some coupling hubs are not mounted flush. Check specific instructions.) If the hub does not slide onto the shaft, check the clearances between the bore and the shaft. In addition, check to assure that there is clearance between the keys at the sides of the coupling keyway and on top of the key.
4. Lock the hub in position (usually with setscrews). Make sure that setscrews have a locking feature such as a Nylok insert, or use locking compound. Some hubs use bolts, nuts, or other means to secure the hub in place. See the specific instructions.

b. *Interference Fit*. This type of installation is the same as that for the straight shaft, with the exception that the hubs must be heated before they slide onto the shaft. The coupling manufacturer usually supplies information as to how to heat the hub and to what temperature. For steel hubs, 160°F is required for every 0.001 in. of interference per inch of hub diameter (0.001 in./in.). For example, a steel hub with a 4-in. bore with an interference of 0.003 or 0.00075/0.001 × 160 = 120°F. Therefore, if the shaft temperature is 80°F, the hub temperature must be 200°F. This does not account for human factors such as cooling due to handling time, errors in measurements, and so on. As a general rule, add 50 to 75°F to the calculated expansion temperatures to account for these factors. The hub should be heated in an oil bath or an oven; a torch or open flame should not be used. This could cause localized distortion or softening of the hub material. It could also cause an explosion in some atmospheres. Oil bath heating is usually limited to approximately 350°F, or under the flash point of the oil used. Special handling devices are required: tongs, threaded rods placed in taped holes in the hub, and so on. Oven heating offers some advantages over oil. Parts can be heated to higher temperatures (usually not exceeding 600°F), and the parts can be handled with heat-resistant gloves. In any event, extreme care must be exercised when handling heated hubs to avoid injury to personnel.

It is also important when mounting interference hubs to make sure that clearance exists over the top of keys; otherwise, when the hub cools, it will rest on the key and produce high stresses in the hub that could cause it to fail.

2. Straight Shafts With Intermediate Bushings. Intermediate bushings come in two basic configurations: with and without

flanges. To assemble, insert the bushing into the hub without tightening the screws or bolts; then slide the hub and bushing onto the shaft. Since the bushing is tapered, tighten the screws or the shaft. Once the hub is at the correct position, the screws should be tightened gradually in a crisscross pattern to the specific torque. Bolts are tightened on a coupling similar to the way in which lugs are tightened on the wheel of an automobile. Refer to the specific instructions for further recommendations and the correct torquing value.

3. Taper Shafts. Tapered shafts have the advantage that the interference between the hub and the shaft can be accomplished by advancing the hub on the shaft. Depending on the amountof interference, the hub may be drawn up with nuts or heating. Removal of the hub is usually easier on tapered shafts than on straight shafts.

Applications using tapered bores require more attention than those using straight shafts because it is easier to machine two cylindrical surfaces that match than two tapered surfaces. The hub can be overstressed if it is advanced too far on the shaft. Dirt and surface imperfections can restrain the hub advance and give the false impression that the desired interference has been reached.

To determine the draw-up required to obtain the desired interference, use the following equation:

$$\text{Draw-up (in.)} = 12 \times \frac{i}{T}$$

where

i = diametral interference (in.)
T = taper (in./ft)

The area of contact between the bore of the hub and the shaft should be checked with machinist's bluing. Fifty to 80% contact is the range of acceptability; usually, 70% is the most desirable. If less than the required contact is achieved, the contact can be increased by lapping the bore and/or the shaft with a plug or ring made from a master plug and ring gage. It is not recommended that the master gages or shaft be used to lap the hub, as the gages could end up with ridges. Ridges in the hub or shaft will prevent proper hub installation and could cause the hub or shaft to fail because of stress concentrations.

1. *Light interference* (under 0.0005 in./in.): When the interference is less than 0.0005 in./in., the hub can usually be advanced without heating. Although heating the hub is the most common method, the hub can usually be advanced by tightening the retaining nut or plate on the shaft. It is also common practice when light interference is used with a combination of keys and a retaining nut or plate to use a light grease or antisieze compound between the hub, shaft, and threads on the shaft and nut. This should help facilitate installation and future removal and help prevent shaft and/or bore gauling.
2. *Medium interference* (usually 0.0005 to 0.0015 in./in.): When the interference is over 0.0005 in,/in., the force required to advance the hub could become too large for manual assembly. When this occurs, the hub *must* be heat mounted or hydraulically mounted. Heating hubs for mounting is the most common method. Regardless of the method used, the amount of draw-up must be measured.
3. *Heavy interference* (usually over 0.0015 in./in.): When the interference is over 0.0015 in./in., hubs are usually heat mounted and removed hydraulically. Some users prefer to both mount and remove hydraulically.

The following is recommended as a general guide when installing hubs on an equipment shaft:

1. Install the hub on the shaft, assuring that the parts mate properly and are burr-free and clean. Using a depth gage or dial indicator (see Figure 4.1A), measure and record the initial reading.
2. Remove the hub and lubricate the bore or shaft if hydraulic assist is to be used; if not, heat in oil or an oven. When using a heating method for mounting hubs, it is best to provide a positive step, such as a clamp on the shaft, to assure proper draw-up (see Figure 4.1B). The reason for this is that a hub advanced too far may not be removable (too much force required or not enough hydraulic pressure available to remove the hub) and normally requires that the hubs be cut off.
3. The hub is installed and advanced the required amount.
4. The shaft nut is then properly tightened and locked in place. Locking is done with a tab washer or setscrew.

4. Rough-Bored Couplings. Most coupling manufacturers will supply couplings with rough bores. This is usually done with a spare

coupling so that as its equipment shafts are remachined, the spare coupling can be properly fitted. This type of coupling also helps to reduce inventory requirements.

It is important that the user properly bore and key these couplings; otherwise, the interface torque-transmission capabilities can be reduced or the coupling balance (or unbalance) can be upset. Recommendations should be obtained from the specific coupling manufacturer. As a general guide, the hub must be placed in a lathe so that it is perpendicular and concentric to its controlling diameters. On rigid hubs the pilot and face are usually the controlling diameter and surface that should be used to bore (see Figure 4.2A). On flex hubs (gear and chain) the gear major diameter (OD) and hub face act as the controlling diameter and surface. [*Note*: Some manufacturers use the hub barrel as the controlling diameter (see Figure 4.2B).]

Some coupling manufacturers supply semifinished bore couplings. In this case, the finished bore should be machined using the semifinished bore as the controlling diameter. Indicate the bore-in, for concentricity and straightness.

Most couplings must have one or two keyways cut in the hub. These should be cut according to the tolerances listed in Section IV of Chapter 3. Particular attention should be given to the following:

1. Keyway offset (the centerline of the keyway must not intersect with the centerline of bore)
2. Keyway parallelism
3. Keyway lead
4. Keyway width and height

5. Keys. The fitting of keys is important to assure the proper capacity of the interface. Refer to Section IV of Chapter 3 and the AGMA standards on keyways and keys. As a general rule, three fits must be checked:

1. The key should fit tightly in the shaft keyways.
2. The key should have a sliding fit (but not be too loose) in the hub keyway.
3. The key should have a clearance fit radial with the hub keyway at the top of the key (see Figure 4.3A).

The key should be chamfered so that it fits the keyway without riding on the keyway radii (see Figure 4.3B and C). A sloppily fitted key can cause the keys to roll or shear when loaded. The results of a sloppy key fit are shown in Figure 4.3D. The forces generated by torque are at distance S, and this movement tends to roll the key and can cause very high loading at the key edges. On the other hand, too tight a fit will make assembly very difficult and increase the residual stresses, which could cause premature failure of the hub and/or shaft.

A key in the keyway that is too high could cause the hub to split (see Figure 4.3D). When there is too much clearance at the top or sides of a key, a path is provided for lubricant to squeeze out. Lubricated coupling clearances between keys and keyways must be sealed to prevent loss of lubricant and thus starvation of the coupling.

III. ALIGNMENT OF EQUIPMENT

Once the hubs have been assembled, the next step is usually to align the equipment, although some people prefer to assemble the coupling and then check and align the equipment with the coupling fully assembled. The danger here is that if the alignment is too far off, the static misalignment imposed on the coupling while assembling it may cause damage to the coupling, which could result in a premature failure. Therefore, it is suggested that at a minimum, the equipment should be rough-aligned before the coupling is installed.

Types of alignment and the steps required are covered in general in the following discussion. The subject of equipment alignment could fill a book by itself and is a very common subject of debate. Alignment tools vary from the simple straightedge to very sophisticated laser methods.

You may be wondering why we need to worry about alignment since flexible couplings are designed to allow some flexibility in alsignment. One important reason is that the coupling's useful life depends on the amount of misalignment. If it were possible to align equipment perfectly and we could assure that it would stay, a flexible coupling would not be required. A second reason for accurate initial alignment is that alignment changes over time due to bearing wear, foundation settling, and thermal changes. When alignment is carefully done initially, generally within one-eight to one-third of the coupling's rated misalignment capacity, there should be enough capacity left in the coupling for it to handle foreseeable increases in misalignment with time.

Two rotating shafts can have the combinations of position shown in Figure 4.4, although conditions A and C are unlikely to occur in real life. A flexible coupling is usually employed as a double flex (except for elastomeric couplings). A gear-type coupling incorporating a spacer is shown in Figure 4.5. Each coupling end (flex element) will accommodate only angular misalignment α_1 at one end, and α_2 at the other; a double flex coupling can accommodate $\alpha = \alpha_1 + \alpha_2$ angular. It is important to remember that *there must be two flex elements* to accommodate any offset misalignment (this does not apply to elastomeric couplings). The amount of offset misalignment a double flex coupling can accommodate is determined as follows:

$$\text{Offset (S)} = L \times \tan \alpha$$

where

α = angular misalignment (deg)
L = center distance between flex points

Example 1

If a gear coupling is rated at $\alpha_1 = 1\ 1/2°$ per mesh and the distance between flex elements (flex points) is 6 in., the amount of offset capacity is

$$6 \times \tan 1.5° = 0.157 \text{ in.}$$

A flexible coupling can accommodate more offset if it is used in conjunction with a spacer; the longer the spacer, the larger the distance between flex points and the greater the allowable offset misalignment allowed.

A. Straightedge, Feeler Gage, and Caliper Methods of Measuring Misalignment

The simplest way to align a coupling is by using a straightedge and a set of feeler gages and/or calipers (see Figure 4.6).

1. Offset Misalignment (Figure 4.6A and C). This type of misalignment is measured by aligning a straightedge on one hub (using the same diameters) and measuring the gap between the straightedge and the other hub. Usually, the gap can be measured with a set of feeler gages. Remember that the angular misalignment created by this offset is

$$\alpha = \tan^{-1} \left(\frac{S}{L}\right)$$

Example 2

If L is 6 in. and S = 0.020 in.,

$$\alpha = 0.19^\circ$$

2. Angular Misalignment. This type of misalignment is measured by taking measurements at maximum openings (between hubs) and a minimum opening (180°) directly with a feeler gage on close-coupled couplings (Figure 4.6A) and with a set of calipers on coupling where there is some distance between hubs. The formula is

$$\tan \alpha_1 = \frac{t_1 - t_2}{D}$$

where

D = diameter at which readings are taken (in.)
t_1 = top gap reading (in.)
t_2 = bottom gap reading (in.)

Example 3

If t_1 = 0.250 in., t_2 = 0.225 in., and D = 4 in.,

$$\alpha_1 = 0.358^\circ$$

B. Dial Indicator Method

The most common method used to align a flexible coupling is with a dial indicator, ruler, vernier, or micrometer (see Figure 4.7). The dial indicator(s) must be mounted on one or both hubs (clamped, magnetically held, or bolted). Both the clamping means and rigidity of setup must be considered. A setup that is too thin or flimsy could give false readings.

To obtain an accurate picture of misalignment, two readings must be taken: one with the indicator attached to the driving shaft and another with the indicator attached to the driven shaft. This procedure is called the *reverse indicator method.*

Figure 4.7A shows why only one reading can be misleading. Reading A makes it appear that there is perfect alignment, whereas reading B shows that there is an angular misalignment present. Figure 4.7B shows that when two readings are the same but of opposite sign, only parallel misalignment exists. [*Note*: This offset misalignment is equal to one-half the total indicator reading (TIR).] Figure 4.7C shows that when two shafts have both offset and angular misalignment, only the angular misalignment can be calculated. Figure 4.8 shows a method of measuring misalignment when there are large distances between pieces of equipment. This method gives the angular misalignment at each flex element.

C. How to Correct Misalignment

Attempting to correct the misalignment between two shafts through trial and error can be a lengthy and frustrating procedure. On the other hand, if a thorough procedure is followed, the job can be done with ease. The equipment required for measuring was described in the preceding section. The tools required for correcting misalignment are the same as those used to measure it, plus a pencil, paper, a good head for calculations, or a calculator, which would make it simpler and possibly more accurate.

There are three important steps to follow when attempting to align equipment:

1. Rotate the shaft to which the indicator is attached and lock or hold the other shaft stationery.
2. Correct the misalignment by moving the piece of equipment with the stationary shaft. Do not adjust the equipment to which the indicator is attached.
3. Make sure that the shafts have no axial play; lock or hold in one position so that no false readings are taken.

1. Shaft Separation. The distance between the shafts (or flanges) is usually determined by the system designer or may be dictated by the type of coupling used. The required shaft separation is usually given on the coupling assembly drawing and with other instructions, which are usually packaged with the coupling. Axial positioning of the equipment should be the first part of an alignment procedure because it usually involves the greatest equipment movement.

a. Rough Aligning. If the shafts are close together, use a straightedge. If they are some distance apart, attach to the shaft a dial indicator, which can easily be rotated, then move the other piece of equipment until the shafts are roughly in line. Following this, try to reduce the horizontal misalignment. Bring the indicator tip into a horizontal plane and set it at zero. Rotate the shaft 180° (one-half the revolution) and take a reading. If the reading is positive (an increase in the reading), move the equipment away from the indicator (move the stationary equipment); if the reading is negative, move the equipment toward the indicator.

Example 4

The reading went from zero to 0.036 (clockwise); therefore, we move the equipment away from the indicator by 0.036/2 = 0.018.

Example 5

The reading went from zero to 0.092 (counterclockwise); therefore, we move the equipment toward the indicator by (0.100 − 0.092)/2 = 0.004.

2. Angular Misalignment

a. Face Readings (Figure 4.9). If the space between shafts allows the indicator to read on the face of the coupling hub, the angular misalignment can easily be corrected. However, there is always the possibility that the face of the hub is not square to the shaft. If so, even when the shafts are perfectly aligned, one would obtain a false reading. To eliminate this "false" reading, the shaft is rotated. If there is some wobble, record the value and mark with chalk (or any other marker) the position of the maximum reading. Turn the shaft until the mark is on top. Now rotate the shaft that holds the indicator and zero the indicator when it lines up with the chalk mark.

Example 6

The diameter D on which the tip of indicator is rotated is 6 in., and the distance Y between the legs of equipment is 15 in. If the face runout is 0.006 and the TIR is 0.040 when the indicator shaft is rotated, shim under the rear legs of equipment by

$$\frac{0.040 - 0.006}{2} \times \frac{15}{3} = 0.085 \text{ in.}$$

If the indicator moved from zero to 0.060 (counterclockwise), the TIR is still 0.040. In this case, shim under the front legs of the equipment by

$$\frac{0.040 + 0.006}{2} \times \frac{15}{3} = 0.115 \text{ in.}$$

b. Horizontal Direction (Figure 4.10). If the distance between the shafts is too small for taking face readings, the angular misalignment can be measured and corrected through two successive readings in two locations w inches apart, starting with the indicator in a horizontal position.

Example 7

In the first location the indicator moves from zero to 0.010 in. (clockwise); in the second location the indicator moves from zero to 0.016 in (clockwise). If the distance w is 4 in. and y = 24 in., move the rear end of the machine

$$\frac{0.016 - 0.010}{2} \times \frac{24}{4} = 0.018 \text{ in.}$$

away from the last position reading on indicator 2.

Example 8

In the first location the indicator moves from zero to 0.082 in. (counterclockwise); in the second location the indicator moves from zero to 0.092 in. (counterclockwise). Move the rear end of the machine

$$\frac{0.092 - 0.082}{2} \times \frac{24}{4} = 0.030 \text{ in.}$$

toward the last position reading on indicator 1.

Note that when both readings are in the same direction, the small number is subtracted from the large one; when the readings are in opposite directions, they should be added.

c. *Vertical Direction* (Figure 4.11). Basically, the same method is used as that described in the preceding section, except that the dimension y is the distance between the machine legs. The indicator must always be zeroed when it is on top of the shaft.

Example 9

In the first location the indicator moves to 0.006 in., in the second location it moves to 0.021 in. If the distance w is 4 in. and y = 24 in., a

$$\frac{0.021 - 0.006}{2} \times \frac{24}{4} = 0.045 \text{ in. shim}$$

should be placed under the front legs of the equipment.

Example 10

In the first location the indicator moves to 0.087 in.; in the second location it moves to 0.098 in. A

$$\frac{0.098 - 0.087}{2} \times \frac{24}{4} = 0.033\text{-in. shim}$$

is required under the rear legs of the equipment.

3. Parallel Misalignment (Offset) (Figure 4.12). First repeat the rough alignment (see Section II.B.4). To avoid inducing any angular misalignment, the machine should be moved parallel to itself. For this, install two indicators at the machine legs and move the machine in such a manner that both indicators always read the same.

Second, zero the indicators on top of the shaft and read after half a revolution (180°). If the indicator reads 0.062 in. (clockwise), shim all four legs of the other machine by 0.062/2 = 0.031 in. If the indicator reads 0.068 in. (counterclockwise), remove

$$\frac{0.100 - 0.068}{2} = 0.016 \text{ in.}$$

from under every leg of the other machine. If a shim cannot be removed, a 0.016-in. shim can be added under the legs of the machine on which the indicator is mounted.

4. Final Check. As you have probably noticed, until now all the indicator readings were done with the indicator attached to only one shaft. Remember that this can lead to false alignment and that the reverse indicator method is best. Now it is time to verify the accuracy of the alignment and perform a reverse indicator reading (Figure 4.7C).

If both readings show TIR = 0.018 in. (but one reading is clockwise and the other is counterclockwise), we have a pure parallel misalignment of 0.009 in. between the shafts. Compare your reading with the value allowed by the manufacture of the coupling or equipment. Most coupling manufacturers give two misalignment capacities: one that is the maximum at which the coupling can operate and one that is acceptable for initial alignment. This is usually one-eighth to one-third of the operating capacity if no guide is given; then a value of less than one-fourth is usually acceptable.

D. Hot Alignment

Machines are usually aligned when cold. Under normal operating conditions, some machines can become hot and the alignment can change significantly due to thermal expansion of shafts and/or support structures. This thermal movement must be measured by operating the equipment, then shutting it down and checking the alignment with the coupling installed, or the information can be obtained from the machinery manufacturer, or an alignment monitoring coupling can be used.

In any case, if the thermal movement is significant, zero cold alignment may not be beneficial. When thermal movements are large and have been accurately calculated or measured, the coupling's initial parameters should be such that its alignment is as close to zero as possible when the equipment is in operation rather than when it is cold. This means that during cold alignment the coupling is usually purposely misaligned to compensate for this thermal effect during operation.

IV. COUPLING ASSEMBLY

Once the hubs are installed on the shafts and the equipment is aligned, the coupling can be assembled. Each coupling type is assembled differently; however, many of the couplings have similar procedures. The following discussion covers some general procedures for several of the most common types available. For more specific instructions and guidelines, consult the specific coupling manufacturer's instructions, which are usually shipped with the coupling.

A. Lubricated Couplings

There are four types of lubricated couplings:

1. Gear type (Figure 2.9)
2. Chain type (Figure 2.10)
3. Steel grid type (Figure 2.11)
4. Universal joint (Figure 2.12)

The gear, chain, and steel grid are very similar and are covered together. The first step is the application of a coating of lubricant to all moving parts, such as gear teeth and chains or steel grids, as well as to the O-rings and their mating surfaces. Then the coupling can be closed by bringing the sleeves, or the covers together. Most lubricated couplings use gaskets or O-rings between the flanges so that the lubricant is sealed in. The importance of properly sealing a coupling cannot be overemphasized, because the amount of lubricant contained in a coupling is small. Even the smallest leak would soon deplete the coupling of lubricant, resulting in rapid wear and ultimate failure.

To ensure proper sealing, a few rules should be followed: The flange surfaces must be clean, flat, and free of burrs or nicks. Any imperfections on the surfaces of the flange must be removed by using a honing stone or fine-grit sandpaper.

The gaskets are made of special material, as oil can seep through regular cardboard or paper. The gasket should be in one piece and be free of tears or folds. To facilitate assembly, some people glue the gasket to one of the flanges with grease. This procedure should be avoided, as it can prevent proper tightening of the bolts. It is suggested that a thin oil be applied to the gasket, which will help hold it in place.

Once the flanges are brought together, the bolts are inserted in the holes. The bolts must be assembled from the flange side, shown on the assembly drawing. The holes in the flanges must be oversized and else countersunk to accept the bolt-head radii. Care should be taken in aligning the holes in the gaskets with those in the flanges. If this is not done, the gasket is torn by the bolts and pieces of material can stick between the flanges, preventing proper sealing.

Once all the bolts are in place, slide on the lockwashers and install the nuts. If self-locking nuts are used, lockwashers are not used. If room permits, the nuts (not the bolts) should be tightened to specifications, which means that a torque wrench should be used. The nuts should be tightened in two or more steps in a crisscross fashion, similar to the way automobile lugs are tightened.

If the sleeves or covers incorporate lube plugs, the plugs on the two sleeves should be diametrically opposite. If the coupling was dynamically balanced, match marks may indicate the relative position not only between the two sleeves (or covers) but also between the sleeves and the hubs, unless the coupling was component-balanced; then alignment may not be required.

Care should be exercised not to mix parts of similar couplings unless the coupling manufacturer's instructions say that it is acceptable to do so. Particular attention should be paid to balanced couplings and special couplings, where parts mixing may affect not only balance but the functional operating characteristics of the coupling as well.

Universal joints are simple to install. They are usually shipped with their center section assembled and ready to be placed between mounted hubs or flanges on the equipment. Caution should be exercised in slinging and handling universal joints. The joints should be slung from yokes; to prevent damage to sliding parts, do not lsing around the shaft or splined protector. Joints should always be transported and handled in the horizontal position; *do not hang or transport in the vertical position*. Prior to installation, the universal joint, flanges, and mating flanges should be carefully cleaned. All traces of rustproofing should be removed. Check the key fit in the shaft's rigid hubs. Install the rigid hubs on the shaft; for shrink-fits, apply heat to the hubs uniformly. A check of mating flange for face runout and pilot runout should be made. Pilot and faces should be checked for nicks and dents. Nicks and dents should be stoned. Check the universal shaft to assure that match marks line up. Check the shaft-to-shaft or flange-to-flange dimensions between pieces of equipment to assure that it conforms to the dimensions supplied on as the drawing. Bolt to the mating flanges. Lubricate the bearing and/or length compensation unless this was done at factory. Refer to specific instructions to verify this.

B. Flexible Membrane Couplings

Flexible disk couplings come in two basic configurations:

1. *Low speed* (Figure 2.15A): The disks are usually taped or wired together. The disks are alternately fastened to the hub and spacer by bolts, nuts, and sometimes stand-off spacers. If stand-off washers are used, they must be assembled with the correct surface against the disk (usually curved). Sometimes, washers of different thicknesses are used. The washers must be placed in the correct position. See the specific instructions.
2. *High speed* (Figure 8): The disks are usually assembled as a pack, and the pack is bolted alternately to the hub and spacer.

Diaphragm couplings come in two basic configurations:

1. *Flex spacer section* (Figure 2.16A): This coupling just drops out between the two installed rigids. The center section is one piece: a subassembly or welded assembly. The center section should not be disassembled unless so instructed by the coupling manufacturer's instructions. Some couplings have pilot rigids, which requires that the center section be compressed for assembly.

2. *Flex subassembly* (Figure 2.16C): On this type of diaphragm coupling the center of the diaphragm coupling section consists of three separate pieces or assemblies. This construction is common where equipment is separated by a long distance, as it helps to reduce the cost of spares.

C. Elastomeric Couplings

The rubber donut coupling (Figure 2.14D) is assembled with radial bolts, and in the process of tightening the bolts, the rubber "legs" are precompressed. Precompression should be uniform around the coupling. The split insert should be the first to reach the final "seated" position in the hub's notch.

Rubber tire couplings (Figure 2.13A) are similar to donut couplings. They have a split element which is wrapped around the hubs. However, the fasteners are arranged axially, and they clamp the side walls of the tire between two plates.

Pin and bushing couplings, as well as jaw-type couplings (Figure 2.14A and F), have an elastomeric element trapped between the hubs. To install or remove the element, usually at least one of the hubs slides on the shaft; if not, the equipment must slide axially out of the way. The shaft separation must be larger than the thickness of the element to allow for installation; however, some elements are split so that close shafts can be used.

Most elastomer-type couplings can be obtained in a "spacer" configuration (Figure 7.27), in which the two hubs and the flexing element can be "dropped out" as a unit. This unit is attached to the two shafts through two "rigid" hubs with axial bolts. To install the couplings between the rigid hubs, the element has to be squeezed so that it will fit between the rabbets of the rigid hubs.

D. Bolt Tightening (Flange Connections) (see Figure 3.13)

Most instruction sheets provided with couplings give information on bolt tightening. Unfortunately, these instructions are not always followed; many mechanics tighten the bolts by feel.

Couplings resist misalignment, and the resulting "forces and moment" put a strain on the equipment and connecting fasteners. If the fasteners are loose, they are subjected to alternating forces and may fail through fatigue. A bolt that is not properly tightened can become loose after a short period of coupling operation.

Few bolts work only in tension. Most coupling bolts also work in shear, which is caused by the torque transmission. Usually, only some of the torque is transmitted through bolt shear; part of load is transmitted through the friction between the flanges. Depending on the coupling design, as much as 100% of the torque can be transmitted through friction. If the bolts are not tightened properly, there is less clamping force, less friction, and more of the torque is transmitted through shear. Because of the combined shear and tensile stresses in bolts, recommendations for bolt tightening vary from coupling to coupling. Coupling manufacturers usually calculate bolt stresses, and their tightening recommendations should always be followed. If recommendations are not available, it is strongly suggested that a value be obtained from the coupling manufacturer rather than by guessing. *Find out what the specific coupling requires.*

Bolts should be tightened to the recommended specification in at least three steps. First, all bolts should be tightened to one-half to three-fourths of the final value in a crisscross fashion. Next, they should be tightened to specifications. Finally, the first bolt tightened to the final value should be checked again after all the bolts are tightened. If more tightening is required, all the bolts should be rechecked. Also, the higher the strength of the bolt, the more steps that should be taken:

Grade 2: two or three steps
Grade 5: three or four steps
Grade 8: four to six or more steps

If an original bolt is lost, a commercial bolt that looks similar to the other coupling bolts should not be substituted. It is best to call the coupling manufacturer for another bolt, or they may suggest an alternative.

If room permits, always tighten the nut, not the bolt. This is because part of the tightening torque is needed to overcome friction. The longer the bolt, the more important it is to tighten the nut rather than the bolt. As there is additional friction when turning the bolt, more of the effort goes into friction than in stretching the bolt.

Some couplings use lockwashers; others use locknuts. Whereas a nut-lockwasher combination can usually be used many times, a locknut loses some of its locking properties every time it is removed from the bolt. If not instructed otherwise, it is best to replace nuts after five or six installations. Some couplings use aircraft-quality bolts and nuts, which can generally be used 12 to 15 times before they lose their locking features.

V. LUBRICATION OF COUPLINGS

Since a large number of flexible couplings require lubrication, a separate section is devoted to lubrication. The group of couplings classified as mechanically flexible require lubrication. These couplings are:

1. The gear coupling
2. The chain coupling
3. The grid coupling
4. The universal joint

These four types of lubricated flexible couplings, although different, have a very similar mechanism of lubrication. The gear coupling is generally used as an example in this section; however, most discussions apply to the first three types of couplings.

The gear coupling has four to five major parts: two hubs, two sleeves, and in many applications a spacer. It also has bolts and nuts and must have some type of seal to assure that the lubricant stays in.

Each gear mesh acts like a spline. Some gear teeth are straight, both internal and external, and some couplings have straight internal external teeth with crowns. The mesh has clearance (backlash). Misalignment results from sliding or rolling these loosely fitted parts. Because of this relative motion, the need for lubrication is clear. In some flexible couplings, motion is only a thousandths of an inch, but in some it may be several inches (4 to 6 in.). From this range of motion a coupling can experience sliding velocities from less than 1 in. per second (ips) to as much as 200 ips.

It is important to remember the effect on lubricants of centrifugal forces. In a gear coupling the lubricant is trapped in the sleeve, which at high speeds acts as a perfect centrifuge. It tends to separate the grease and for continuously lubricated couplings it will separate foreign particulars out of lubricant, which tend to build up in the coupling (sludge). Chain and grid couplings are much like gear couplings, except that the "sleeve" in one case is the chain and cover and in the other case is the grid and cover.

The universal joint acts slightly differently. If a telescoping spline is used, the spline acts like a gear coupling and must be so lubricated. In the yoke heads there are bearings (needle or roller type) which are lubricated by the rolling action of these bearings. Since most universal joints operate at relatively low speeds, centrifugal force usually has very little effect on lubrication.

A. Types of Seals

For lubricated couplings the seals are almost as important as the lubricant itself. Almost as many coupling failures can be attributed to seal failure as to lube failure.

There are two basic types of seals used to retain lubricants: the metallic labyrinth seal and the elastomeric seal. Again we will use the gear coupling to discuss specifics, but most of these sealing devices are incorporated in the chain and grid couplings and the universal joint.

1. Metallic Labyrinth Seals. When gear couplings were introduced about 75 years ago, synthetic positive seals, which can withstand repeated flexing motion between the hub and the sleeve, and the deterioration caused by the lubricant were not available. Thus, as shown in Figure 4.13A, the all-metal labyrinth seal plate with its center of contact in line with the center of the gear flex point was used. Metallic seals usually require clearance and are not considered to be a "positive" seal. Lubricant can leak out of a coupling if the coupling is stopped and or is constantly being reversed. Also, this seal does not provide adequate protection in keeping out of couplings contaminates which could cause the coupling or its lubricant to deteriorate. This type of seal is still used very successfully in many applications.

2. Elastomeric Seals. There are four basic elastomeric seals used in couplings:

1. The O-ring
2. The H or T cross-section seal
3. The lip seal
4. The boot seal

The O-ring seal shown in Figure 4.13B was developed by the Air Force during World War II and is now very common in mechanical equipment. Most couplings requiring lubrication use O-rings, which are one of the most positive means of retaining lubricant. The major disadvantage with O-rings is that they usually allow relatively small amounts of motion between parts. To account for the oscillatory motion, seals can only be squeezed through approximately 10% of their cross section before they start to take a set and no longer function.

The H or T cross-section seal (Figure 4.13C) functions and operates very similarly to an O-ring but has more surface area in contact with the sealing surface and for relatively small motion will usually do a slightly better job. For many applications that require moderate amounts of oscillatory motion together with axial movement, th cross-section seal tends to roll and twist. When this occurs, a spiral path is created through which the lubrication can exit the coupling.

The lip seal (see Figure 4.13D) is another very common seal used with couplings. This seal can be designed with "extended" lips that can accommodate high eccentricities and thus a large degree of misalignment. The lip seal can also be designed with spring retainers and steel inserts to prevent the lips from lifting off at high speed.

The boot seal (see Figure 4.13E) is not in very common use, but it is a very effective means of retaining lubricant. This type of seal is usually limited to operation at low speed, but with a special fabric reinforcement material (see Figure 4.13F) it has been used on couplings that operate at 7 1/2° of misalignment and 6000 rpm.

B. Methods and Practices of Lubrication

1. Grease Lubrication. There are two methods for lubricating flexible couplings with grease: lubrication before closing coupling (pack lubrication) or lubrication after closing coupling. To pack-lubricate a coupling, an appropriate quantity of grease is applied manually to each half coupling, making sure that the teeth and slots are coated. Then the coupling is bolted up or assembled. If couplings are supplied with lube plugs, before the coupling is assembled the working surfaces are wiped with a light coating of lubricant, then assembled. Usually, couplings are provided with two lube plugs in their periphery; both should be removed before attempting to fill. Only the amount specified by the coupling manufacturer should be used, as too much lubricant could cause seals to be damaged or cause coupling to bind up in operation. After filling, the lube plugs are replaced.

Periodic maintenance of grease-lubricated couplings is necessary if the coupling is to give satisfactory service. A coupling should be relubed at regular intervals, usually at least every 6 to 12 months. The coupling manufacturer's suggestions should always be followed. Most relubrication cycles are established by experience, so records should be kept of relubrication history.

It is also suggested that the couplings be disassembled and cleaned at regular intervals to get rid of foreign materials and to check for wear and deterioration. Parts should be washed and dried and inspected. If parts appear to be worn, they should be replaced, relubricated, and reassembled.

Note: When disassembling and relubricating a coupling, it should be inspected for the condition of the lubricant. If the grease is soapy or separated into oil and soap, it is evident that the lubricant previously used is not suitable for this application. It would be wise to contact the coupling manufacturer and obtain a better lubricant recommendation.

2. Oil Lubrication

a. Continuous Flow. In this method oil is injected into the coupling by an oil jet(s) directed toward the collecting lips (fixed or removable) or rings. Oil may be injected at one end of the coupling and exit at the other, or each end is lubricated separately. Oil collectors may be provided on the hubs, sleeves, or spacer, depending on the coupling design. Oil jets are made from tubes that have been capped and then drilled to provide the correct quantity of oil and exit velicity based on supply oil pressure and the size of the supply line. The oil flow for a given type of design is function coupling size, transmitted horsepower, speed, and possible coupling misalignment. Figure 4.14 shows a curve for a high-speed gear coupling which gives the required lube flow requirements as a function of horsepower and misalignment. Such curves differ for different types of couplings and are also from one manufacturer to another. Use whatever is recommended by the specific coupling manufacturer.

There are several things that influence the required lubricant flow. Material hardness is one; another is whether the coupling incorporates positive dams or is damless. Damless couplings usually require much more oil. Usually, continuously lubricated couplings use the same oil as that used on the connected machinery bearings. It is mandatory that a sufficient quantity of clean oil (filtered not contaminated with water or corrosive media) be supplied to the coupling. It is suggested that a separate filtration be used, usually 5 micron in size or smaller and the oil properties and contamination be monitored and a centrifuge be used.

Sludge accumulation (Figure 4.15) may be detrimental to equipment and the coupling. It can produce high moments and forces on the equipment and the coupling which could result in catastrophic failure. Sludge may affect a coupling in several ways: it can reduce axial movement between coupling parts, tends to accelerate wear, stop the flow of oil, causing the coupling to operate without sufficient lubricant, and in contaminated oil systems cause or encourage corrosion. Sludge can be controlled by filtration to remove foreign particles and to prevent lube contamination (water and corrosive media) the use of separators and constant sampling of the lube is required, and/or the coupling housing system can be sealed from contaminated atmospheres.

Most couplings are designed to minimize sludge (see Figure 4.16). Many people believe that the solution to sludge accumulation is to remove the coupling dams. Unless you can drastically increase the coupling lubricant flow and assure a reliable lube source, the problems associated with damless couplings usually far exceeds the possible benefits. Figure 4.17 shows what happens to the oil in a damless coupling. Unless the lubricant flow is substantially increased, most of the working portion of the teeth will be starved. If couplings are properly maintained, the sludging problem can be minimized. It is suggested that at regular intervals and every time the equipment is stopped the coupling be opened and inspected for sludge accumulation. The sludge should be removed mechanically and/or chemically and the parts thoroughly cleaned and inspected for damage (tooth distress, corrosion, breakage, etc.).

b. *Confined and Self-Contained.*

Confined lubrication. Whenever a coupling is confined in a housing together with other lubricated components, such as gears, it may use the oil from the housing for its lubrication. When the coupling operates above the oil surfaces, lubrication is provided either by an oil pump or through splashing. In either case the oil flow to the coupling is a function of operating speed. If the working surfaces of a coupling are partially or totally submerged in oil, the lubrication requirements are satisfied by the flow of oil from the housing into the coupling. To ensure adequate lubrication, submerged couplings are provided with holes in their covers or sleeves, or the coupling may incorporate pickup scoops or other arrangements to aid in assuring adequate lubrication. Because the centrifugal action of these types of couplings tend to retain dirt from the system, it is advisable to clean the coupling when the equipment is down or during scheduled periodic cleaning intervals.

Self-contained lubrication. This type of coupling is supplied with seals and can be filled at assembly and operated for an extended period without further attention. The usual method of filling this type of coupling is to remove the lube plugs and pour the correct amount of recommended oil into the coupling. It is recommended that the working portions of the coupling be coated with lubricant before assembly. This will prevent bare metal contact between mating parts at assembly and at startup before oil properly distributes. Maintenance of an oil-filled coupling consists basically of preventing the oil from being lost from the coupling. It is therefore important to maintain flange surfaces, seals, and gaskets in good condition. Lube plugs should be properly seated (or sealed) so that no leakage occurs. A regular lubrication schedule should be established and maintained following all of the manufacturer's assembly, lubrication, maintenance, and cleaning instructions.

C. Types of Lubrication

Couplings are usually lubricated with oil or grease. Grease-lubricated couplings are usually lubricated by one of two methods. They are either self-lubricated or are lubricated from an external supply. For the self-contained method, oil or grease can be used and covers and/or seals are required. For the externally lubricated coupling, the lubricant is oil. The oil is supplied with a specific flow rate or it may be dipped or intermittently lubricated.

Which type of lubrication is best? Sealed lube or continuous? Gear, chain, and grid couplings may be filled with a specific quantity of lubricant and sealed. There are many factors governing which method should be chosen, and each coupling application must be evaluated with respect to its particular requirements. Some of the factors that should be considered are listed next.

Sealed lubrication:

1. Affords the opportunity to choose the best lubrication available.
2. Outside contaminants are effectively prohibited from entering working surfaces.
3. Case design is simplified—not required to be an oil-tight enclosure.
4. Coupling may operate for extended periods of time without servicing except for normal leakage checks.
5. Seals are required in connected equipment.
6. Seals tend to wear and age, and replacement is not always easy.
7. High ambient temperatures may adversely affect the lubricant.

Continuous lubrication:

1. Permits continuous operation but requires periodic sludge removal.
2. Removes generated heat effectively and increases coupling life significantly in applications subject to high ambient temperatures.
3. May eliminate the need for costly seals in conencted equipment.
4. Eliminates seals in the coupling itself.
5. Does not permit the choice of the best lubricating oils.
6. Requires oil supply filtration to 5 to 10 μm or less absolute particle size.
7. Requires an oil-tight case.

D. Grease Lubrication

Because gear, chain, and grid couplings are similar as to how they accommodate misalignment and transmit torque, a grease that works satisfactory in one should be good for the others. Various manufacturers have many recommendations for lubricants to use in their couplings, and it is very difficult to find a common denominator. Many tests and studies have been undertaken for gear couplings and the conclusion should also apply to chain and grid couplings. The most important discovery of many of these studies is that the wear rate of a coupling is greatly influenced by the viscosity of the base oil of the grease; the higher the viscosity, the lower the wear rate.

One study shows that centrifugal force is not only important in helping to lubricate a coupling but can cause grease to deteriorate and create serious problems. The sliding parts of a lubricated coupling would run dry if it were not for the centrifugal forces created by the rotation of a coupling. Centrifugal force is generated by the coupling's rotation and is a function of the coupling's diameter and the square of the rotational speed. Centrifugal force generates pressure in the lubricant. This pressure helps the coupling in two ways. First, it forces the lubricant to assume an annular form, flooding the coupling's teeth (or chain, or steel grid); and second, it forces the lubricant to fill all the voids rapidly.

Figure 4.18 can be used to calculate the magnitude of the centrifugal forces in a coupling. It can be seen that even in a coupling operating at motor speeds, the centrifugal force can exceed $500g$, Because the centrifugal force is a function of the square of the rotational speed but the frequency of the hub tooth oscillatory motion increases directly proportional with the speed, couplings get better lubrication when operating at high speeds. The reverse is true when a coupling operates at very low speeds. The centrifugal force decreases rapidly and below a given level can no longer force a thick lubricant, such as grease, between the coupling teeth, with rapid wear as a result.

Lubricant selection depends on the type of coupling and particularly on the application. Even the best lubricant for one application can cause rapid wear if used in another application. This fact cannot be overemphasized. A coupling user who lubricates all the couplings in the plant with whatever grease is available cannot expect good performance from the couplings. In reality, a substantial reduction in maintenance costs can be realized through the use of proper lubricants.

The best approach is to read and then follow the coupling manufacturer's recommendations. If lubricant recommendations are not available, the following can be used as a guide:

Steel grid coupling manufacturers generally recommend an NLGI No. 1 to 3 grease.
Gear and chain coupling manufacturers recommend an NLGI No. 0 to 3 grease.

Normal applications can be defined as those where the centrifugal force does not exceed 200*g*. In general, these couplings can operate at motor speeds:

$$\text{rpm} \leqslant 3600 \text{ rpm}$$

Also, normal applications are those where the misalignment at each hub is less than 3/4° and where the peak torque is less than 2 1/2 times the continuous torque. For these applications, an NLGI No. 2 grease with a high-viscosity base oil (preferably higher than 198 centistokes at 40°C) should be used.

Low-speed applications can be defined as those where the centrifugal force is lower than 10*g*. If the pitch diameter (d) is not known, the following formula can be used:

$$\text{rpm} \leqslant \frac{200}{\sqrt{\text{d}}}$$

The same conditions for misalignment and shock torque as those above are valid. For these applications, an NLGI No. 0 or No. 1 grease with a high-viscosity base oil (preferably higher than 198 centistoke at 40°C) should be used.

High-speed applications can be defined as those where the centrifugal force is higher than 200*g*, the misalignment at each hub is less than 1/2°, and where the torque transmitted is fairly uniform. For these applications, the lubricant should have a very good resistance to centrifugal separation. This information is seldom published and can be obtained from either the lubricant or the coupling manufacturer.

High-torque, high-misalignment applications can be defined as those where the centrifugal forces are lower than 200*g*, the misalignment is larger than 3/4°, and the shock loads exceed 2.5 times the continuous torque. Many such applications also have high ambient temperatures (e.g., 100°C), at which only a few greases can perform satisfactorily. Besides the characteristics of a grease for "normal applications," the grease should have antifriction and antiwear additives (such as molydisulfide), extreme pressure (EP) additives, a Timken load larger than 40 lb, and a minimum dropping point of 150°C.

E. Oil Lubrication

Oil is seldom used as a lubricant for chain and grid couplings. For gear couplings, oil lubrication is very common. The coupling can be oil filled or continuous, dipped or submerged. For oil-filled couplings the oil used should be of a high-viscosity grade, not less than 150 SSU at 100°C. The higher the viscosity, the better. For high-speed applications an oil viscosity of 2100 to 3600 SSU at 100°F can be used very successfully. Generally, an oil that conforms to MIL-L-2105 grade 140 is acceptable. The only problems with an oil of such high viscosity is that it is very difficult to pour and fill a coupling, but the time it takes is usually well worth the good operating characteristics obtained.

Continuously lubricated couplings use the oil from the system. This lubricant is seldom a high-viscosity oil. Because of this, it is best to cool the lubricant before it enters the coupling. This will help supply this lubricant to the coupling at its highest lubricity (viscosity). The viscosity should be a minimum of 50 SUS at 100°C. An oil that conforms to MIL-L-17331 is usually satisfactory.

VI. COUPLING DISASSEMBLY

A. Opening the Coupling

A visual inspection of couplings before opening them can sometimes help in planning the required maintenance procedure. For example, loose or missing bolts indicate poor installation practices; discolored (usually blue) coupling parts indicate that the coupling probably ran dry and lubrication procedures should be reviewed. If sleeves, covers, or seal carriers have to be slid off the hubs, the exposed hub surface should be cleaned with emery paper to remove rust or corrosion. Elastomer parts should be examined for wear, which indicates that they came in contact with stationary components; for localized melting or permanent set; and for signs of hardening or chemical attack. It is important to observe the state of a coupling even if it is damaged beyond repair. Understanding how or why a coupling failed can help to prevent similar occurrences in the future.

Before opening a coupling it is advisable to have on hand a container to use to store the parts. It is important to remember that most couplings use "special bolts" which should not be replaced with bolts of lesser quality, different grips, or different thread length. The coupling should be checked to see if there are match marks, particularly for some balanced couplings. If not and the coupling is to be reused, match marks should be scribed *before* the coupling is opened. Because many couplings wear-in, it is best to ensure that match marks be made even if the coupling was not balanced, so that the coupling parts can be reassembled in the same relative position they were in when the coupling was running. To shorten downtime, it is also best to have on hand all the parts that might need replacing, such as gaskets, O-rings, lubricant, and possibly a spare flex element or an entire coupling.

After visual inspection, the bolts should be removed. Make sure that wrenches of the proper size are used; this is very important. Otherwise, the heads of the bolts or the nuts could be damaged. Do not use vise-grips or adjustable wrenches. If lockwashers are used, it is suggested that the bolt be loosened rather than the nuts. This method should keep the lockwasher from digging into the flange and nut surface.

Once the bolts have been removed, one would expect the coupling to open up easily, but this is seldom the case. Parts usually stick together, particularly when gaskets are used. To separate two flanges it is best to use the jacking holes (if provided). Tapping the parts with a soft-face hammer can also help in separating parts. Prying the parts apart by hammering, chiseling, or by prying with a screwdriver is not recommended, as it will damage the surfaces of the flanges and the surfaces become difficult to seal when reassembled.

Be prepared for a mess, as some of the lubricant (if a lubricated coupling) will leak out of coupling when it is opened. All the lubricant should be removed from the coupling and discarded, as lubrication properties may deteriorate when a coupling is opened.

When the connected machines are removed from their location or when couplings are left disconnected for long periods of time, it is best to bolt a piece of cardboard or plywood to the half coupling. There are two advantages to this procedure: the coupling is protected from dirt and contamination, and the coupling bolts that are used to hold the sleeves or cover will not get lost.

B. Removing the Hubs from the Shafts

Most installed hubs need some force to be removed. The hubs can be removed by applying a continuous force or through impact.

Hammering will always damage the coupling, which would be all right only if the coupling is to be replaced. However, hammering is not recommended because it can damage the bearing in the equipment and even bend shafts.

The first thing to do before removing the hub is to remove the locking means, such as tapered keys, setscrews, intermediate bushings, or the shaft or nut. Make sure that all retaining rings, bolts, setscrews, and so on, are loosened.

The most common way to remove the hubs is by using the puller holes (Figure 4.19). If the hubs have no puller holes when received, two holes can be drilled and tapped before the hub is installed on the shaft. It is best to buy them equal to one-half the hub wall thickness to ensure proper sizing. To facilitate removal, use oil to lubricate the threads and use penetrating oil between the hub and the shaft. For straight shafts use a spacer between the shaft end and the flat bar, as force has to be applied for as long as the hub slides on the shaft. If there are no puller holes, a "wheel puller" can be used; however, care should be taken not to damage the coupling. Figure 4.20 illustrates some don'ts for this procedure.

If the hub does not move even when maximum force is applied, apply a blow with a soft-face hammer. If nothing happens, heat must be applied to the hub. It is important to use a low-temperature flame (not a welding torch) and to apply the heat uniformly around the hub (localized temperature should not exceed 600°F for most steel). Heat does not have the same effect in this case as it has at installation, when only the hub is heated. At removal the hub on the shaft and the heat will also expand the shaft. The secret is to heat only the outside diameter of the hub, shielding the shaft from the heat. If possible, the shaft should be wrapped with wet rags. *The heat should be applied while the pulling force is being applied.*

Caution: If heat has been applied to remove the hub, hardness checks should be made at various sections before using the hub again, to assure that softening of the material has not occurred. If there is any doubt, the coupling should be replaced. Rubber components should be removed during heating. If not removed, they will have to be replaced if they come into contact with excessive heat.

The force required to remove a hub is a function of its size (bore and length) and of the interference used at installation. For very large couplings, pressure can be used in two ways for hub removal. The first is to use a portable hydraulic ram such as the puller press shown in Figure 4.19. The force that can be applied is limited only by the strength of all the threads and rods.

There are many cases where nothing helps to remove the hub; then the hub has to be cut. This usually cheaper than replacing the shaft. The place to cut is above the key, as shown in Figure 4.21. The cut should be made with a saw, but a skilled welder could also flame-cut the hub, taking precautions that the flame does not touch the shaft. After the cut is made, a chisel is hammered in the cut, spreading the hub and relieving its grip on the shaft. If the hub is too thick, it can be machined down to a thin ring, then split.

C. Intermediate Bushings

To loosen intermediate bushings, remove the bolts (usually more than two) and insert them in the alternate holes provided, making sure to lubricate the threads and the bolt (or setscrew) points. When tightened, the bolts will push the bushing out of the hub and relieve the grip on the shaft. The bushing does not have to be moved more than 1/4 in. If the bushing is still tight on the shaft, insert a wedge in the bushing's slot and spread it open. Squirting penetrating oil in the slot will also help in sliding the hub-bushing assembly off the shaft. Do not spread the bushing open without the hub on it, as this will result in the bushing breaking or yielding at or near the keyway.

D. Returning Parts to Service

Any part that is to be reused should be thoroughly cleaned and examined before using. The part should be dimensionally inspected to assure that it is within the coupling manufacturers limits. The part should be examined for dents, burrs, corrosion. It should be non-destructively tested (magnaflux or dye penetrant tested) for surface cracks or other distress. If there is any question about the coupling integrity, *throw it away* and use a new one or return it to the coupling manufacturer to determine its suitability for further service. *Using damaged or deteriorated parts could result in serious failure of the coupling or the equipment.*

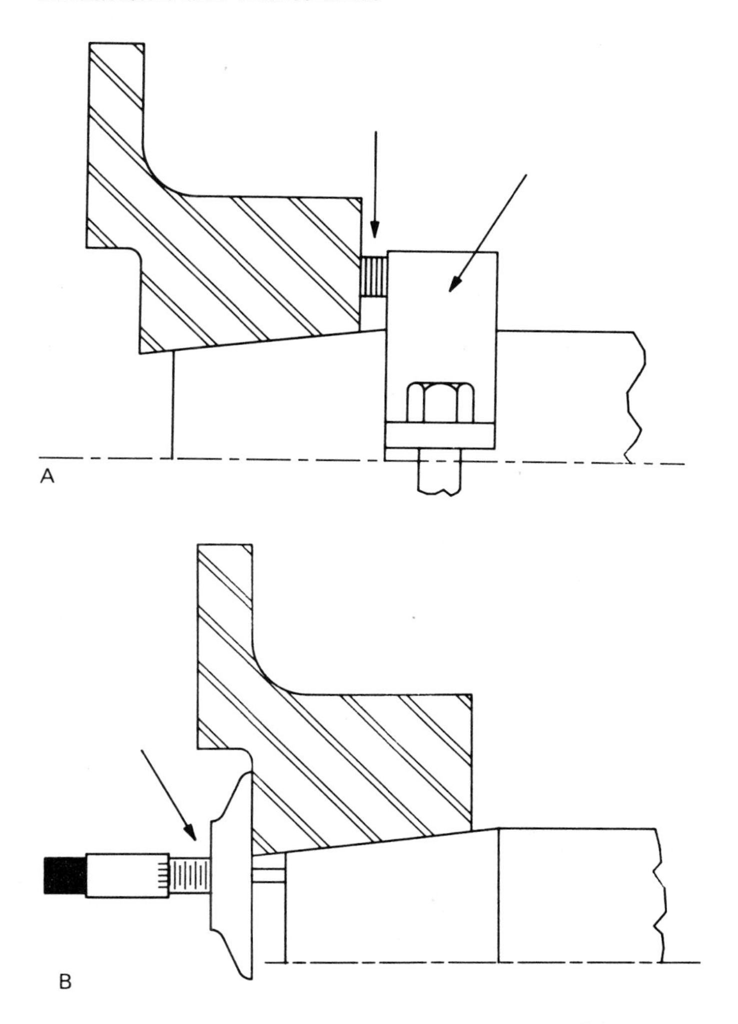

Figure 4.1 Positioning hubs: (A) hub position measurement with depth gage; (B) positive stop used when mounting a hub.

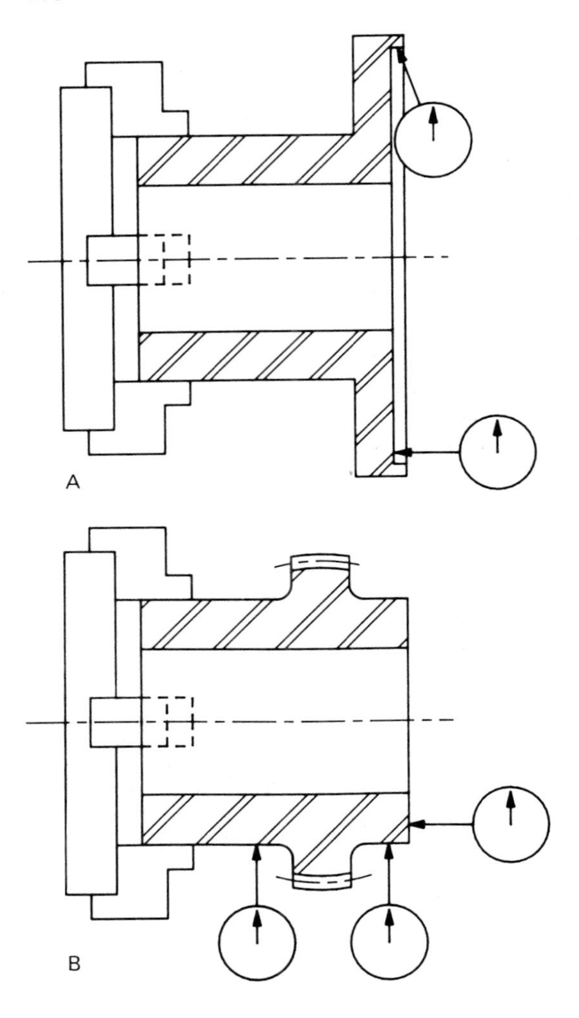

Figure 4.2 Setup for reboring hubs: (A) rigid hub setup; (B) flex hub setup.

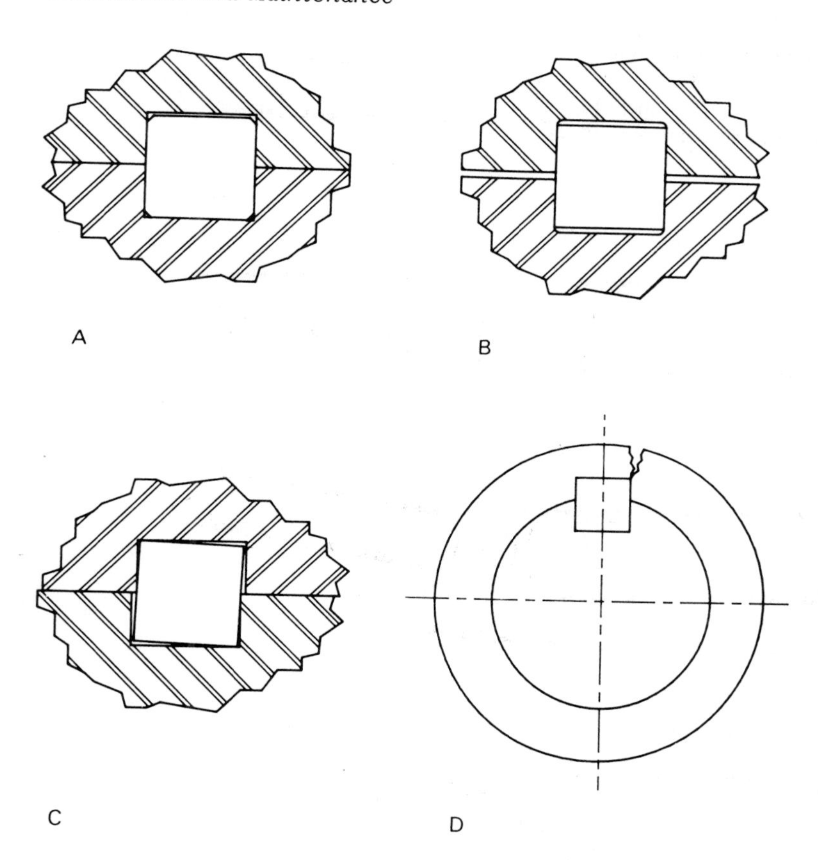

Figure 4.3 Improperly fitted keys: (A) proper key fit; (B) improper fit of key with no chamfers; (C) rolling of a sloppy key; (D) a split hub from an improperly fitted key.

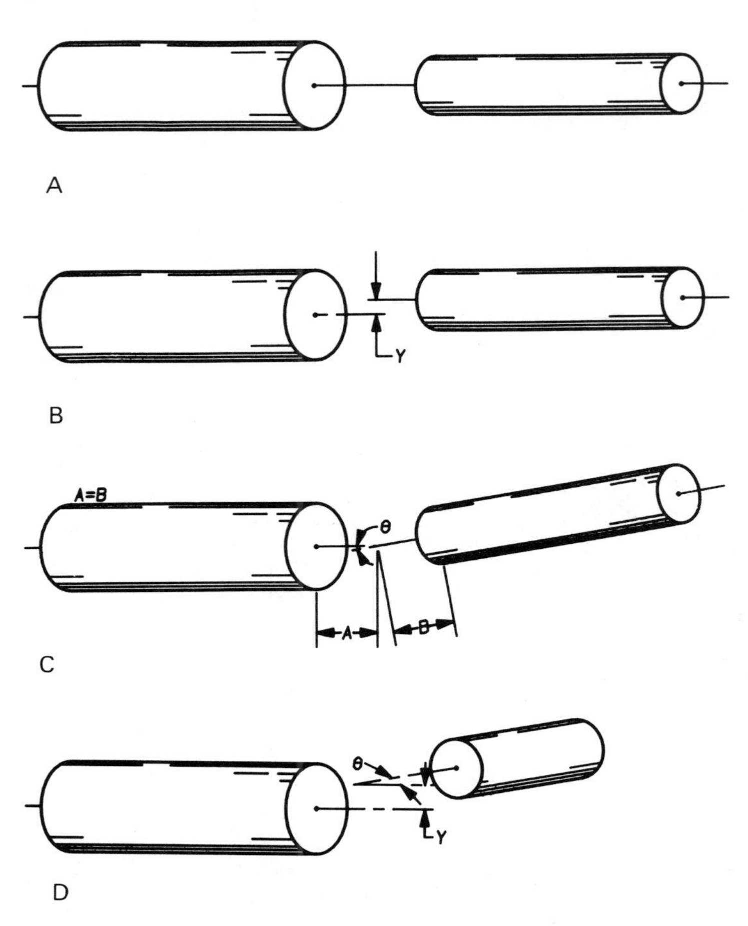

Figure 4.4 Shaft alignment conditions: (A) alignment; (B) parallel offset misalignment; (C) symmetrical angular misalignment; (D) combined angular and offset misalignment.

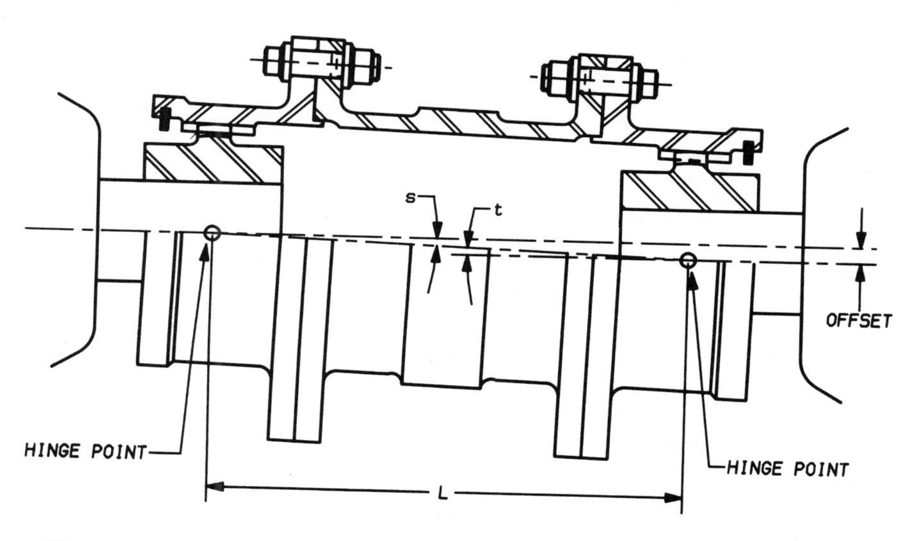

Figure 4.5 Misalignment of a spacer-type gear coupling.

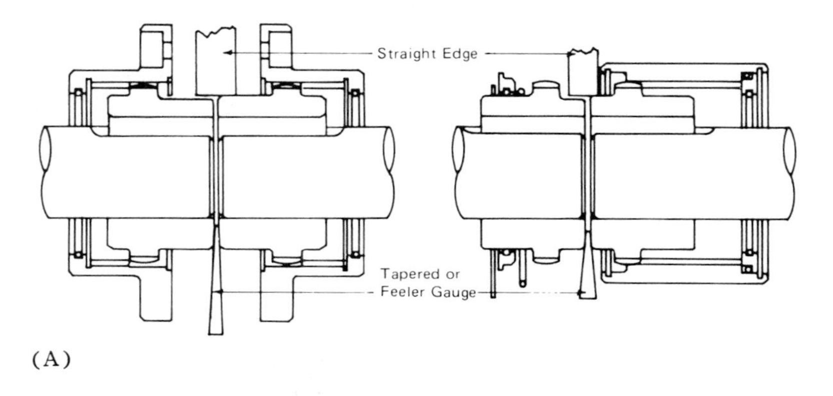

(A)

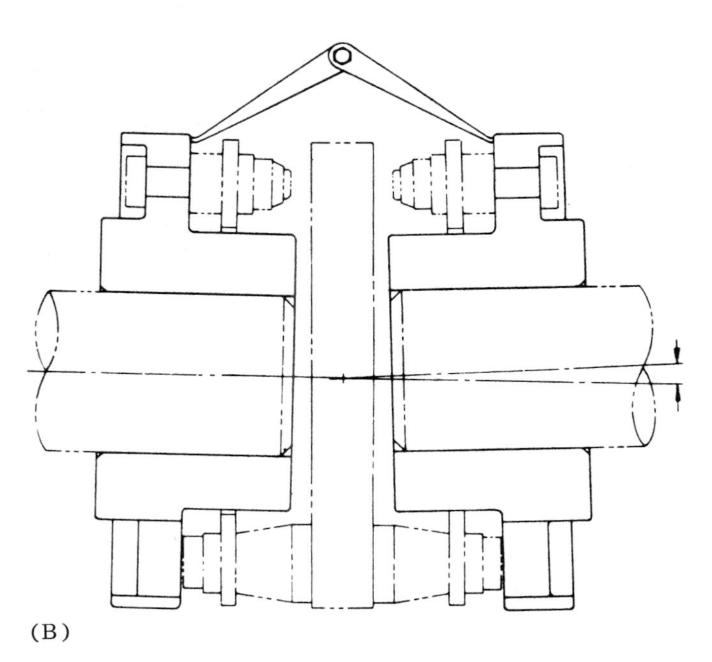

(B)

Figure 4.6 Alignment methods: (A) straightedge and feeler gage alignment of a gear coupling; (B) caliper alignment of a disk coupling; (C) straightedge alignment of a disk coupling.

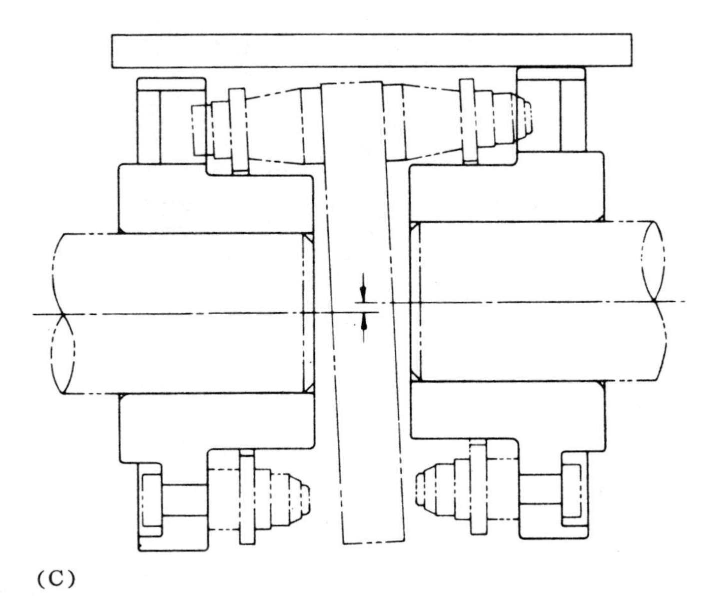

(C)

Figure 4.6 (continued)

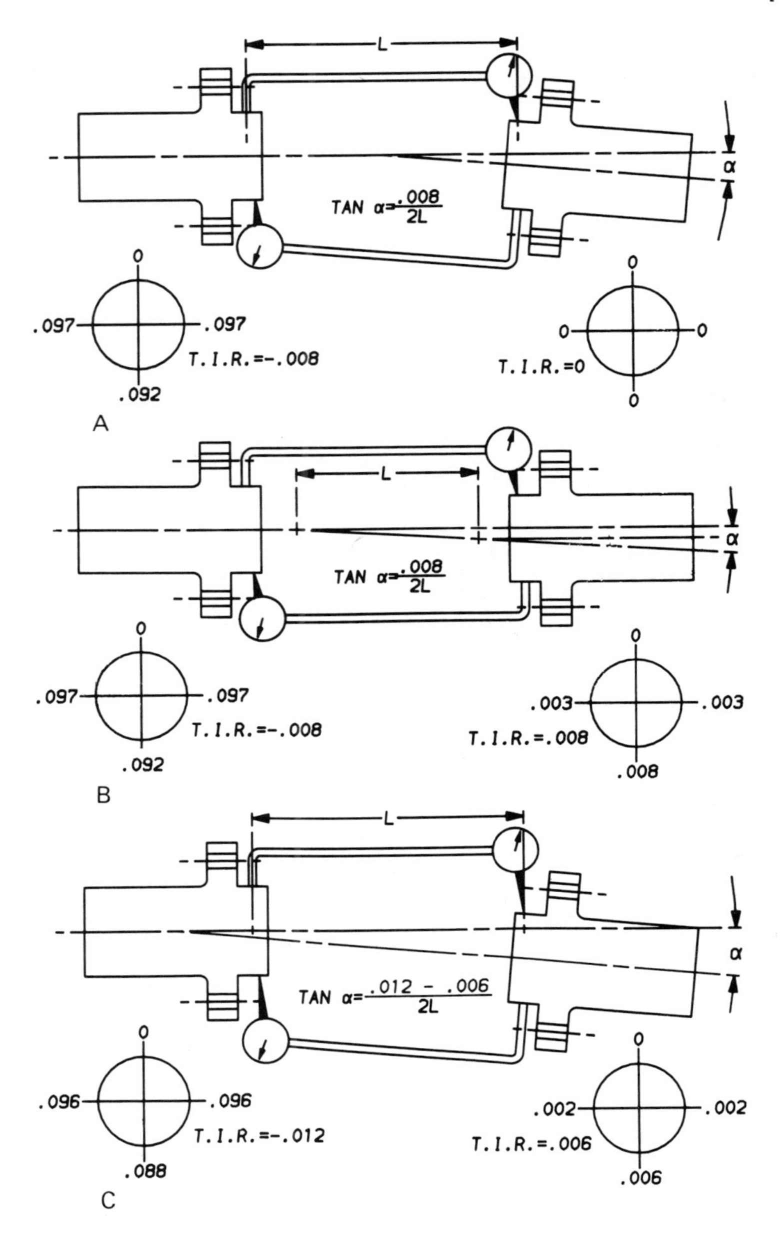

Figure 4.7 How to measure misalignment: (A) measuring angular misalignment; (B) measuring offset misalignment; (C) measuring combinations of misalignments.

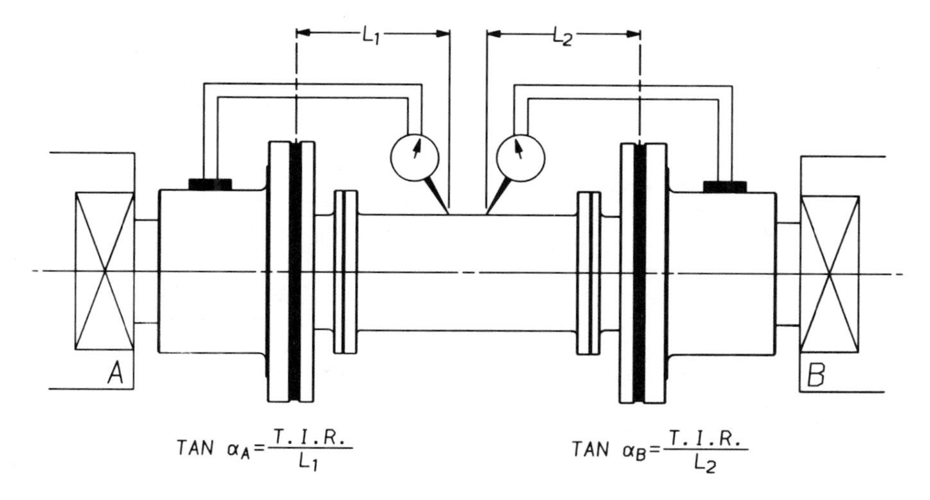

Figure 4.8 Measuring misalignment on an assembled coupling.

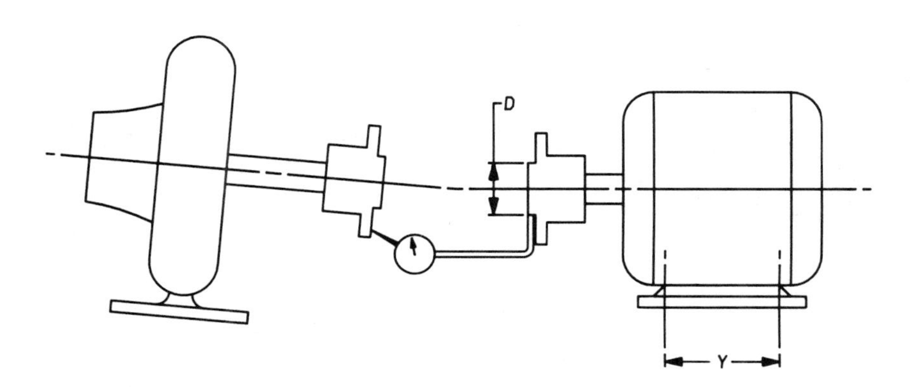

Figure 4.9 Correction of face runout.

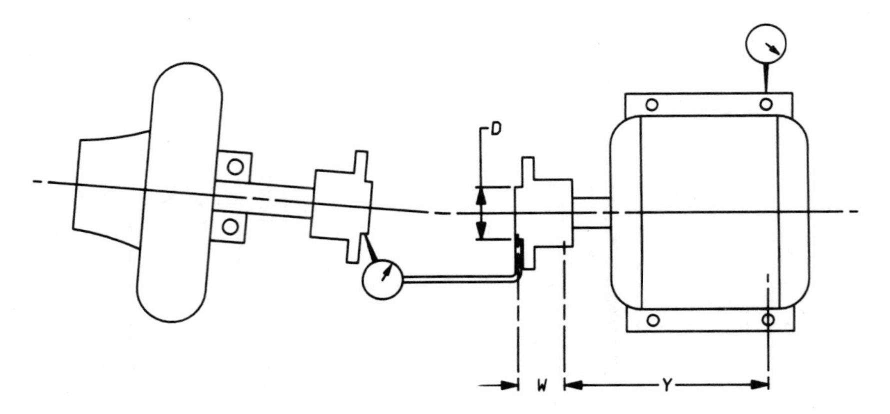

Figure 4.10 Correction for horizontal misalignment.

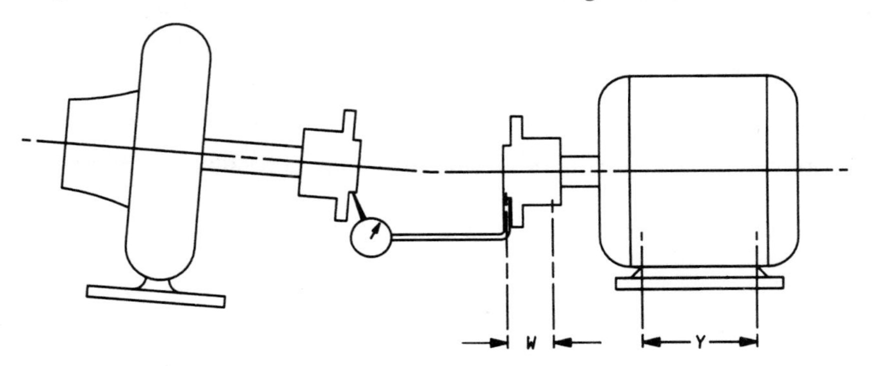

Figure 4.11 Correction for angular misalignment.

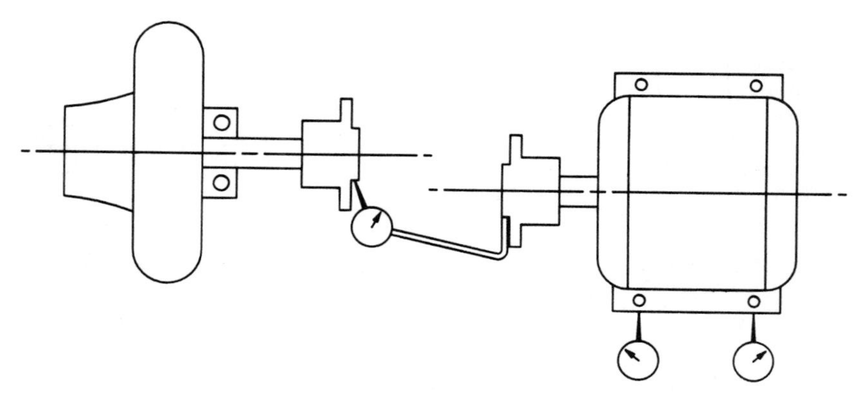

Figure 4.12 Correction for offset misalignment.

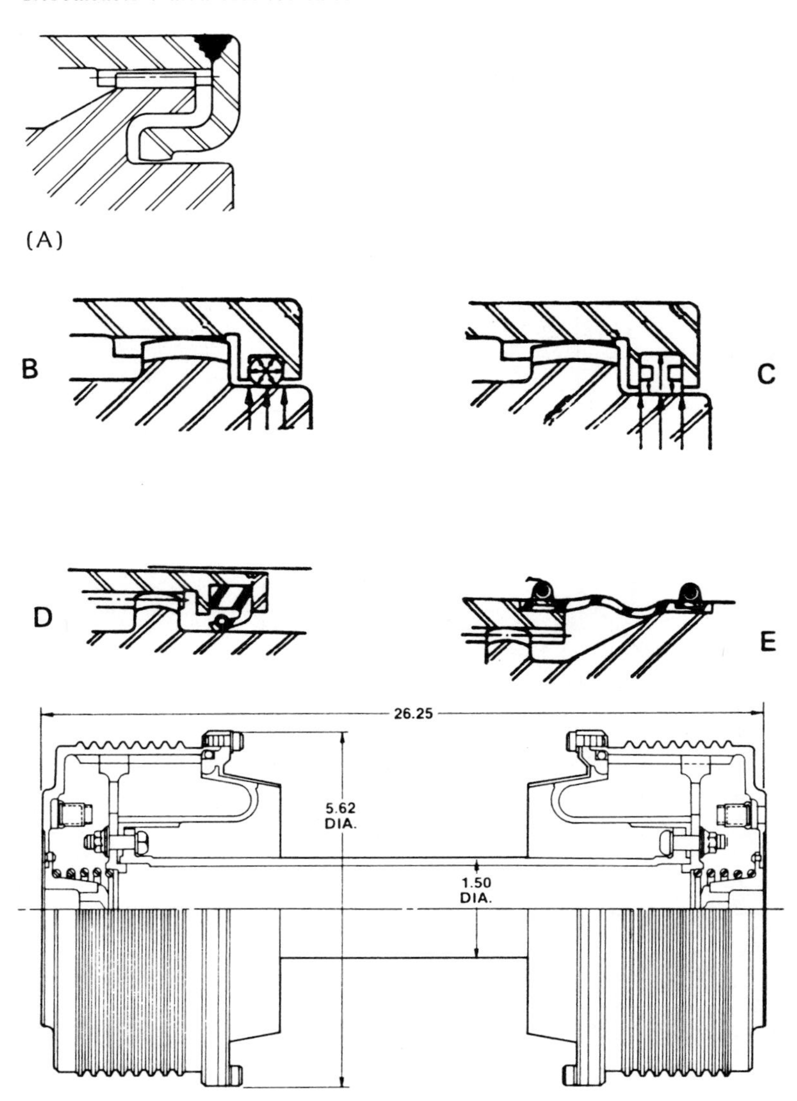

Figure 4.13 Types of seals: (A) metal labyrinth seal; (B) O-ring; (C) H or T cross-section seal; (D) lip seal; (E) boot seal; (F) high-speed boot seal.

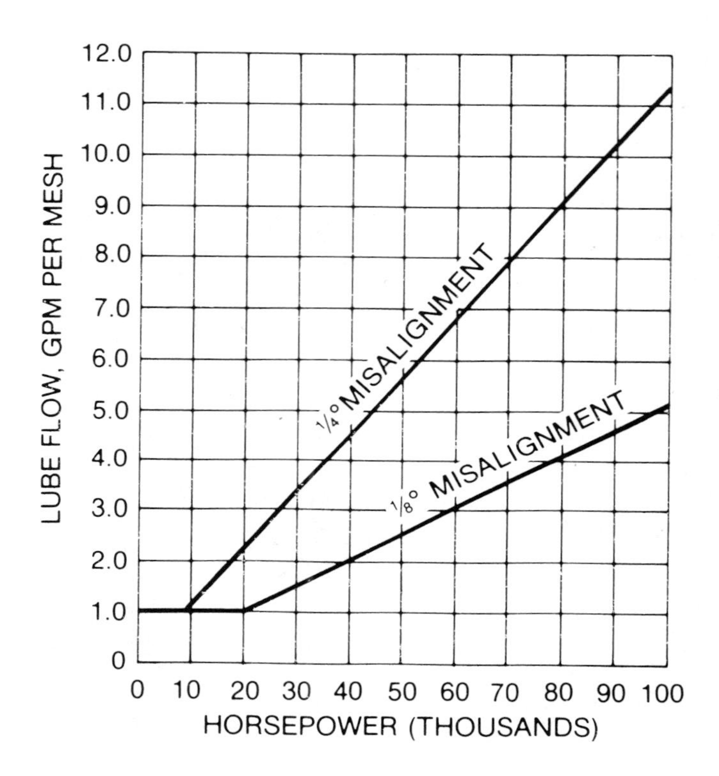

Figure 4.14 Amount of lube flow required for gear couplings.

Figure 4.15 Sludge-contaminated gear coupling.

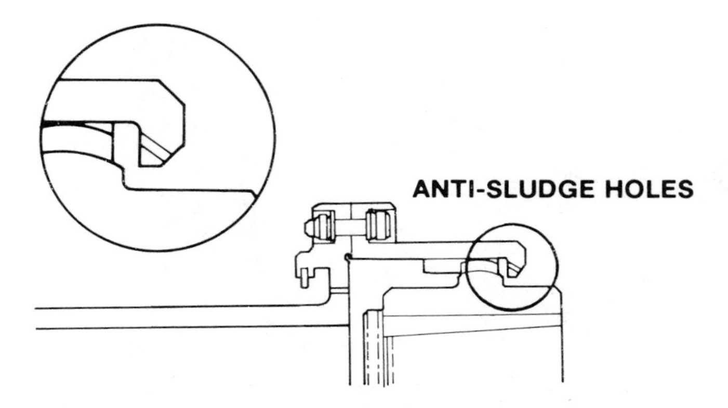

Figure 4.16 Design to help minimize sludge buildup: flow-through design.

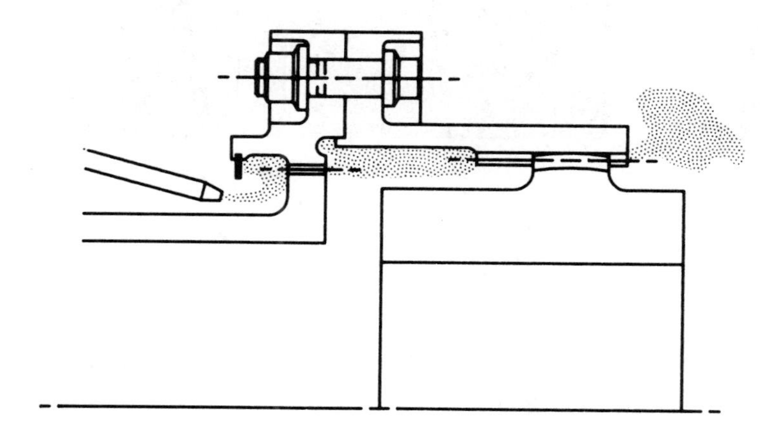

Figure 4.17 Oil flow in a damless gear coupling.

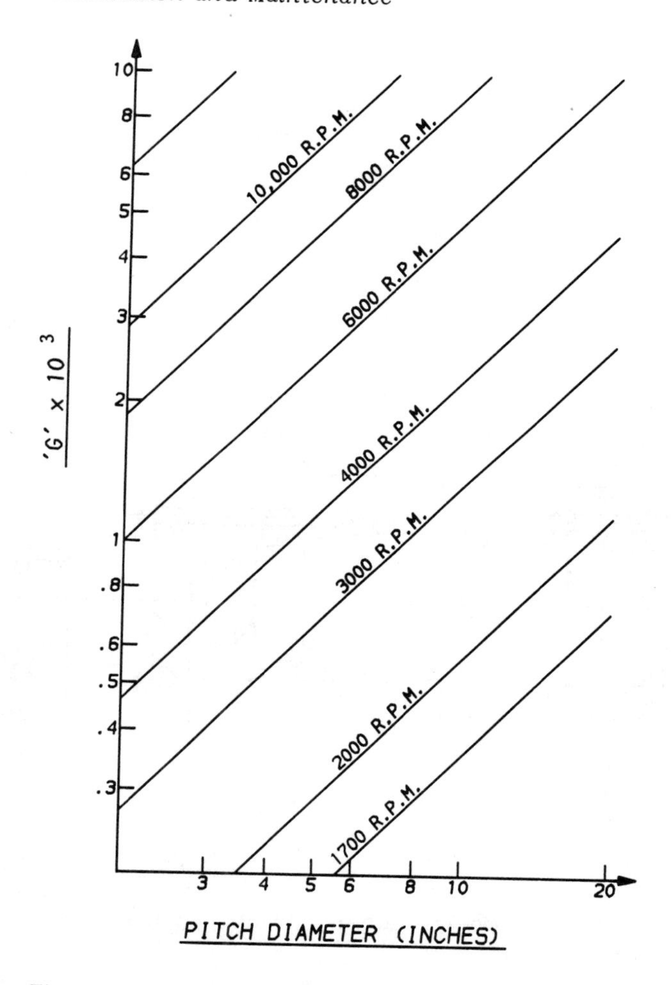

Figure 4.18 Centrifugal force developed in a coupling.

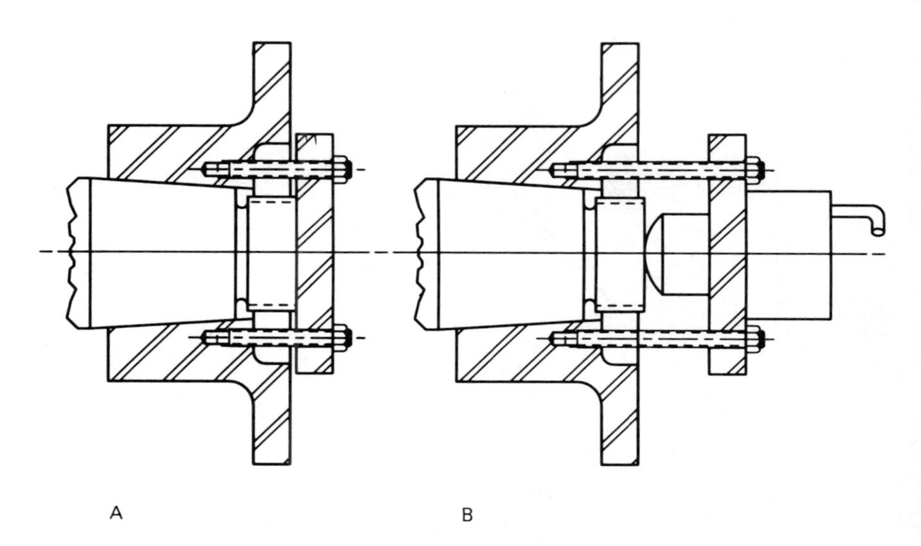

Figure 4.19 Methods of removing hubs: (A) using puller holes; (B) using a hydraulic press.

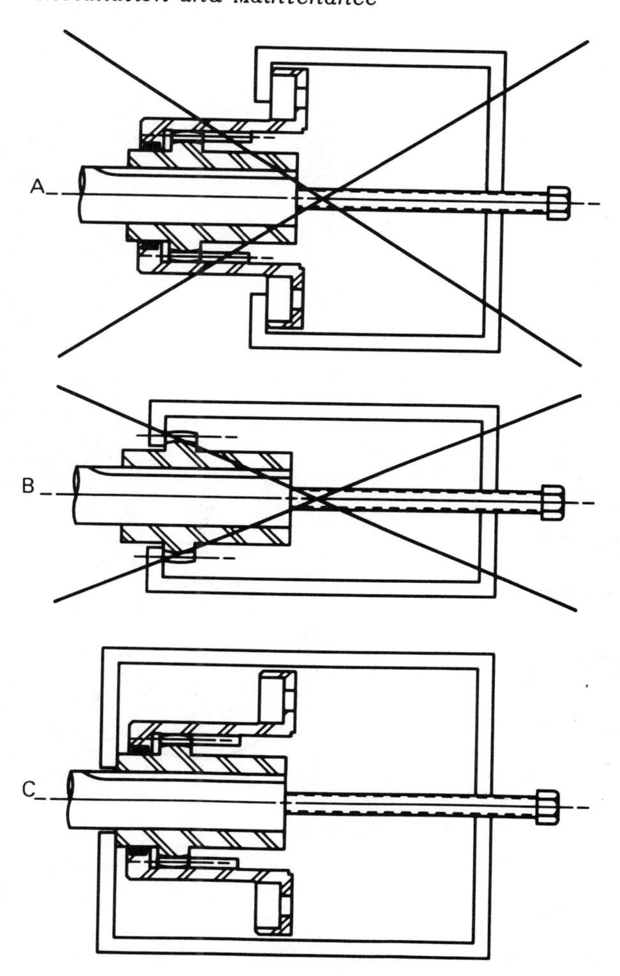

Figure 4.20 Removing hubs with a wheel puller: (A), (B) improper positioning of puller; (C) proper positioning of puller.

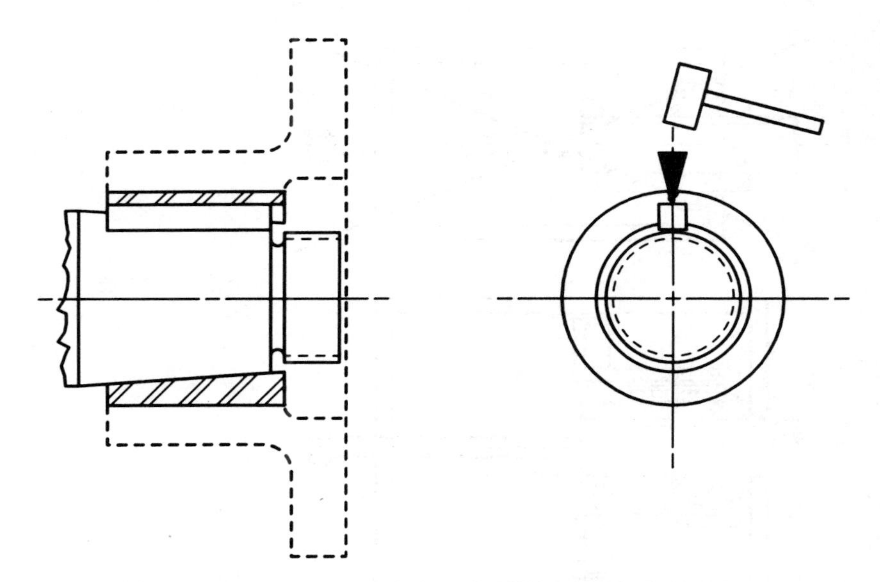

Figure 4.21 Splitting a hub for removal from a shaft.

5
Rigid Couplings

I. INTRODUCTION

Whenever two pieces of rotating equipment need to be connected, the question arises as to whether they should be directly coupled or indirectly coupled. Equipment can be indirectly coupled by belts or chains (see Figure 5.1A), but indirectly coupled equipment is usually less efficient than directly coupled equipment. This is normally due to the frictional loses from the belts and chains slipping and/or from sliding when they transmit power. Therefore, it is usually preferable to directly couple two pieces of rotating equipment (Figure 5.1B).

The main problem associated with directly coupled equipment is how to accommodate for the usual misalignment that can occur between two pieces of rotating equipment. Rigid couplings are used when misalignment does not exist or when it is very small. If this is the case, rigid couplings are a very effective means of connecting equipment. There are four basic types of rigid couplings:

1. The flanged rigid coupling (Figure 2.7A)
2. The ribbed rigid coupling (Figure 2.7B)
3. The sleeve rigid coupling (Figure 2.7C)
4. The quill shaft rigid coupling (Figure 2.7D)

In general, rigid couplings connect two pieces of rotating equipment. They allow for the transfer of power from one piece of equipment to the other. They also allow equipment that has different size shafts to be connected. It is important to stress the fact that rigid couplings should be used only when the equipment has virtually no misalignment and/or when the shafts of the equipment or the rigid coupling (quill shaft rigid coupling) are long and slender enough so that they can flex and accept the forces and moments produced by the mechanical deflection on these parts due to the misalignment imposed by the connected equipment.

II. FLANGED RIGID COUPLINGS

The flanged rigid coupling is probably the most common type of rigid connection used. This coupling is available from almost every flexible coupling manufacturer in several standard series. One of the most common is interchangeable with standard flanged gear couplings. Figure 5.2 lists some of their dimensions. Rigid couplings are usually made from carbon steel 1035 to 1050 either from bar stock or from forgings. They can handle large amounts of torque for their sizes; see Figure 5.2, which also shows the typical torques that can be handled by the various sizes.

There are many other configurations and sizes available (see Figure 5.3). They can be supplied with replaceable bushings (see Figure 5.4); this allows the rigid to be easily removed so that equipment bearings and seals can be replaced or serviced. The flanged rigid is also available in various materials: gray iron, mallable iron, various carbon steels, and in alloy steels. The design is usually limited by the number, size, and type of bolts used (see Section II of Chapter 3) for the various stress calculations that are usually considered. The ratings should consider the stresses in bolts, hubs, and the flange itself. It is suggested that you check with specific coupling manufacturers for information on material used, bore sizes allowed, number of bolts used, and torque and speed ratings.

The flanged rigid coupling can be used any place where no misalignment is present or virtually none. Some applications include pumps (vertical and horizontal) and crane drives (see Figure 5.5).

III. RIBBED RIGID COUPLINGS

The ribbed rigid coupling is used where ease of assembly and disassembly is required. The coupling clamps on the shafts. The shaft and the coupling hubs are usually keyed. The two halves are held together by radial bolts at the split. The quantity of bolts can vary depending on the size of the coupling; typically, four to eight are used. Torque is transferred from one half to the other by the frictional force produced by the bolt rather than by direct loading of the bolts themselves.

These couplings are usually made of material having a strength compatible with AISI 1018 steel shafts (relatively low torque capacities). Many coupling manufacturers make a standard line of ribbed couplings. There is no industrial standard. Typically, these couplings are available in bore sizes from 1 to 7 in. Figure 5.6 shows some typical information and dimensions.

These couplings are usually used for low-speed, low-torque applications. The ribbed coupling is typically used on vertical pumps, agitators, winches (see Figure 5.7), and in many other types of applications.

IV. SLEEVE RIGID COUPLINGS

This form of a coupling is probably one of the simplest forms of a coupling available. On small drives (usually fractional horsepower) where the equipment shafts are of the same diameter, a sleeve rigid coupling can be slid onto the shaft of one piece of equipment, the equipment put in place, and the sleeve rigid coupling slid onto the other piece of equipment's shaft. The sleeve is usually locked to the shafts with two setscrews, one for each piece of equipment. What could be simpler? There is no industrial standard for this type of coupling. Figure 5.8 shows some information for a standard line offered by one coupling manufacturer. For larger sizes the sleeve rigid coupling can be supplied with a replaceable bushing for ease of assembly and disassembly. See Figure 5.9 for some typical dimensions.

For high-torque applications a more sophisticated sleeve rigid coupling has been used. The coupling consists of two sleeves of high-quality steel (see Figure 5.10), a thin inner sleeve and a thick outer sleeve. The outer surface of the inner sleeve is tapered and the bore of the outer sleeve has a corresponding taper. The inner sleeve bore is somewhat larger than the diameter of the shafts so that the sleeve can be passed over them with ease. The outer sleeve is then driven up the tapered inner sleeve using the hydraulic unit incorporated in the coupling. This action compresses the inner sleeve onto both shafts. To allow this drive-up, the friction of the matching tapered surfaces is first overcome by injecting oil at high pressure between them, where it forms a load-carrying film separating the two components. When the outer sleeve has reached its correct position, the injection pressure is released and the oil is drained off, restoring normal friction between the sleeves. Dismounting the coupling is equally simple. Oil is injected between the coupling sleeves to overcome the friction. As a result of the taper, the compressive force has an axial component which causes the outer sleeve to slide down the taper, forcing the oil out of the hydraulic unit. By controlling the flow of this oil, the sleeve can be prevented from sliding too quickly.

Sleeve rigid couplings with setscrews are used on motor-driven pumps. The hydraulic sleeve rigid coupling is used on high-torque applications such as marine propulsion shafting (see Figure 5.11).

V. QUILL SHAFT RIGID COUPLINGS

The quill shaft rigid coupling gets its name from the fact it looks like the quill shafts that go through some large gears in gearboxes; they are usually long and slender. These rigids do accommodate for some misalignment; they do this through the flexing of their long slender shafts. Quill shaft rigid couplings are usually much smaller than flexible couplings. Generally, they are 25 to 50% smaller. For example, a 40-in. quill shaft rigid coupling is approximately equivalent to a 60-in. flexible coupling.

Quill shaft rigid couplings are usually made of high-grade alloy steels. This is because the stresses imposed on them when they are misaligned (flexed) are usually quite high and they must be designed for the cyclic loads imposed on them when they are flexed due to misalignment. These couplings are usually used on large, high-horsepower steam and gas turbines (see Figure 5.12).

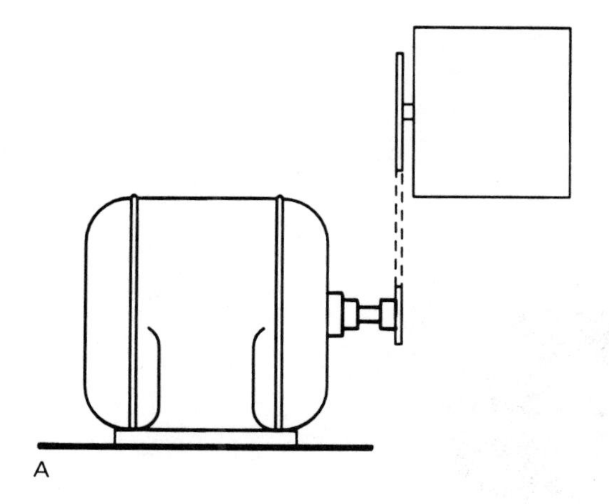

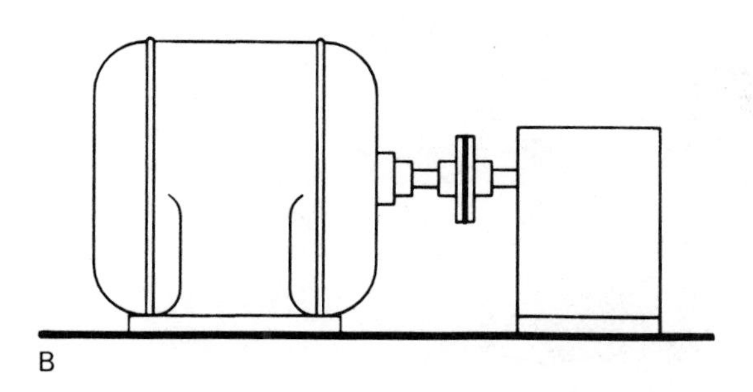

Figure 5.1 Equipment connection methods: (A) indirectly coupled equipment; (B) directly coupled equipment.

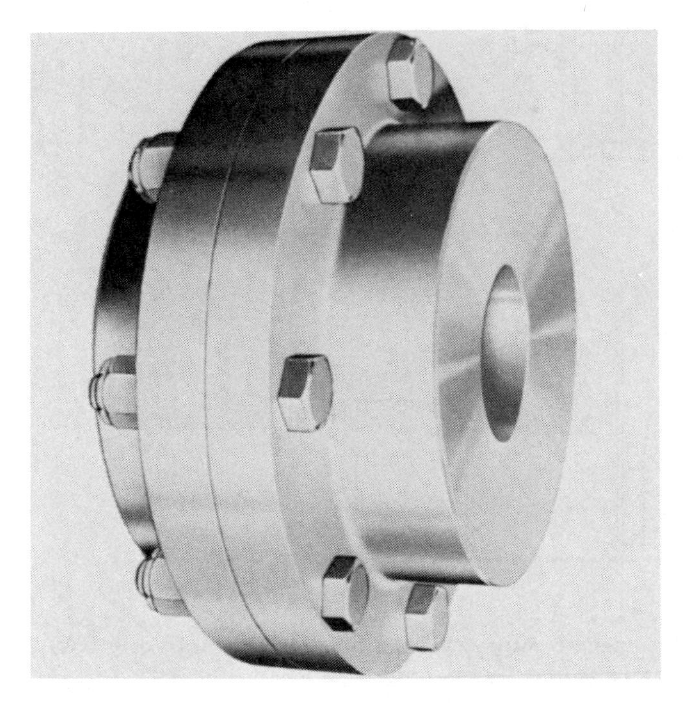

Figure 5.2 Typical rigid coupling dimensions (courtesy of Dodge Division of Reliance Electric).

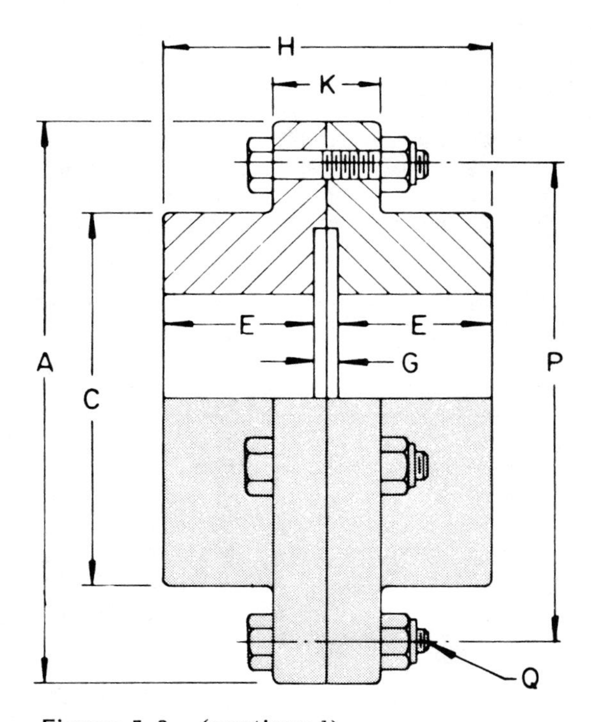

Figure 5.2 (continued)

Size	Max. rpm	Torque (lb.-in.)	A diameter	C diameter	E	G
1W	10,000	8820	4 9/16	3	1 9/16	3/16
1 1/2W	7400	22,680	6	3 13/16	1 27/32	3/16
2W	5900	44,100	7	4 13/16	2 9/32	3/16
2 1/2W	5000	75,600	8 3/8	5 23/32	2 29/32	3/16
3W	4300	132,300	9 7/16	6 23/32	3 13/32	3/16
3 1/2W	3900	189,000	11	7 3/4	3 31/32	3/16
4W	3500	299,250	12 1/2	8 31/32	4 7/16	3/8
4 1/2W	3200	409,500	13 5/8	10 1/8	5	3/8
5W	2900	576,450	15 5/16	11 3/8	5 3/4	3/8
5 1/2W	2700	756,000	16 9/16	12 9/16	6 1/8	3/8
6W	2500	1,008,000	18	13 7/8	7 1/4	3/8
7W	2200	1,575,000	20 3/4	15 3/4	8 11/16	1/2

Size	H	K	P bolt circle	Q bolts Number	Q bolts Size
1W	3 9/32	1 1/8	3 3/4	6	1/4—28 × 1 1/2
1 1/2W	3 27/32	1 1/2	4 13/16	8	3/8—24 × 2
2W	4 23/32	1 1/2	5 7/8	6	1/2—20 × 2 1/4
2 1/2W	6	1 7/8	7 1/8	6	5/8—18 × 2 3/4
3W	7	1 7/8	8 1/8	8	5/8—18 × 2 3/4
3 1/2W	8 5/32	2 1/4	9 1/2	8	3/4—16 × 3 1/4
4W	9 3/16	2 1/4	11	8	3/4—16 × 3 1/4
4 1/2W	10 11/32	2 1/4	12	10	3/4—16 × 3 1/4
5W	11 27/32	3	13 1/2	8	7/8—14 × 4 1/4
5 1/2W	12 19/32	3	14 1/2	14	7/8—14 × 4 1/4
6W	14 27/32	2	15 3/4	14	7/8—14 × 3 1/4
7W	17 13/16	2 1/4	18 1/4	16	1—14 × 3 1/2

Figure 5.2 (continued)

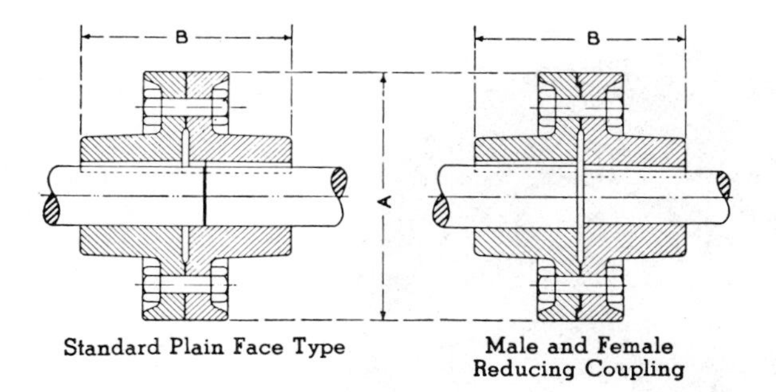

Shaft Size	In-cluded Sizes	Weight lbs.	A	B	Shaft Size	In-cluded Sizes	Weight lbs.	A	B
1 3/16	1 1/4	15	6	4 5/8	3 15/16	4	130	12 1/2	10 5/8
1 7/16	1 1/2	19	6 3/4	5 1/8	4 7/16	4 1/2	181	13 1/2	13 5/8
1 11/16	1 3/4	26	7 1/4	5 3/4	4 15/16	5	251	16 1/4	14 3/8
1 15/16	2	34	8	6 3/8	5 7/16	5 1/2	310	17 1/2	15 1/8
2 3/16	2 1/4	40	8 1/2	6 7/8	5 15/16	6	400	19	16 1/8
2 7/16	2 1/2	48	9	7 3/8	6 7/16	6 1/2	525	20	16 3/4
2 11/16	2 3/4	57	9 3/4	7 7/8	6 15/16	7	575	21	17 1/2
2 15/16	3	74	10 1/2	8 3/8	7 7/16	7 1/2	800	22 1/2	18 1/2
3 7/16	3 1/2	93	11 1/4	9 3/8	7 15/16	8	875	24	19 1/2

Figure 5.3 Other rigid coupling configurations (courtesy of Dodge Division of Reliance Electric).

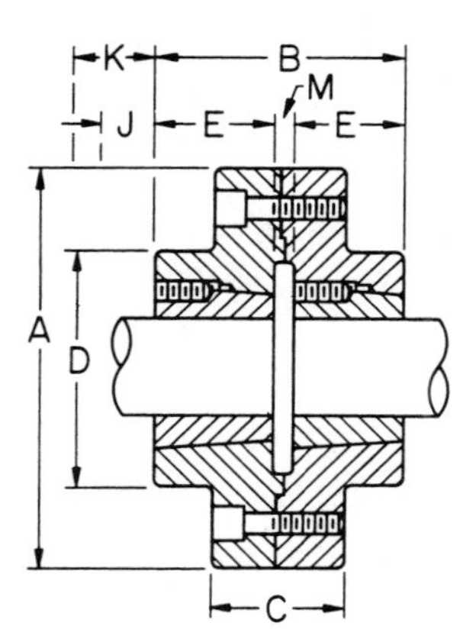

Figure 5.4 Rigid couplings with replaceable bushings (courtesy of Dodge Division of Reliance Electric).

Figure 5.5 Shop crane application of a flanged rigid coupling (courtesy of Dodge Division of Reliance Electric).

Shaft size	Maximum bore	Maximum rpm	Outside diameter	Length	Number and size of bolts
1, 1 3/16, 1 1/4	1 1/4	5360	3 5/8	5 1/4	4–3/8
1 3/8, 1 7/16, 1 1/2	1 1/2	4130	4 5/8	6 3/16	4–1/2
1 11/16, 1 3/4	1 3/4	3965	4 13/16	7 1/16	4–1/2
1 7/8, 1 15/16, 2	2	3635	5 1/4	7 15/16	4–1/2
2 3/16, 2 1/4	2 1/4	3180	6	8 5/8	4–5/8
2 7/16, 2 1/2	2 1/2	2965	6 7/16	9 11/16	6–5/8
2 11/16, 2 3/4	2 3/4	2830	6 3/4	10 9/16	6–5/8
2 15/16, 3	3	2545	7 1/2	11 3/8	6–3/4
3 3/16	3 1/4	2315	8 1/4	12 1/4	6–7/8
3 7/16	3 1/2	2165	8 13/16	13 3/16	6–7/8
3 15/16	4	1900	10 1/16	15 1/4	6–1
4 7/16	4 1/2	1775	10 3/4	18 3/16	6–1 1/8
4 15/16	5	1625	11 3/4	19 5/8	6–1 1/8
5 7/16	5 1/2	1390	13 3/4	20 3/8	8–1 1/8
5 15/16	6	1365	14	20 3/4	6–1 1/4
7	7 1/16	1230	15 1/2	21 15/16	8–1 1/4

Figure 5.6 Ribbed rigid coupling: typical dimensions (courtesy of Dodge Division of Reliance Electric).

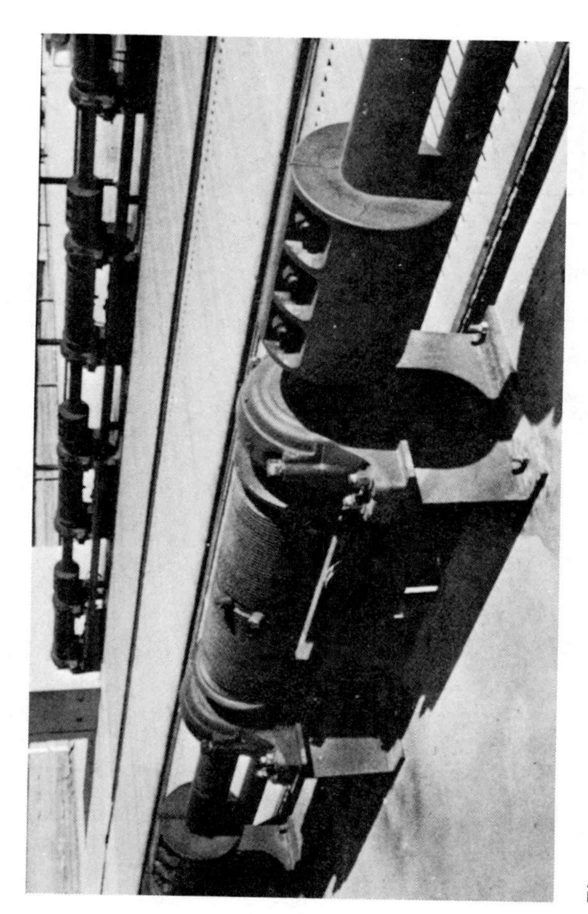

Figure 5.7 Ribbed rigid coupling application (courtesy of Dodge Division of Reliance Electric).

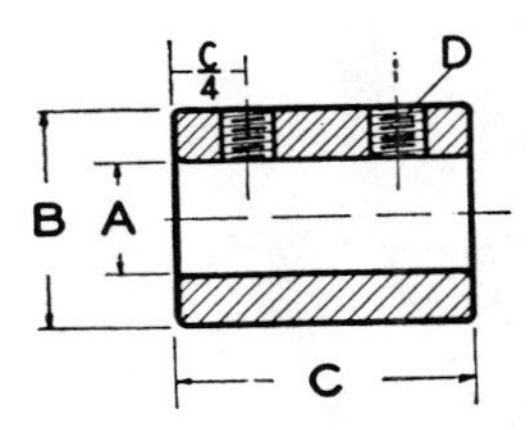

Part number	B (in.)	C (in.)	A (in.)	C/4 (in.)	Weight
CS−04	1/2	3/4	1/4	3/16	0.06
CS−05	5/8	1	5/16	1/4	0.06
CS−06	3/4	1	3/8	1/4	0.1
CS−08	1	1 1/2	1/2	3/8	0.2
CS−10	1 1/4	2	5/8	1/2	0.5
CS−12	1 1/2	2	3/4	1/2	0.8
CS−14	1 3/4	2	7/8	1/2	1.0
CS−16	2	3	1	3/4	1.9
CS−18	2 1/8	3	1 1/8	3/4	2.1
CS−20	2 1/4	4	1 1/4	1	3.1
CS−22	2 1/2	4 1/2	1 3/8	1	4.3

Figure 5.8 Sleeve rigid coupling: typical dimensions.

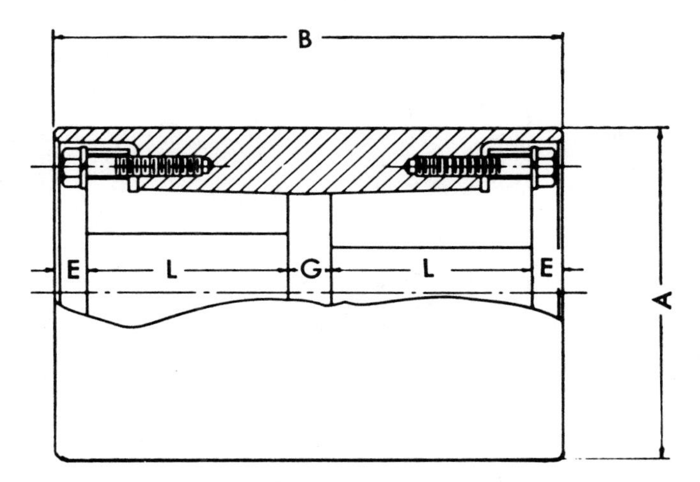

Figure 5.9 Sleeve rigid coupling with replaceable bushings. Data on the coupling appears on the next page (courtesy of T. B. Wood's Sons Company).

Coupling number	Maximum bore		Bushing	Dimensions					Weight (including bushings)
	Light loads	Heavy loads		A	B	E	G	L	
44-SD	1 3/4	1 1/2	SD	4	4 5/8	3/8	1/4	1 13/16	11
44-SF	2 1/2	2 1/4	SF	5 1/2	5 1/4	1/2	1/4	2	22
44-E	3 3/16	2 3/4	E	6 7/8	6 3/4	5/8	1/4	2 5/8	54
44-J	4	3 5/8	J	8 1/4	11	3/4	1/2	4 1/2	122
44-M	5 1/4	4 11/16	M	10	16	1	1/2	6 3/4	270

Figure 5.9 (continued)

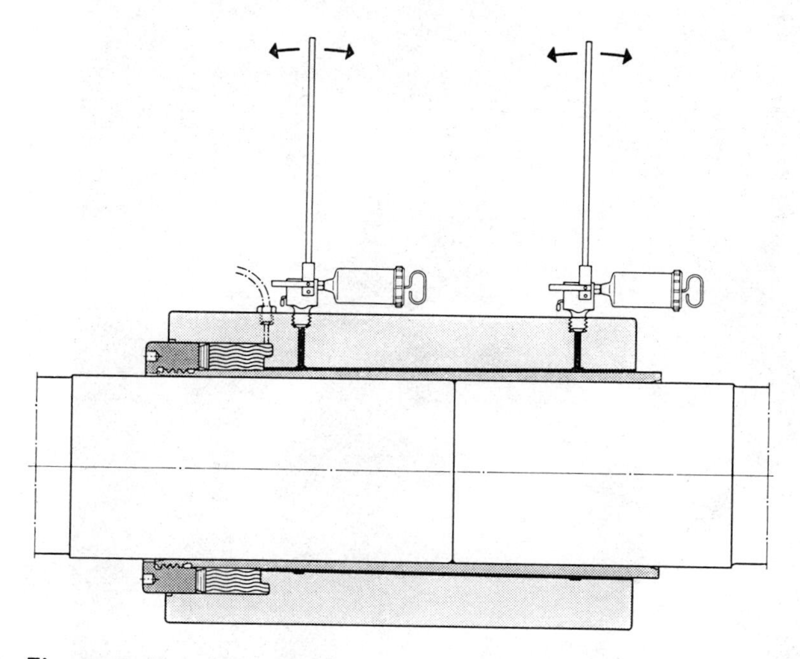

Figure 5.10 High-torque sleeve rigid coupling (courtesy of SKF Steel, Coupling Division).

Figure 5.11 Application of a sleeve rigid coupling for a propeller shaft (courtesy of SKF Steel, Coupling Division).

Figure 5.12 Quill shaft rigid coupling (courtesy of General Electric Schenectady, Gas Turbine Division).

6

Mechanically Flexible Couplings

I. INTRODUCTION

The mechanical flexible coupling accommodates the three basic requirements of a flexible coupling: they transmit power, accommodate for misalignment, and compensate for shaft end float. The couplings in this group obtain their flexibility from loose-fitting parts and/or rolling or sliding of mating parts. They usually require lubrication unless one of the moving parts is made of a material that supplies its own lubrication, such as nylon (chain on chain couplings and sleeves on gear couplings).

There are four basic types of couplings in this group:

1. The gear coupling
2. The chain coupling
3. The grid coupling
4. The universal joint

Gear couplings have been applied to bores to 36 in., misalignment to 7 1/2°, speeds to 40,000 rpm, and torques to 40,000,000 in.-lb. *Chain couplings* have been applied to bores to 10 in., misalignment to 2°, speeds to 6500 rpm, and torques to 1,500,000 in.-lb. *Grid couplings* have been applied to bores to 20 in., misalignment to 1/3°, speeds to 4000 rpm, and torques to 4,000,000 in.-lb. *Universal joints* have been applied to bores to 20 in., misalignments to 20°, speeds to 8000 rpm, and torques to 25,000,000 in.-lb.

Note: The foregoing information is based on present published information. Also, it is unlikely that all the maximum conditions could be handled in any or all combinations listed.

The couplings in this group are probably the most commonly used in the industry. Although these couplings must be lubricated, when properly installed, lubricated, and maintained, they have been known to last 25 years or more. Normally, the life span is 3 to 5 years. The most common mode of failure for this group of couplings is that they wear out. This type of failure mode is usually predictable and detectable at maintenance checks and/or from the symptoms associated with a roughly running piece of equipment. This type of failure mode usually allows for replacement of coupling before a disasterous failure occurs. This does not preclude other failure modes, such as bolt breakage, hub splitting, spacer breakage, or other broken parts, but wear-out of flexing elements is the most common. Mechanical flexible coupling failures can usually be attributed to sloppy installation, poor lubrication, and maintenance practices.

In the following section we discuss in detail how some of the mechanical flexible couplings work, where they are used, various types, what is considered in their design, how they affect a system, and failure modes.

II. GEAR COUPLINGS

A gear coupling is one of the simplest and most common types of couplings in use today. It is also one of the most difficult to design and evaluate, the reason being the number of variables that can affect its successful operation. Some of these variables are:

1. Tooth design
    a. Straight teeth
    b. Type of crown and amount of crown
    c. Pressure angle of tooth
    d. Amount of backlash
    e. Accuracy of tooth spacing
2. Material
    a. Medium-carbon steel, alloy steels
    b. Type of core heat treatment: as rolled, normalized, heat treated
    c. Type of surface hardening: none, induction, nitrided, and carburized

3. Lubrication
   a. Oil
   b. Grease
   c. Sealed lubrication
   d. Continuous lubrication

These are just some of the important variables and indicate why no general criteria have been established. Experience and success are the real criteria and proof of a design. Section II of Chapter 3 gives a method of comparing various stresses in coupling components.

A. How Gear Couplings Work

A gear coupling usually consists of four major components, as illustrated in Figure 6.1. When two shafts are to be connected by this type of coupling, a coupling half (consisting of one hub and its mating sleeve) is keyed to each shaft, the shaft ends brought together, and the sleeves bolted together securely. Since the hubs have external teeth that mesh with the internal teeth on the sleeves, the connection is in effect an internal spur gear drive wherein the gear ratio is 1:1.

Figure 6.1 indicates that there is a gap between the two hubs and that the sleeve teeth are somewhat longer than the hub teeth. It is therefore possible for axial movement of the hubs to occur without interfering with the operation of the coupling. If axial movement were the only type of misalignment encountered, the design of flexible couplings would be a relatively simple task. However, due to the fact that perfect initial alignment of machinery is difficult, and since operating conditions will impose forces on the shafts which tend to throw them out of line radially and angularly, it is necessary for the coupling to compensate for misalignments other than just axial movement.

Pure angular misalignment is illustrated in Figure 6.2A. The misalignment between shafts A and B is designated by the angle α. Figure 6.2B shows the case where the shafts are displaced radially. This type of misalignment may be resolved into pure angular misalignment, as in Figure 6.2A, by use of a floating shaft E between shafts C and D. Thus it can be seen that the two types of misalignments shown in Figure 6.2 are basically the same. Henceforth, the term "misalignment" will apply to angular displacement and will be designated by the angle α.

To accommodate for misalignment, it is necessary to build considerable backlash into the teeth of the coupling. This is usually done by making the sleeve tooth thinner than its standard thickness. However, if the teeth on both the hub and sleeve are of conventional form, the teeth would be loaded at its end during misalignment. This is illustrated in Figure 6.3, which shows the entire tooth being applied at points P and Q. The tooth is loaded at these points during operation. At high loads and misalignments this may be an undesirable condition, as the ends of the hub teeth could dig into the faces of the sleeve teeth, and high bending moments would be developed at the point of contact due to the high coefficient of friction. One method of relieving end loading is to crown the teeth on the hub. The resultant form in Figure 6.4A may be compared with the standard tooth form shown in Figure 6.4B.

Cutting a crowned tooth is done more readily on a hobbing and shaping machine. It is a relatively simple matter to put a cam in the machine which will make the hob or shapper travel through a curved path rather than a straight line as it advances across the gear. The path of the hob and shapper is represented by the arrows in Figure 6.4.

The effect of crowning is illustrated in Figure 6.5. Figure 6.5A shows a coupling conenction two shafts whose angular misalignment is represented by the angle α. The relative position of the hub teeth and sleeve teeth at point A is shown in Figure 6.5B. Figure 6.5C shows the same for the teeth at point B. By comparing Figure 6.5C with Figure 6.3, it may be seen that the end-loading condition has been minimized. This does not mean that the hub tooth does not carry the load at or near its end under extreme misalignment for which the coupling was designed, but due to the curvature of the hub tooth at some misalignment less than designed, the point of application of load is closer to the center of the tooth, and therefore the moments would be less. It may also be seen by inspection of Figure 6.5C that a crowned hub tooth allows more misalignment than a straight tooth by virtue of the fact that the ends (where contact would first occur) are rounded off so that the sleeve tooth in effect pivots about point C.

B. Where Gear Couplings Are Used

Gear couplings are used on the following types of equipment: centrifugal pumps, conveyors, exciters, fans, generators, blowers, mixers, hydraulic pumps, compressors, steel mills and auxiliary equipment, cranes, hoist, mining machinery, and others.

Figure 6.6 shows a No. 28 Poole coupling of approximately 80 in. OD, weight 60,000 lb, with a 28-in.-diameter shaft, in a steel mill drive rated at 11,425,000 in.-lb at 19.3 rpm. Figure 6.7 shows a hot strip finishing mill. This 20 in. and 35 in. × 36 in. six-stand hot strip finishing mill is equipped with spindles designed to deliver 1,178,000 lb-in. of torque continuously and 2,060,000 lb-in. momentary at 2 1/2° misalignment. Figure 6.8 shows a 25-ton 100-ft-tall yard crane. Figure 6.9 shows a typical gear coupling/pump application. Figure 6.10 shows a typical gear coupling/fan application.

C. Types of Gear Couplings

Gear couplings can be classified in many ways. Two of the most common are: by the type of tooth form and by their functionality.

1. Typed by Tooth Form.

In all gear couplings the internal teeth are straight-sided teeth. They are involute formed teeth, usually at a 20° pressure angle. Generally, most of backlash required to misalign a gear coupling is cut into the internal tooth.

1. *Straight hub tooth form*: This is the same form as that of the internal teeth described above (see Figure 6.11A).
2. *Crowned tooth with pitch diameter on a curve*: These teeth are crowned hobbed or shaped with a cutter on a cam. This produces a pitch diameter that is on a curve (see Figure 6.11B).
3. *Crowned tooth with pitch diameter on a straight line*: The external teeth are straight hobbed or shaped, then a crown is shaved on the tooth (see Figure 6.11C).
4. *Variable crowned teeth*: The external teeth are generated with a cam that has multiple curvatures (see Figure 6.11D).

It is difficult to say whether a crown is better, and if so, which crown, given equal pitch diameter, diametral pitch, and tooth spacing, since gear couplings rely mostly on tooth deflection to carry load.

1. At small angles of 1/8° or less, the straight tooth coupling works almost as well as the crowned tooth coupling. For small angles a tooth shaved or relieved by 0.0015 to 0.003 in. works like a hobbed or shaped crowned tooth. For small

angles it is also probable that a straight tooth will wear itself into its required crown.

2. Over 1/8°, all crowned teeth (curvatures of over 0.003 in.) work about the same, assuming that we are comparing the same curvatures produced by different methods.

2. Typed by Functionality. This method is probably the best way to classify gear couplings. There are five functional classes for gear couplings:

1. The general-purpose gear coupling
2. The high-speed gear coupling
3. The high angle/low speed gear coupling
4. The high angle/high speed gear coupling
5. The special-purpose gear coupling

a. General-Purpose Gear Couplings. There are two basic designs in this group: straight tooth couplings (Figure 6.12) and crowned tooth couplings (Figure 6.13). Guston Fast was the inventor of this coupling and his basic design is still available from Kopper-Fast and from Poole.

The general-purpose gear coupling is one of the few couplings that has some degree of standardization. Basically, flange, OD, and bolt circles are standard and most half couplings are interchangeable. Torques and bore capacities are not standard. Figure 6.14 provides a comparison of most commercial gear couplings.

General-purpose gear couplings are also available in larger sizes, but standards and interchangeability data do not exist for these sizes. Figure 6.15 shows one manufacturer's typical dimensions for these larger sizes.

There are many variations of general-purpose gear couplings, two of which are:

Gear coupling with spacer (Figure 6.16)
Gear couplings for large shaft separation (Figure 6.17)

b. High-Speed Gear Couplings. These couplings are usually available in sealed lubrication and continuous lubrication types in four basic styles:

1. Standard configuration (Figure 6.18)
2. Reduced moment configuration (Figure 6.19)
3. Continuously lubricated (Figures 6.19 and 6.20)
4. Marine style or flanged connected (Figure 6.20)

These couplings are usually made from alloy steels and operate at speeds usually in excess of normal motor speeds. Two typical applications are:

Gas turbine application using high-speed gear couplings (Figure 6.21)
Centrifugal compressor application using high-speed gear couplings (Figure 6.22)

c. *High Angle/Low Speed Gear Couplings.* These gear couplings are very common on steel mill drives (see Figure 6.23). There are two basic types: standard lines (flange types) similar to those shown in Figure 6.24 and designed styles, (Figure 6.25) which are used for primary mill roll drives.

The flange-type flexible spindle (Figure 6.24) is similar to a tandem arrangement using flange-type couplings except that the gearing is cut to accommodate higher misalignment. The gear teeth are heat treated to provide higher torque ratings, and special molded high-angle lip-type seals are used.

This spindle is for medium-duty, medium-torque applications with relatively high misalignment capacities required. It is used on applications where equipment is not subjected to frequent disconnecting of drive components—used primarily on auxiliary equipment such as floating shafts for pinch rolls, tension bridles, continuous casting equipment, plastics and rubber calenders, rotary side guides, paper mill equipment, and auxiliary equipment applied on electrolytic cleaning, pickling, and galvanizing lines.

The custom-designed gear spindle (Figure 6.25) is used on heavy-duty, high-torque applications requiring rugged strength, such as in metal rolling mill main drives and similar heavy equipment. Mill-type flexible spindles deliver maximum torque capacity at relatively high operating angles and under severe shock load conditions on all types of equipment for the metals, rubber, plastic, paper, automotive, and chemical industries.

Variations and Features of Spindle Gear Couplings

1. *Splined hubs* (Figure 6.26A): Male gear hubs are connected to the intermediate shafts through snug-fitting splines employing a side bearing fit held concentric with the spindle shaft by an accurately machined pilot diameter on both ends of the spline. Thus the hub is held in a fixed position, able to resist radial load. Because of these pilot diameters, a long spline is used only to transmit torque. Relative movement

between the male and female spline is minimized, ensuring that fretting and spline wear cannot take place. Also, the spline requires less radial space than do keys, thus permitting the use of large shafts.

Whenever possible, spindles are manufactured with hubs retained externally. This is done through the use of a split retainer which fits in a groove in the shaft and is through-bolted into the back of the hub. This eliminates all internal bolting and the possibility of broken capscrews working into the gear teeth, causing premature failure.

2. *Insert gear rings* (Figure 6.26B): This component is connected to the respective sleeve or adapter usually through a snug-fitting spline employing a side bearing fit. The gear ring is usually held concentric with the adapter by a pilot, usually at both ends of the splined engagement.

3. *Spring-loaded thrust button* (Figure 6.26C): Many spindles are designed to incorporate a spring-loaded thrust button in the roll end coupling which holds the roll end coupling in line with the spindle shaft during roll change. This button is actuated by a spring force when the roll neck is removed from the coupling. The force of the spring drives the sleeve assembly forward until the sleeve assembly locks against the hub element for a full 360° contact, thrusting it into a horizontal position, ready to receive a new roll.

Most spindles are designed so that the contact point of the thrust surface is located on the centerline of the gear mesh. This provides ball-and-socket action rather than the sliding motion that occurs if the contact point is located back on the thrust plate.

4. *Splined replaceable sleeve* (Figure 6.26D): Many spindles are designed with splined replaceable sleeve on one or both ends. Basically, this is an adapter that fits onto the roll or the pinion shaft and intermediate sleeve, is splined to the adapter, and has an internal gear that meshes with the hub. This type of design is desirable when a quick-disconnect feature is required.

5. *Lip seals* (Figure 6.26E): The most common seal for high angularity is the lip seal. The positive seal is an integral one-piece molded lip seal, usually designed specifically for the specific flexible spindle application.

6. *Roll end bores* (Figure 6.26F): Many spindles are furnished with either developed bores or replaceable wear keys. Keys can be hardened or use standard commercial steels,

depending on the application. Roll end bores of the developed bore configuration are also used with or without heat treatment. The heat-treatment possibilities are directly related to the type of material of which the adapter or sleeve is made.

d. High Angle/High Speed Gear Couplings. This type of gear coupling is typically used on helicopters and on traction drives for subway cars. Helicopter couplings include accessory drive couplings, main drive couplings, and tail rotors. Figure 6.27 shows drawings and schematics of some of these; Figure 6.28A, a main drive helicopter coupling; Figure 6.28B, the helicopter that uses this coupling; and Figure 6.29, a traction coupling design and application.

Turbine Helicopter Drives. Figure 6.27A is a simplified drawing of the drive train of a typical turbine-powered helicopter utilizing three types of high-performance gear couplings. These types are listed and described briefly below.

1. *Main drive* (Figure 6.27B): connects the engine to the rotor gearbox. Fins on the sleeve aid in heat dissipation and stiffening. Rated at 2020 hp at 6360 rpm with 2° of angular misalignment per gear mesh.
2. *Accessory drive* (Figure 6.27C): connects the turbine-engine-driven main gearbox to the accessory gearbox. Rated at 130 hp at 6600 rpm with ±2° angular misalignment per gear mesh.
3. *Tail rotor drive* (Figure 6.27D): Rated at 185 hp at 3130 rpm with 2° angular misalignment per gear mesh. A series of these single engagement couplings are used in the tail rotor drive system to compensate for air frame deflection.

e. Special-Purpose Couplings. Gear couplings have been designed to do many things and accommodate many situations. Some of these are:

Brake drum incorporated into gear coupling (Figure 6.30)
Insulated gear coupling: driver and driven insulated from stray electric currents (Figure 6.31)
Limited-end-float-position motor rotors that do not have thrust bearings (Figure 6.32)
Shear pin coupling for overload protection (Figure 6.33)
Disconnects (Figure 6.34)

1. *Low-speed disconnect* (Figure 6.34A): provides manual or automatic disengagement of driver and driven, usually done at 0 rpm
2. *High-speed disconnect* (Figure 6.34B): provides automatic disengagement of driver and driven, usually done at 0 rpm or at very low speeds

D. Materials and Heat Treatment for Gear Couplings

1. Materials. The various materials used for gear couplings include the following:

a. *As rolled or normalized, AISI 1035 to 1050*: This material is used on standard general-purpose gear couplings. Couplings using this material are generally lightly torqued and cut for less than 1 1/2° up to 14 PD, over 14 PD they are usually cut for less than 3/4° misalignment (Figure 6.35A).
b. *Flank-hardened, AISI to 1035 to 1050 (usually induction hardened), 50 Rc*. This is used to improve wear. Couplings using flank-hardened teeth are usually rated for light torque, low speed, and high angle (up to 6°) (Figure 6.35B).
c. *Fully contoured induction hardened, AISI 1035 to 1050, 50 Rc*: This material is used for steel mill applications. It not only improves wear but increases the bending strength of teeth. This process is limited to medium-crowned teeth (usually under 4°) and tooth size between 5 and 2.5 DP. Couplings using this material are usually rated for moderate torque, low speed, and high angle (Figure 6.35C).
d. *Heat-treated AISI 4100 and 4300 steels, core 28 Rc* (usually 4140 and 4340): This material usually increases capacity slightly but does not affect wearability. Couplings of this material are usually applied to medium-torque, medium-speed, low-angle applications (Figure 6.35D).
e. *Flank-hardened (induction hardened) AISI 4100 and 4300 steels, surface 50 Rc, core 28 Rc* (usually AISI 4140 or 4340): This material is used on steel mill applications to improve wear. Couplings using flank-hardened teeth of this material are usually applied to moderate-torque, low-speed, high-angle applications (Figure 6.35E).
f. *Fully contour (induction hardened) AISI 4100 and 4300 steels, surface Rc, core 28 Rc* (usually AISI 4140 or 4340): This material, used for steel mill applications, not only improves wear but increases the bending strength of teeth. This process is limited to medium-crowned teeth (under 4°) and tooth

sizes between 5 and 2.5 DP. Couplings using this material are usually applied to high-torque, low-speed, high-angle applications (Figure 6.35F).

g. *Nitrided AISI 4100 and 4300, surface 54 to 60 Rc, core 35 Rc* (usually 4140 or 4340); *case 10% tooth thickness and 0.035 in. maximum depth, maximum tooth size 5 DP*: This gives a coupling increased strength in the root. Couplings of this material are usually used for small steel mill couplings for high torques, medium speed, and high angles (Figure 6.35G).
h. *Nitrided AISI 4100 and 4300, surface 54 to 60 Rc, core 28 Rc minimum* (usually AISI 4140 or 4340); *case depth 0.010 to 0.035 in. depending on tooth size*: This material takes heat treating well and produces little distortion. Couplings of this material are usually used in a wide range of applications: light to moderate torque, low to high speeds, and low to high angles (Figure 6.35H).
i. *Nitralloy (135 or N), surface 60 Rc, core 32 to 35 Rc; case depth 0.010 to 0.035 in. depending on tooth size*: This material provides excellent wear resistance and produces very little distortion. Couplings of this material are usually of special-purpose type: aircraft, traction coupling, and others (Figure 6.35I).
j. *Carburized AISI 8620, 4320, 3310, and others, surface 58 Rc, core 32 to 38 Rc; case depth 0.062 to 0.187 in. depending on tooth size*: These materials are used in many steel mill applications. Couplings made of these materials usually require grinding or lapping after carburizing to correct for tooth distortion caused by the carburizing process (Figure 6.35J).

Figure 6.36 shows the relative strength of various materials and heat/treatments compared to as-rolled (1035 to 1050) steel at 1 1/2° of misalignment.

2. Heat Treatment. Basically, heat treatment of steel consists of raising the material some specified temperature. The process is performed to change certain characteristics of steel to make them more suitable for a particular kind of service. Some of the reasons for heat treating are:

1. To soften a part so that it can be machined more easily
2. To relieve internal stresses so that a part will maintain its dimensional stability (will not warp or fail prematurely due to locked-in stresses)

3. To refine the grain structure so that the part will be less apt to fracture abruptly (or to toughen)
4. To throughly harden a part so that it will be stronger
5. To case harden a part so that it will be more wear resistant

a. Some Heat Treatment Processes.

Annealing: Annealing generally refers to the heating and controlled cooling of a material for the purpose of removing stresses, making it softer, refining its structure, or changing its ductility, toughness, or other properties.

Carburizing: Adding carbon to iron-base alloys by absorption through heating the metal at a temperature below its melting point in contact with carbonaceous materials. Such treatment followed by appropriate quenching hardens the surface of the metal. This is the oldest method of quenching.

Flame hardening: In this method of hardening, the surface layer of a medium- or high-carbon steel is heated by a high-temperature torch and then quenched.

Induction hardening: A hardening process in which the part is heated above the transformation range by electrical induction.

Nitriding: Adding nitrogen to solid iron-base alloys by heating at a temperature below the critical temperature in contact with ammonia or other nitrogenous material.

Hardening: (as applied to heat treatment of steel): Heating and quenching to produce increased hardness and increased strength, respectively.

Normalizing: Heating to about 100°F above the critical temperature and cooling to room temperature in still air. Provision is often made in normalizing for controlled cooling at a slower rate, but when the cooling is prolonged, the term used is "annealing."

Stress relieving: Reducing residual stresses in a metal by heating to a suitable temperature for a certain period. This method relieves stresses caused by casting, quenching, normalizing, machining, cold working, or welding.

Tempering: Reheating after hardening to a temperature below the critical temperature and then cooling.

a. The Importance of Heat Treatment. The best heat treatment for a coupling gear tooth gives the correct combination of core hardness versus case depth and hardness be used. If the core is too soft, the wear and load-carrying capabilities of a coupling become limited because the core will cause the coupling to fail in subsurface shear instead of compression.

1. *Normalizing and/or annealing*: It is important that the contoured or forged bar materials used in couplings be normalized or annealed. This assures a refined grain structure which helps throughout the manufacturing cycle. The refined grain structure helps produce a good finish and minimizes localized cold working, which can produce stress risers that can cause distortion in parts after final heat treating or machining and premature failure from cracks produced by the stress risers.

2. *Stress relieving*: This process is used for material that is heat treated to a high core hardness and used in a high-torque, high-misalignment, or high-speed coupling. Parts are stress-relieved after rough machining and/or before case hardening. This reduces internally locked-in stresses caused by core hardening and machining. Stress relieving helps prevent distortion or cracking after final heat treating and machining.

3. *Hardening and tempering*: Steel is usually hardened and tempered to obtain the desired level of strength. This process is used to increase the strength capacity not only of gearing but also of such torque-transmission parts as spacers, hubs, and rigid coupling. With the increase in core strength there is a reduction in ductility or impact resistance. This is one reason why parts are not hardened to maximum hardness throughout. Depending on the material, typical heat-treated-material core hardnesses range from 28 to 38 Rc. Beyond 38 Rc, not only does ductility decrease but normal machining becomes almost impossible.

4. *Case hardening of gear teeth*: Why surface harden gear teeth? The main purpose is to impart a more highly wear resistant surface. Increased wearability can be obtained by hardening the gear flank surface by one of the following methods:

a. Flame or induction hardening
b. Fully contoured induction hardening
c. Nitriding
d. Carburizing

An additional benefit can be derived from the surface hardening process if a uniform, adequate increase in tensile strength can be obtained. The hardened surface and the ductile core must be properly balanced to provide a high-strength ductile tooth for bending.

A rectangular cantilever beam subject to bending will result in the stress distribution shown in Figure 6.37A. The greatest stress occurs at the outer fiber and decreases linearly to zero at the neutral axis or center. The maximum stress without permanent deformation is the yield stress of the material. Figure

6.37B illustrates the same cantilever surface hardened. The stress distribution is similar, but the maximum stress without permanent deformation is not the yield of the core material but the yield of the surface and the ductile core. A gear tooth is like a cantilever beam. A properly balanced tooth is one in which the extreme fiber stress reaches the yield strength of the hardened material when the stress at the juncture of the core and the surface reaches the yield strength of the core. It can be shown algebraically that the proper case-hardened depth should be approximately one-tenth to one-fifth of the tooth thickness for surface-hardened alloy gear teeth.

For all case hardening listed below, not only is wear resistance increased but tensile strength is also increased, by approximately a factor of 2.

1. Fully contoured (bending load) induction hardened 5 to 2.5 DP teeth cut for 4° or less of misalignment
2. Nitriding smaller than 5 DP
3. Carburizing all

3. Surface Hardening

a. Hardening of Medium-Carbon Steels. How is the gear tooth hardened? First, the part is heated to above the transformation temperature. This temperature varies with different grades and types of steels, but it is the temperature at which the structure of the steel changes and carbon goes into solution. This simply means that the carbon atom is freed to move about in the steel. For nearly all steels processed, this temperature is about 1500°F.

The next step after heating is quenching or very rapid cooling. In this process excess carbon atoms are trapped and cannot return to their original locations. If the quench is ideally rapid and the part is cooled to a low temperature, there will exist a condition of 100% martensite. This is the strongest of all structures, but it is very difficult to obtain and not at all desirable for couplings because it is very brittle and subject to impact failures.

b. Flame Hardening. Flame hardening is a surface-hardening process in which a steel is rapidly heated above its transformation temperature by a high-temperature flame and then cooled at a rate to produce the hardness desired. Flame-hardened gear teeth can be made to 50 to 56 Rc. Due to the fact that the heat is localized, there is relatively low distortion, but there is a great difficulty in producing a repeatable concentration of heat. There can not only be hardness variation from tooth to tooth but also hardness variation within a single tooth. This process does add wear resistance, but the overall strength of the base material is lowered.

c. *Induction Hardening.* Induction heating takes place when an electrically conducting object (not necessarily magnetic steel) is placed as a varying magnetic field. Induction heating is due to hysteresis and eddy current losses; and although it has been in wide use for about 25 years, this heating effect is not completely understood. Probably the primary reasons for using this type of heating are that it is very fast and is very easy to apply to a particular area.

Since a gear tooth should be case-hardened without affecting its resilient core, it is necessary to apply the heat very rapidly so that the cooler center of the gear does not rob the heat; and since the heat pattern must be controlled very closely, induction heating is ideal for this process.

Conventional flank hardening. Hardening the flanks of a gear tooth by induction hardening, shown in Figure 6.38A, can provide a hardened surface for increased life. The lack of hardness in the root of the tooth usually not only does not improve the strength of the tooth, but can reduce its strength. To understand this, you have to realize that hardening creates stresses. The hardening process is performed by heating a steel part to an elevated temperature, usually around 1500°F, and rapidly quenching the part. The heated and quenched portion actually increases in volume, setting up compressive stresses internally. It becomes a very dense, hard structure, and its strength increases in proportion to its increase in hardness. In conventional processes, there could be sharp stress discontinuities if this tooth is highly loaded at the root area. These discontinuities could be starting points for stress cracks.

Fully contoured hardening. Fully contoured induction hardening is an answer to the need for gear tooth wearability and strength. Figure 6.38B illustrates the fully contoured pattern. Observe that the hardened case is contoured through the root and a transition between the hard case and the soft core occurs at the unstressed tooth tip.

The process is carried out by core hardening the rough-turned forging to the upper machineability limit, and then after cutting the gear teeth, the coupling is hardened one tooth space at a time. The absence of large quantities of heat results in a gear essentially free of overall distortion. The depth of the case can be controlled to permit long wear life and a strong gear tooth in bending for impact loads.

d. Nitriding. Nitriding is a process used for case hardening of alloy steel parts, usually of special composition, by heating them in an atmosphere of ammonia gas and disassociated ammonia mixed in suitable proportions. The process is carried out below the transformation range for steel and quenching is not involved. Nitriding is particularly desirable because of the following features:

1. Exceptionally high surface hardness
2. Very high wear resistance
3. Low tendency to seizing and galling
4. Minimum warpage or distortion
5. High resistance to fatigue
6. Improved corrosion resistance

Nitriding a gear coupling is done selectively. That is, all areas except the gear tooth are blocked off with paint or plating during nitriding. This procedure minimizes part distortion and allows conventional machining in critical areas such as bolt holes and pilots to be done after nitriding.

In nitriding the depth and hardness of the case is a function of two things: the nitriding process time and in some alloys, the hardenss of the core. The maximum case depth is approximately 0.035 in. for material with a core hardness above 36 Rc. With the use of special nitriding steels such as Nitralloy, surface hardnesses as high as 65 and 70 Rc are attainable.

If we apply the rule for depth of case versus tooth size, case depth should be approximately one-tenth of the tooth thickness. We can see that with a maximum surface depth of 0.35 in. we will not obtain any increase in bending capacity at root of the tooth for DP values larger than 5 DP. Normally, nitriding is very advantageous to increase wear resistance of high-speed, high-angle couplings but can supply good fatigue resistance at tooth roots for small DP's (smaller than 5 DP).

e. Carburizing. Carburizing is the ultimate means of hardening a coupling tooth as it can produce the balance needed for wear resistance and strength. Its use results in a very hard deep case (as high as 65 Rc) and a tough ductile core. The uniform hardness pattern at the flanks extends through the root of the tooth. The depth of hardenss can be controlled by the amount of time in the furnace.

Wear has been found to be a function of hardness and carbon content. Figure 6.39 shows the wear resistance of several materials as a function of hardness.

The major problem with carburizing is that the entire part must be placed in a furnace at a high temperature. A fully machined gear tooth is subject to approximately 1700°F for many hours and is then quenched in oil. As one would expect, a great amount of distortion results. Gear coupling designs require uniform tooth loading to survive the rigors involved in many applications. The distorted carburized gear cannot produce uniform tooth loading. The few teeth carrying the load must "wear in" to allow other gear teeth to share in the load. This early wear can result in reduced gear life.

E. How to Evaluate Gear Couplings

1. How to Compare. One question often asked is how to evaluate or compare a gear coupling. The answer is very difficult because no two gear couplings are exactly alike. What I have attempted to supply is a procedure that can be used to compare couplings, not a design procedure. Design should be left to the coupling manufacturer.

Most gear couplings fail due to tooth distress, breakdown of lubricant, compressive stress, and the motion due to misalignment. If an overload occurs, the gear coupling will fail at its weakest link:

1. Tooth shear or bending
2. Bolts
3. Shaft or spacer
4. Flanges
5. Hubs
6. Key or keyways

Items 2 through 6 are covered in Section II of Chapter 3.

Failures of gear teeth in shear or bending are uncommon. The stress equations vary due to the tooth stress used. The stresses are a function of:

Bending stress:

$$S_b = K_b \frac{T}{(P.D.)^2 F} = \text{psi} \qquad (6.1)$$

Shear stress:

$$S_s = K_s \frac{T}{(P.D.)^2 F} = \text{psi} \tag{6.2}$$

where

T = torque (in.-lb)
P.D. = pitch diameter (in.)
F = face width (in.)

K_b and K_s are a function of:

1. The amount of crown
2. The number of teeth in contact
3. Misalignment
4. Other geometric variables

2. Criteria for Comparison. Obtain the following information:

T = raw motor torque per spindle (in.-lb)
P.D. = diameter of pitch circle of gear tooth (in.)
h = active tooth height (in.)
D = diameter of curvature of tooth face (in.)
D.P. = diameter pitch of gear tooth
N = maximum speed of spindle
α = full-load misalignment angle (deg)
F.W. = gear tooth face width (in.)
R = P.D./2 (in.)
n = number of teeth in coupling

For crowned teeth (based on the compressive stress equation in MIL-C-23233):

$$S_c = 2290\sqrt{\frac{T}{(P.D./2) \times (D/2) \times c \times h \times (D.P. \times P.D.)}} = \text{psi} \tag{6.3}$$

where c represents the percentage of teeth in contact (see Figure 6.40). It is advisable to make the comparison at 0°, so that c = 1.

If tooth curvature is expressed in amount of crown:

K = amount of crown = (in.) (see Figure 6.41)

$$D = \frac{(FW/2)^2}{K} = \text{(in.)} \tag{6.4}$$

$$h = \frac{1.8}{D.P.} = \text{(in.)}$$

$$S_c = 3600\sqrt{\frac{T}{(P.D.)^2 \times D \times c}} = \text{psi} \tag{6.5}$$

For straight-tooth couplings:

$$S_c' = \frac{T}{RnA} = \text{psi} \tag{6.6}$$

$$A = \frac{1.8 \times F.W.}{D.P.} \tag{6.7}$$

Maximum sliding velocity (V) of gear coupling:

$$V = \pi \times P.D. \times \sin\alpha \times \frac{N}{60} = \text{ips} \tag{6.8}$$

S_cV criteria can be used to compare equivalent couplings and applications:

$$S_cV = S_c \times V = \text{psi} - \text{ips} \tag{6.9}$$

Limits on S_cV depend on the type of lubrication and the material used for a particular coupling. Some typical values based on testing and field experience are given in the following table. It should be remembered that the allowable values change with material, tooth geometry, tooth finish, and type of lubrication.

Material	AISI 1045
Lubrication method	Grease packed
Misalignment	3/4°
S_cV (psi − ips)	1,500,000
Material	AISI 4140 with nitrided teeth
Lubrication method	Continuously lubed
Misalignment	1/4°
S_cV (psi − ips)	750,000

Material	Nitralloy N with nitrided teeth
Lubrication method	Grease packed
Misalignment	4 1/2°
S_cV (psi – ips)	6,000,000
Material	AISI 3310 with carburized teeth
Lubrication method	Grease packed
Misalignment	3°
S_cV (psi – ips)	500,000

For a crowned tooth gear coupling, the number of teeth in contact changes with misalignment and load. For comparison purposes the chart in Figure 40 can be used.

3. Torque Capacity. The torque capacity (T) can be approximated by using equation (6.10) and Figures 6.42, 6.43, and 6.44.

$$T = T_B \times S \times M \tag{6.10}$$

where

T_B = base torque (from Figure 6.42)
M = misalignment factor (from Figure 6.43)
S = material factor (from Figure 6.44)

This equation and the chart are empirical, based on 25 years of experience by Zurn Industries, Inc., Mechanical Drives Division.

F. How Gear Couplings Affect the System

The most significant effect that a gear coupling has not only on itself but also on the system comes from the moments and forces generated when it slides and/or misaligns.

1. Axial Force. The formula for the axial force reacted back onto the thrust bearings is

$$\text{Force} = \frac{T \times \mu}{R} = \text{lb} \tag{6.11}$$

where

T = torque (in.-lb)
μ = coefficient of friction
R = pitch radius of gear (in.)

From test data available from various coupling manufacturers, typical values for coefficient of friction are:

Sealed lubricated gear couplings:

$\mu = 0.05$

Continuous lubricated gear couplings:

$\mu = 0.075$

For equipment design purposes it should be safe to multiply the foregoing values by 2. Values higher than the above can be experienced, but if they are seen for any period of time, the coupling is no longer flexible and more than likely will fail or cause a failure of some portion of the equipment. Experience indicates that for values of μ above 0.15, a failure of some type could occur. In fact, if a coupling is locked (mechanically) from sludge or wear, the forces experienced could be seven to eight times those normally expected.

There is much discussion of how large a value to use when designing a system or, in fact, when designing the coupling. Designing as if the coupling were locked would completely defeat the purpose of a flexible coupling. It would be like trying to design rotating equipment to run with no lubrication in the bearings.

If couplings are lubricated and maintained properly, and operated within the loads and misalignment values used to select them, the coefficient of friction should never exceed 0.15. It is up to the system designer to decide how safe is safe. This must be decided on the basis of what would happen if a failure were to occur.

2. Bending Moment. The moments produced in a gear coupling can load equipment shafts and bearings and change the operating characteristics of equipment. First, let's define a coordinate system for a gear coupling. In Figure 6.45, the line X' − X' line defines a cut through the plane of misalignment. Moments that tend to rotate the coupling around the Z − Z axis will be designated M_p. Moments that tend to rotate the coupling around the Y − Y axis will be designated M_s.

Gear tooth Sweep:

$$\sin \alpha = \frac{x}{R_c} \tag{6.12}$$

$$R_c = \frac{x}{\sin \alpha} \tag{6.13}$$

$$D_c = \frac{2x}{\sin \alpha} \tag{6.14}$$

$$2x = D_c x \sin \alpha \tag{6.15}$$

where

x = distance from the centerline of gear tooth to the point along the flank of the gear contact at $\alpha°$ misalignment (see Figure 6.46)

Average radius for a misaligned coupling (see Figure 6.47):

$$R' = \sqrt{R^2 - \left(\frac{R}{3} \sin \frac{A}{2}\right)^2}$$

where

R = pitch radius of gear (in.)
A = angle for gear tooth contact range (degrees)

There are three basic moments in a gear coupling:

1. Moment generated from transmission torque and angle (rotated around the Z–Z axis)
2. Moment generated from frictional loading (rotated around the Z–Z axis)
3. Moment generated from displacement of load from it condition rotating around the Y–Y axis

Forces on coupling tooth (Figure 6.48):

$$\Sigma F_{X'} = +P \cos \alpha - \mu N - W \sin \alpha = 0 \tag{6.16}$$

$$\Sigma F_{Z'} = N - P \sin \alpha - W \cos \alpha = 0 \tag{6.17}$$

$$N = P \sin \alpha + W \cos \alpha \tag{6.18}$$

Eliminating N from equations (6.16 and (6.17), we have

$$P \cos \alpha - \mu(P \sin \alpha + W \cos \alpha) - W \sin \alpha = 0 \tag{6.19}$$

$$P \cos \alpha - \mu P \sin \alpha = W \sin \alpha + \mu W \cos \alpha \tag{6.20}$$

$$P = \frac{W(\sin \alpha + \mu \cos \alpha)}{\cos \alpha - \mu \sin \mu} \tag{6.21}$$

Moments causing rotation around the Z–Z axis:

$$M_p = P \times 2R' \tag{6.22}$$

$$M_p = \frac{2W(\sin \alpha + \mu \cos \alpha}{\cos \alpha - \mu \sin \mu} \sqrt{R^2 - \left(\frac{R}{3} \sin \frac{A}{2}\right)^2} \tag{6.23}$$

$$M_p = \frac{M_t(\sin \alpha - \mu \cos \alpha) \sqrt{R^2 - [(R/3) \sin (A/2)]^2}}{R(\cos \alpha - \mu \sin \alpha)} \tag{6.24}$$

Moments causing rotation around the Y–Y axis:

$$M_s = M_F\alpha + M_G = 0 + M_G = W(2x) \quad \text{where } W = \frac{M_t}{2R} \tag{6.25}$$

$$M_s = \frac{M_t x}{R} \tag{6.26}$$

Resultant moment: vectorial combination of two moments:

$$M_2 = \sqrt{(M_s)^2 + (M_p)^2} \tag{6.27}$$

at approximately $\theta = \tan^{-1}(2R/Dc)$ from Y in the ZY plane. (6.28)

Final combined moment equation:

$$M_2 = \sqrt{\left[\frac{M_t(\chi)}{R}\right]^2 + \left[\frac{M_t(\sin \alpha + \mu \cos \alpha)}{R(\cos \alpha - \mu \sin \alpha)} \sqrt{R^2 - \left(\frac{R}{3} \sin \frac{A}{2}\right)^2}\right]^2}$$

or (6.29)

$$M_2 = \sqrt{\left[\frac{M_t(\chi)}{R}\right]^2 + \left[\frac{M_t(\sin \alpha + \mu \cos \alpha)}{R(\cos \alpha - \mu \sin \alpha)} (R')\right]^2}$$

3. Reactionary Forces on Equipment Bearings. Since maximum moment occurs at $\theta = \tan^{-1}(2R/D_c)$ from Y in the ZY plane, maximum bearing loads will be perpendicular. Pictorially this is difficult to show, so we will show loads as if they are parallel to the Z axis. This does not effect the magnitude of the load, only the point of application, and this is not very important. We will solve bearing reactions in terms of maximum moment, as if the maximum moment were tending to rotate the coupling around the Y–Y axis.

Considering Figure 6.49 (plan view), with torque and rotation directions shown as indicated and an α° plane of misalignment as given, it may be observed that the force pattern nets a clockwise moment transferral (M) to the shaft in an amount. The bearing nearest the transferral moment would then have a high-pressure condition, B, in the Z plane, and loading with the front bearing down and the back bearing up. This assumes the direction of rotation to be over the top of the shaft. Refer to Figure 6.49 for further details.

Taking the Σ M values about A and setting them equal to zero, we have Σ M about A = $B'L - M = 0$

$$B' = \frac{M}{L} \quad \text{resisting force required of bearing}$$

$$b' = A' \quad \text{equal and opposite in direction}$$

The resulting force pattern for A and B would then be the opposite of C and D, reading from left to right. Figure 6.49 indicates the force pattern that results for the direction of rotation shown. When the direction of shaft rotation is reversed with the same α° misalignment condition, the forces reverse themselves as indicated.

G. Failure Mode for Gear Couplings

The most common type of failure for gear couplings is tooth distress. Common types of failures are (Figure 6.50 shows typical good wear patterns):

1. Failure due to lubricant breakdown (Figure 6.51)
   a. Loose
   b. Lubricant washed out
   c. Lubricant deterioration
   d. Improper lubricant used

2. Failure due to improper contact (Figure 6.52). In this case the matched lapped set was mixed. The resulting failure was not tooth distress itself but increased moment, which broke the coupling spacers.
3. Failure due to over-misalignment, in which the ends of the teeth are broken or cracked (Figures 6.53 and 6.54).
4. Failure due to operating at torsional critical levels at which the lubricant breaks down and cold welding takes place (Figure 6.55).
5. Worm tracking (Figure 6.56). Cold flow or welding usually occurs on continuously lubed couplings. This occurs near the end of the teeth when misalignment approaches the design limit of the crown. At this point the lubrication film breaks down and causes metal-to-metal contact. Sometimes cold flow distress that takes place anywhere on a tooth is described as worm tracking. Cold flow occurs when the lubricating film breaks down. This can result from high localized tooth loading or from lubricant deterioration.
6. Occasionally, an overload can occur which can cause a gear tooth to shear off (see Figure 6.57).
7. Sludge buildup can cause failure of coupling components other than teeth or equipment components such as shafts or bearings (see Figure 6.58). Once a gear coupling locks up due to sludge, anything can break (see Figure 6.59). Sludge also collects corrosive residue, which can corrode coupling parts and act as a source of crack initiations for fatigue-propagated failure of a part (see Figure 6.60).
8. Lock-up can also occur due to tooth wear. The hub tooth embeds (wears) into the internal gear tooth, which causes mechanical lock-up (Figure 6.61).

Other type of gear coupling failures include:

1. Fatigue failures through gear teeth (Figure 6.62)
2. Key shear or roll (Figure 6.63A)
3. Damage to key, shaft, and hubs caused by improper fit (Figure 6.63B)
4. Hub bursting—usually due to overload on couplings that are used with maximum bore and keyway sizes. The hub may be the weak link. Hubs split over corners, and in some cases this wedging can cause a sleeve to split (Figure 6.64).

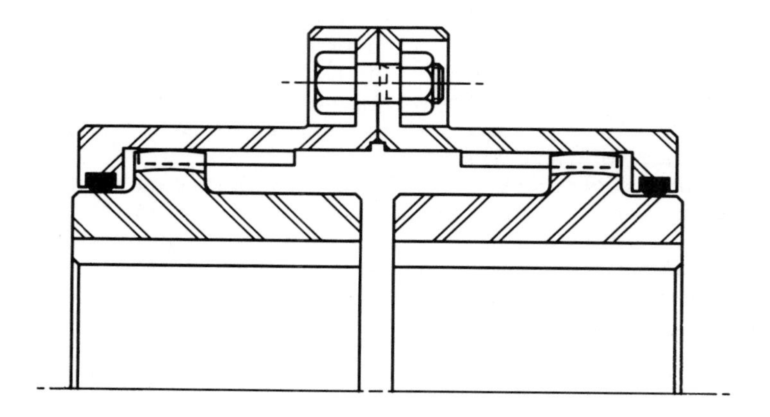

Figure 6.1 Gear coupling.

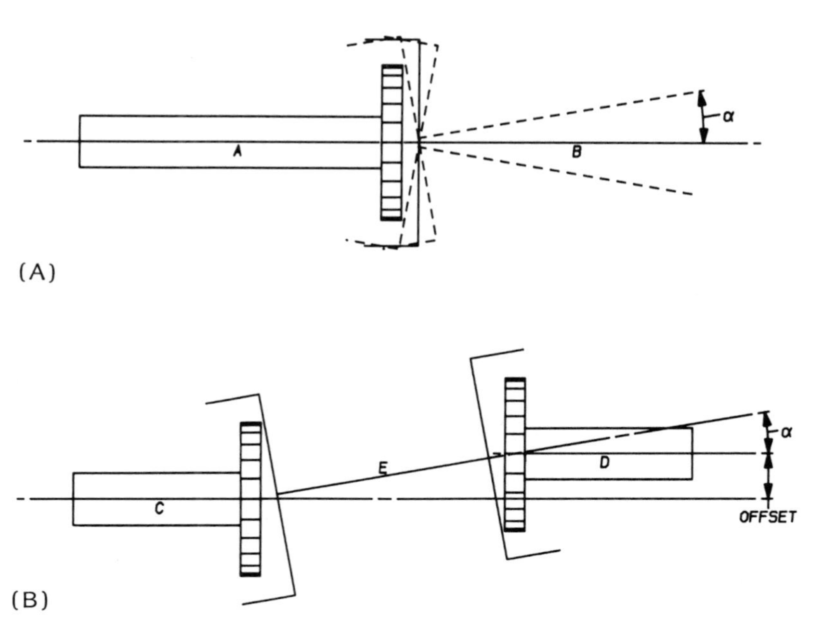

Figure 6.2 (A) Single mesh, angular misalignment; (B) double mesh, angular and offset misalignment.

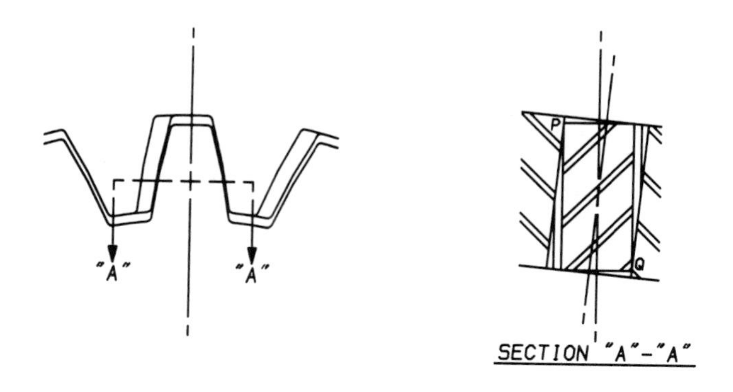

Figure 6.3 Misalignment of a conventional tooth.

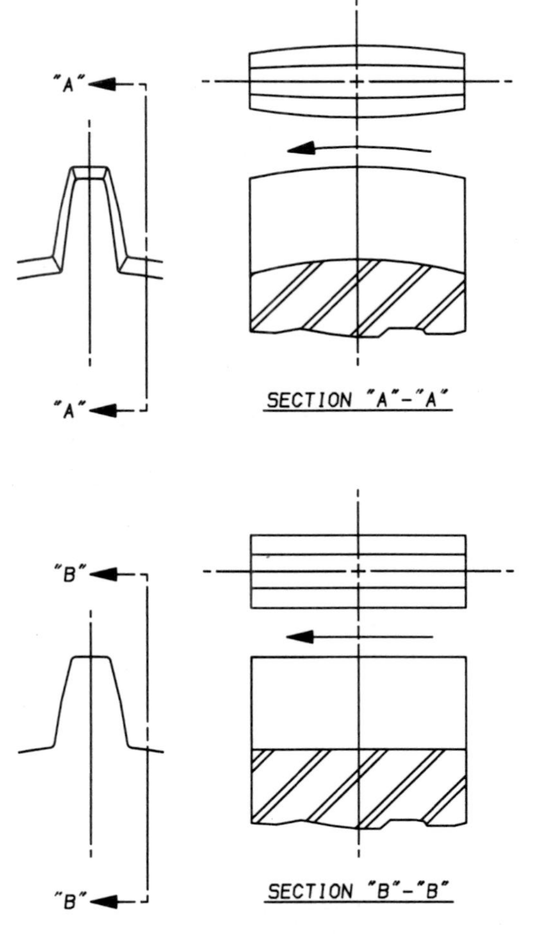

Figure 6.4 Path of hob for (A) crowned tooth, and (B) conventional tooth.

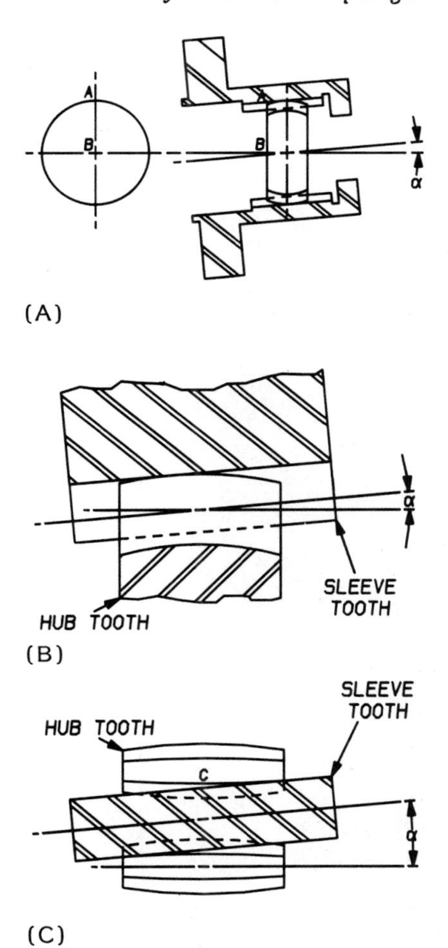

Figure 6.5 Misalignment of a crowned tooth coupling.

Figure 6.6 Large gear coupling (courtesy of Poole Company).

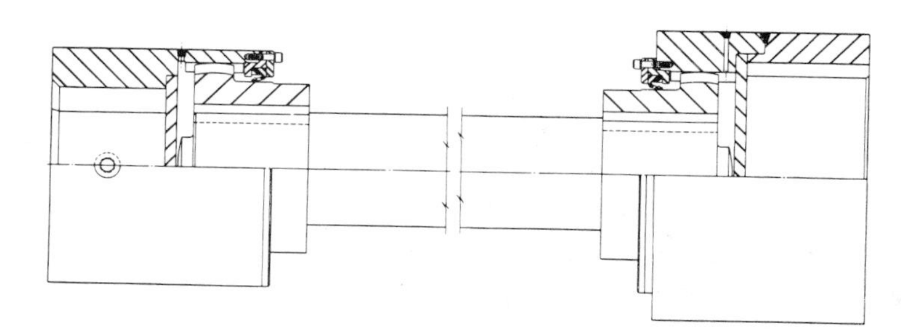

Figure 6.7 Hot strip mill with gear spindles (courtesy of Zurn Industries, Inc., Mechanical Drives Division).

Figure 6.8 Crane drive (courtesy of Morgan Construction Company).

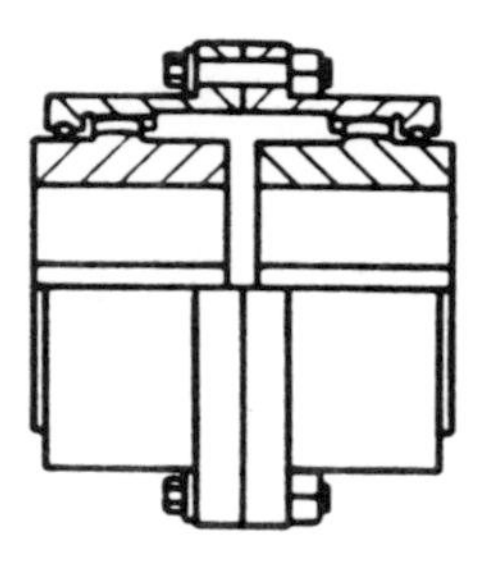

Figure 6.9 Pump drive (courtesy of Falk Corporation).

Figure 6.10 Gear drives (courtesy of Falk Corporation).

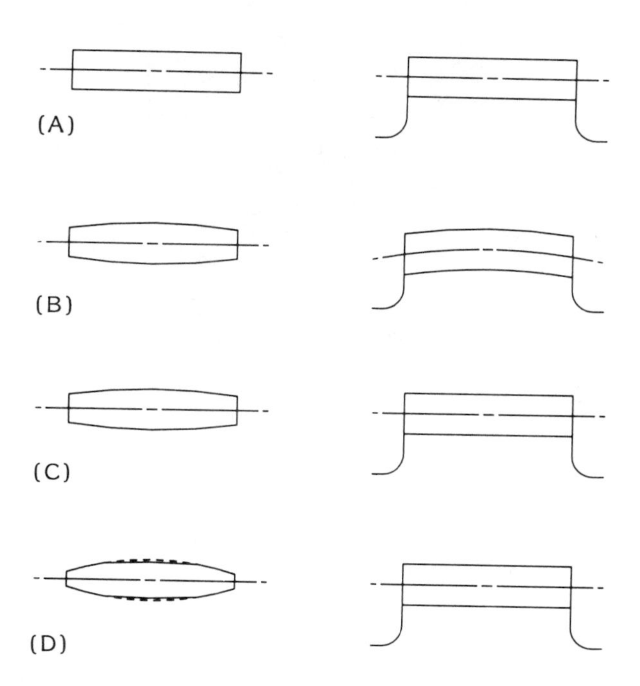

Figure 6.11 Gear couplings: (A) straight-tooth form; (B) crowned tooth with pitch on a curved path; (C) crowned tooth with pitch on a straight line; (D) variable crowned tooth.

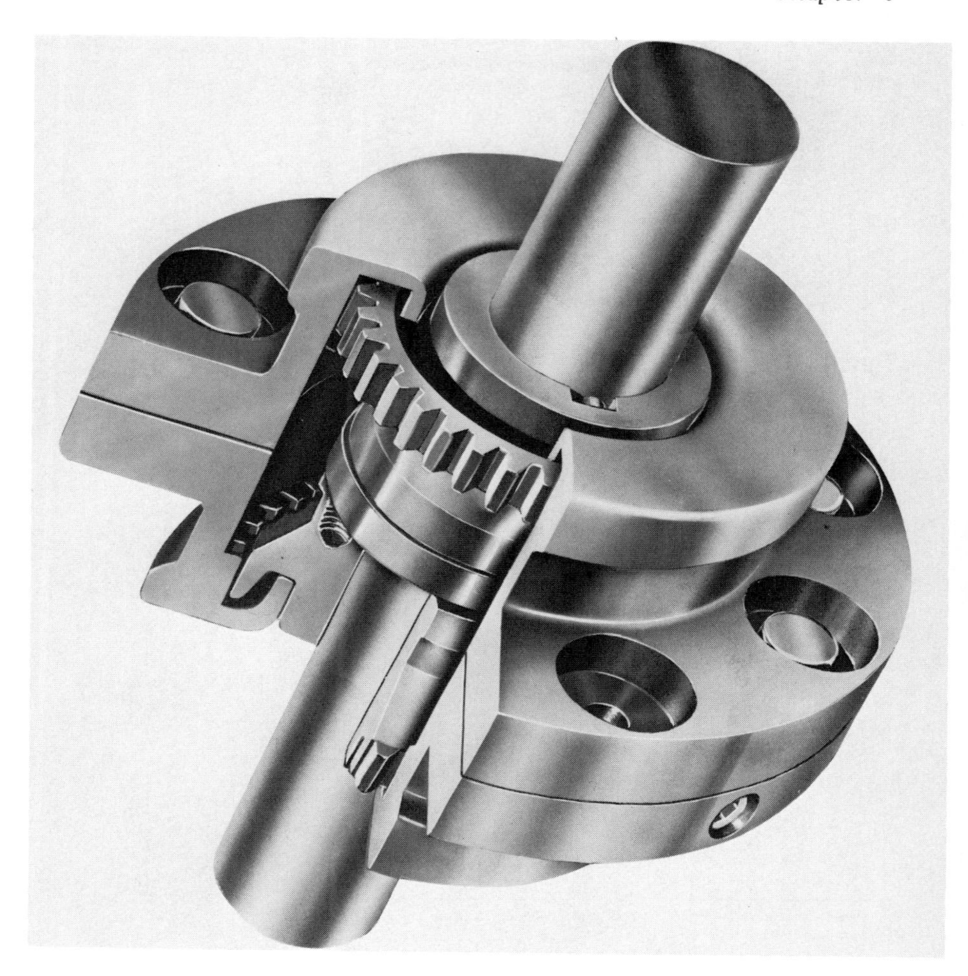

Figure 6.12 Fast gear coupling (courtesy of Koppers Company, Inc., Engineered Metal Products Group).

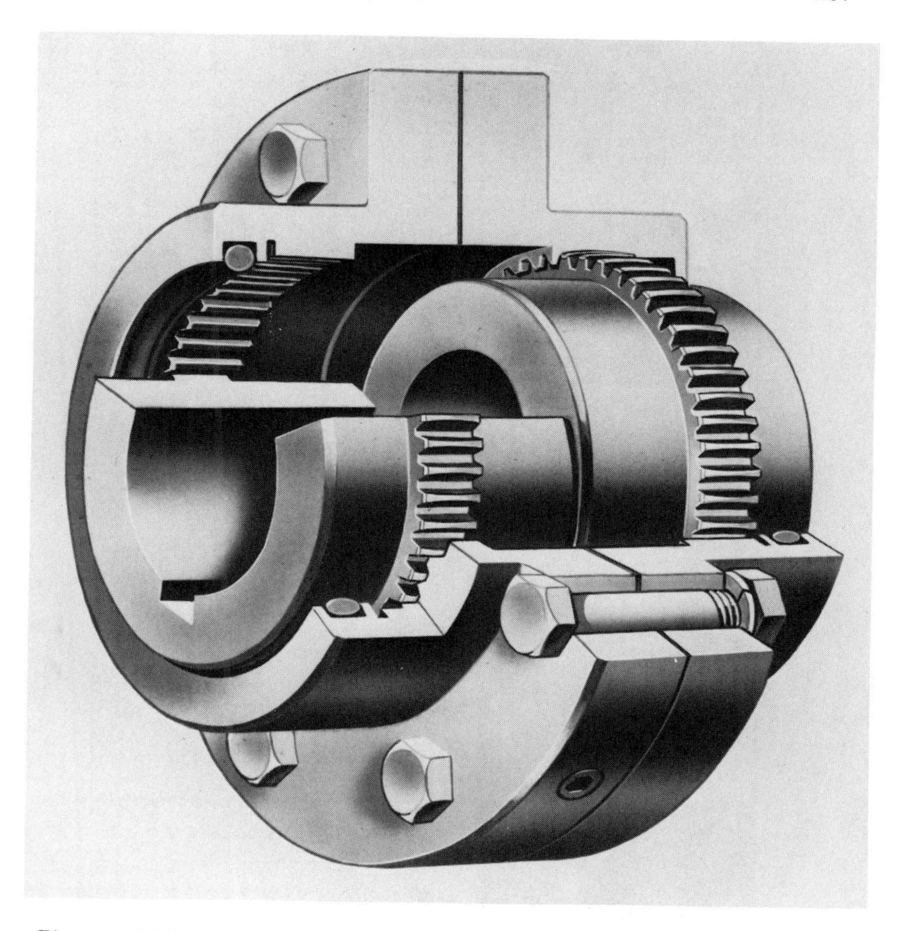

Figure 6.13 Crowned tooth gear coupling (courtesy of Falk Corporation).

SHROUDED BOLTS

DIMENSIONS COMMON TO ALL MANUFACTURERS SHOWN ON THIS CHART

DOES NOT INTERCHANGE WITH EXPOSED BOLT DESIGN.

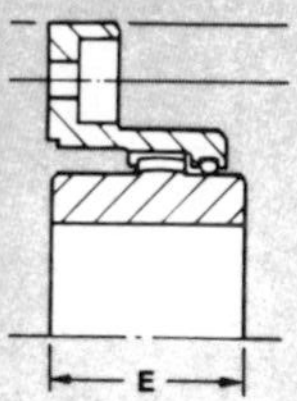

OUTSIDE FLANGE DIAMETER	NUMBER OF BOLTS	BOLT SIZE	BOLT CIRCLE	GAP	SIER-BATH SERIES F VARI-CROWN: COUPLING SIZE	SIER-BATH SERIES F VARI-CROWN: (1) HP/100	SIER-BATH SERIES F VARI-CROWN: MAX. BORE	SIER-BATH SERIES F VARI-CROWN: E DIMENSION	KOPPERS F.S.: COUPLING SIZE	KOPPERS F.S.: E	KOPPERS H: COUPLING SIZE	KOPPERS H: E
4 9/16	6	5/16	3.562	⅛								
4 9/16	6	¼	3.750	⅛	F-1	8	1⅝	1 11/16			1 H	1 11/16
6	8	⅜	4.812	⅛	F-1½	14	2⅛	1 15/16	1½	1 15/16	1½ H	1 15/16
7	10	⅜	5.812	⅛	F-2	38	2¾	2 7/16	2	2 7/16	2 H	2 7/16
8⅜	10	½	7.000	3/16	F-2½	55	3¼	3 1/32	2½	3 1/32	2½ H	3 1/32
9 7/16	12	½	8.000	3/16	F-3	88	4	3 19/32	3	3 19/32	3 H	3 19/32
11	12	⅝	9.281	¼	F-3½	137	4⅝	4 3/16	3½	4 3/16	3½ H	4 3/16
12½	14	⅝	10.625	¼	F-4	196	5⅜	4¾	4	4¾	4 H	4¾
13⅝	14	⅝	11.750	5/16	F-4½	315	6	5 5/16	4½	5 5/16	4½ H	5 5/16
15 5/16	14	¾	13.187	5/16	F-5	536	6½	6 1/32	5	6 1/32	5 H	6 1/32
16¾	16	¾	14.437	5/16	F-5½	618	7⅜	6 29/32	5½	6 29/32	5½ H	6 29/32

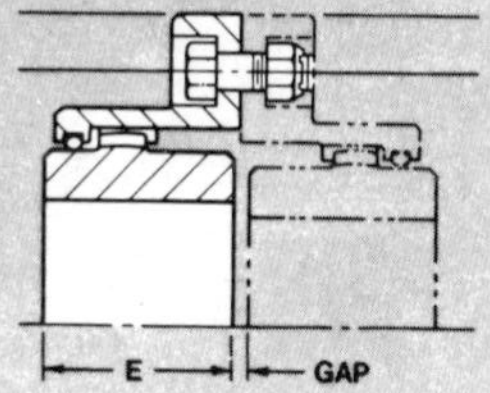

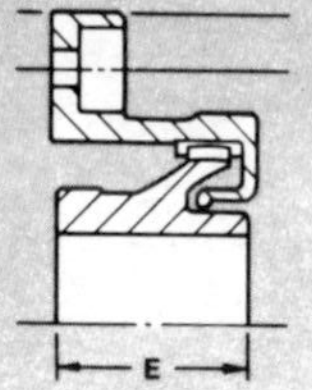

A

WALDRON Series A		WALDRON FLEXALIGN W		AMERIGEAR Series F		FALK G-10 Series 10		FALK* G-10 Series 1000		AJAX Series 6000		POOLE Standard	
COUPLING SIZE	E	COUPLING SIZE	E	COUPLING SIZE	E	COUPLING SIZE	E	COUPLING SIZE	E	COUPLING SIZE	E	COUPLING SIZE	E
1¼ A	1½					10G10	1½			6125	1 11/16		
		1 W	1 11/16					1010G	1 11/16				
2 A	2 1/16	1½ W	2 1/16	F-101½	1 15/16	15G10	2	1015G	1 15/16	6150	1 15/16		
2½ A	2 7/16	2 W	2 7/16	F-102	2 7/16	20G10	2 7/16	1020G	2 7/16	6200	2 7/16	2	2 7/16
3 A	3 1/32	2½ W	3 1/32	F-102½	3 1/32	25G10	3 1/32	1025G	3 1/32	6250	3 1/32	2½	3 1/32
3½ A	3 19/32	3 W	3 19/32	F-103	3 19/32	30G10	3 19/32	1030G	3 19/32	6300	3 19/32	3	3 19/32
4 A	4 3/16	3½ W	4 3/16	F-103½	4 3/16	35G10	4 3/16	1035G	4 3/16	6350	4 3/16	3½	4 3/16
4½ A	4¾	4 W	4¾	F-104	4¾	40G10	4¾	1040G	4¾	6400	4¾	4	4¾
5 A	5⅜	4½ W	5⅜	F-104½	5 7/16	45G10	5 5/16	1045G	5 5/16	6450	5 5/16	4½	5 5/16
5½ A	6⅛	5 W	6⅛	F-105	6 1/32	50G10	6 1/32	1050G	6 1/32	6500	6 1/32	5	6 1/32
6 A	6⅝	5½ W	6⅝	F-105½	6⅝	55G10	6⅝	1055G	6⅝	6550	6⅝		

B

Figure 6.14 Comparison of standard gear couplings (courtesy of Seir-Bath Gear Co., Inc.).

EXPOSED BOLTS

DIMENSIONS COMMON TO ALL MANUFACTURERS SHOWN ON THIS CHART

DOES NOT INTERCHANGE WITH SHROUDED BOLT DESIGN.

					SIER-BATH SERIES F VARI-CROWN				KOPPERS F.S.		KOPPERS H	
OUTSIDE FLANGE DIAMETER	NUMBER OF BOLTS	BOLT SIZE	BOLT CIRCLE	GAP	COUPLING SIZE	(1) HP/100	MAX. BORE	E DIMENSION	COUPLING SIZE	E	COUPLING SIZE	E
4 9/16	6	5/16	3.562	1/8								
4 9/16	6	1/4	3.750	1/8	F-1	8	1 5/8	1 11/16			1 H	1 11/16
6	8	3/8	4.812	1/8	F-1½	14	2 1/8	1 15/16	1½	1 15/16	1½ H	1 15/16
7	6	1/2	5.875	1/8	F-2	38	2 3/4	2 7/16	2	2 7/16	2 H	2 7/16
8 3/8	6	5/8	7.125	3/16	F-2½	55	3 1/4	3 1/32	2½	3 1/32	2½ H	3 1/32
9 7/16	8	5/8	8.125	3/16	F-3	88	4	3 19/32	3	3 19/32	3 H	3 19/32
11	8	3/4	9.500	1/4	F-3½	137	4 5/8	4 3/16	3½	4 3/16	3½ H	4 3/16
12½	8	3/4	11.000	1/4	F-4	196	5 3/8	4 3/4	4	4 3/4	4 H	4 3/4
13 5/8	10	3/4	12.000	5/16	F-4½	315	6	5 5/16	4½	5 5/16	4½ H	5 5/16
15 5/16	8	7/8	13.500	5/16	F-5	536	6½	6 1/32	5	6 1/32	5 H	6 1/32
16 9/16	10	7/8	14.625	5/16	—				—			
17 7/8	12	7/8	16.000	3/8	—							
16 3/4	14	7/8	14.500	5/16	F-5½	618	7 3/8	6 29/32	5½	6 29/32	5½ H	6 29/32
18	14	7/8	15.750	5/16	F-6	880	8	7 13/32	6	7 13/32	6 H	7 13/32
20 3/4	16	1	18.250	3/8	F-7	1425	9	8 11/16	7	8 11/16	7 H	8 11/16

C

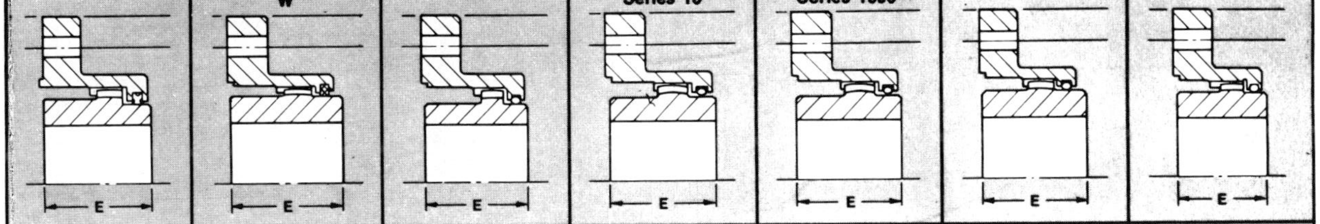

WALDRON Series A		WALDRON FLEXALIGN W		AMERIGEAR Series F		FALK G-20 Series 10		FALK* G-20 Series 1000		AJAX Series 6000		POOLE Standard	
COUPLING SIZE	E	COUPLING SIZE	E	COUPLING SIZE	E	COUPLING SIZE	E	COUPLING SIZE	E	COUPLING SIZE	E	COUPLING SIZE	E
1 1/4 A	1 1/2					10G20	1 1/2						
		1 W	1 11/16					1010G20	1 11/16				
2 A	2 1/16	1 1/2 W	2 1/16	F-101 1/2	1 15/16	15G20	2 19/32	1015G20	1 15/16	6150	1 15/16		
2 1/2 A	2 7/16	2 W	2 7/16	F-102	2 7/16	20G20	2 7/16	1020G20	2 7/16	6200	2 7/16	2	2 7/16
3 A	3 1/32	2 1/2 W	3 1/32	F-102 1/2	3 1/32	25G20	3 1/32	1025G20	3 1/32	6250	3 1/32	2 1/2	3 1/32
3 1/2 A	3 19/32	3 W	3 19/32	F-103	3 19/32	30G20	3	1030G20	3 19/32	6300	3 19/32	3	3 19/32
4 A	4 3/16	3 1/2 W	4 3/16	F-103 1/2	4 3/16	35G20	4 3/16	1035G20	4 3/16	6350	4 3/16	3 1/2	4 3/16
4 1/2 A	4 3/4	4 W	4 3/4	F-104	4 3/4	40G20	4 3/4	1040G20	4 3/4	6400	4 3/4	4	4 3/4
5 A	5 3/8	4 1/2 W	5 3/8	F-104 1/2	5 7/16	45G20	5 5/16	1045G20	5 5/16	6450	5 5/16	4 1/2	5 5/16
5 1/2 A	6 1/8	5 W	6 1/8	F-105	6 1/32	50G20	6 1/32	1050G20	6 1/32	6500	6 1/32	5	6 1/32
*6 A	6 5/8												
7 A	7 1/4												
—		*5 1/2 W	6 5/8	*F-105 1/2	6 5/8	55G20	6 5/8	1055G20	6 5/8	6550	6 5/8		
—		6 W	7 3/8	F-106	7 13/32	60G20	7 3/8	1060G20	7 13/32	6600	7 13/32		
—		7 W	8 11/16	F-107	8 11/16	70G20	8 11/16	1070G20	8 11/16	6700	8 11/16		

D

Figure 6.14 (continued)

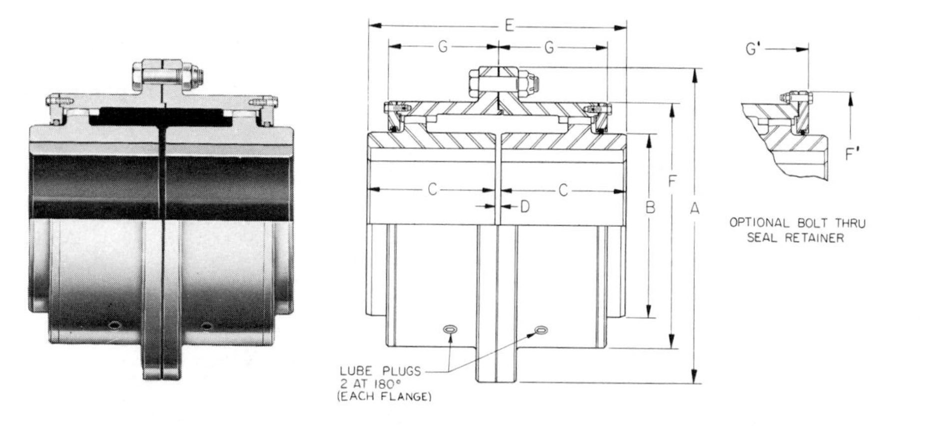

F SIZE	LOAD CAPACITY: HP Per 100 RPM	LOAD CAPACITY: Torque In.-Lbs. x 10^6	Parallel Offset Capacity In.	DIMENSIONS: A	B	C	D	E	F	G	Optional Bolt Through Seal Retainer: Parallel Offset	F'	G'
108	1,700	1.07	.167	23.25	13.50	9.75	.375	19.88	18.00	8.47	.167	20.25	8.47
109	2,500	1.58	.183	26.00	15.00	10.75	.500	22.00	20.00	9.28	.183	22.25	9.28
110	3,300	2.08	.203	28.00	17.00	12.00	.500	24.50	22.00	10.09	.203	24.75	10.09
111	4,600	2.90	.216	30.50	19.00	13.00	.500	26.50	24.75	10.91	.216	26.75	10.91
112	6,000	3.78	.228	33.00	20.50	14.00	.500	28.50	26.75	11.59	.228	28.75	11.59
113	8,000	5.04	.249	35.75	22.00	15.00	.750	30.75	28.75	12.47	.249	30.75	12.47
114	10,000	6.30	.262	38.00	24.00	16.00	.750	32.75	30.75	13.09	.262	32.75	13.09
115	12,000	7.56	.275	40.50	26.00	17.00	.750	34.75	32.75	13.72	.275	35.50	13.72
116	14,000	8.82	.203	44.50	28.00	18.00	1.000	37.00	35.50	11.34	.294	39.50	14.84
118	18,000	11.35	.203	48.50	32.00	20.00	1.000	41.00	39.50	11.47	.347	43.50	16.97
120	23,000	14.50	.203	52.50	36.00	22.00	1.000	45.00	43.50	11.59	.399	48.00	19.09
122	30,000	18.90	.203	58.00	40.00	24.00	1.000	49.00	48.00	11.75	.451	52.00	21.50
124	37,000	23.30	.203	62.88	44.00	26.00	1.000	53.00	52.00	11.91	.504	56.00	23.41
126	44,000	27.70	.203	69.00	48.00	28.00	1.000	57.00	57.00	12.22	.556	61.00	25.72

Figure 6.15 Large gear coupling line (courtesy of Zurn Industries, Inc., Mechanical Drives Division).

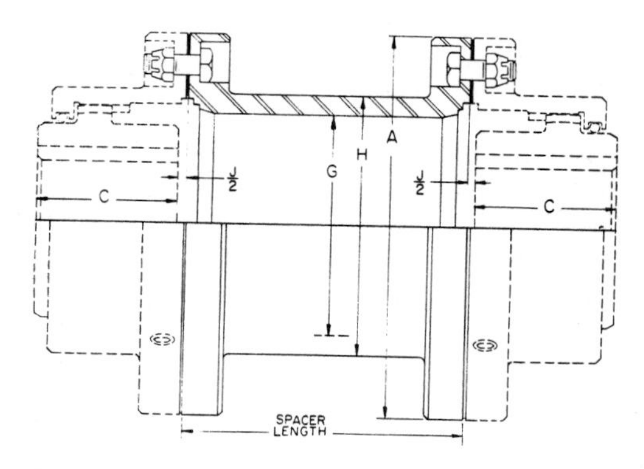

Figure 6.16 Spacer-type gear coupling (courtesy of Zurn Industries, Inc., Mechanical Drives Division).

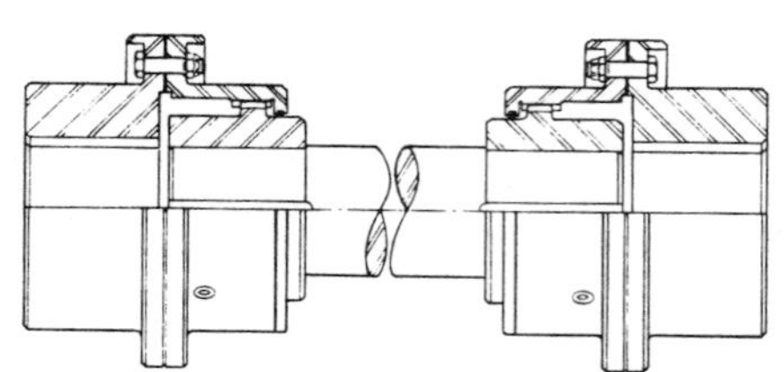

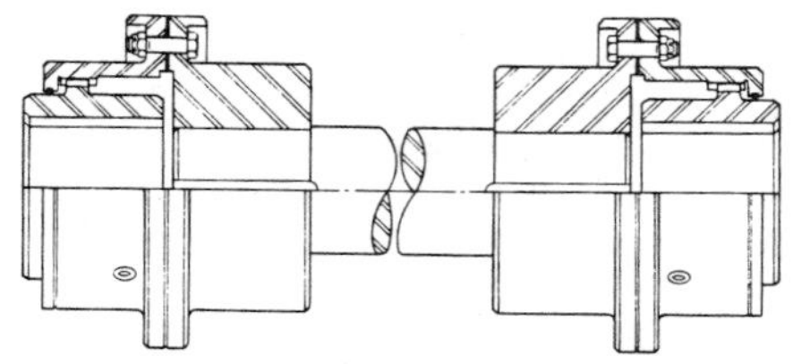

(A)

(B)

Figure 6.17 (A) Solid-shaft gear coupling; (B) solid-shaft steel mill application (courtesy of Zurn Industries, Inc., Mechanical Drives Division).

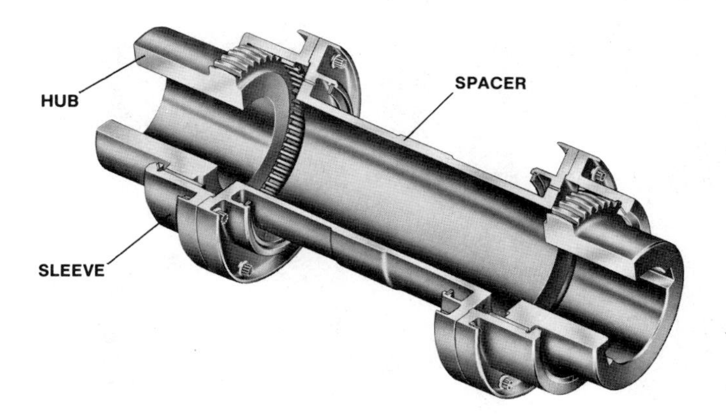

Figure 6.18 High-speed gear coupling (courtesy of Zurn Industries, Inc., Mechanical Drives Division).

Figure 6.19 High-speed continuously lubed reduced moment gear coupling (courtesy of Koppers Company, Inc., Engineered Metal Products Group).

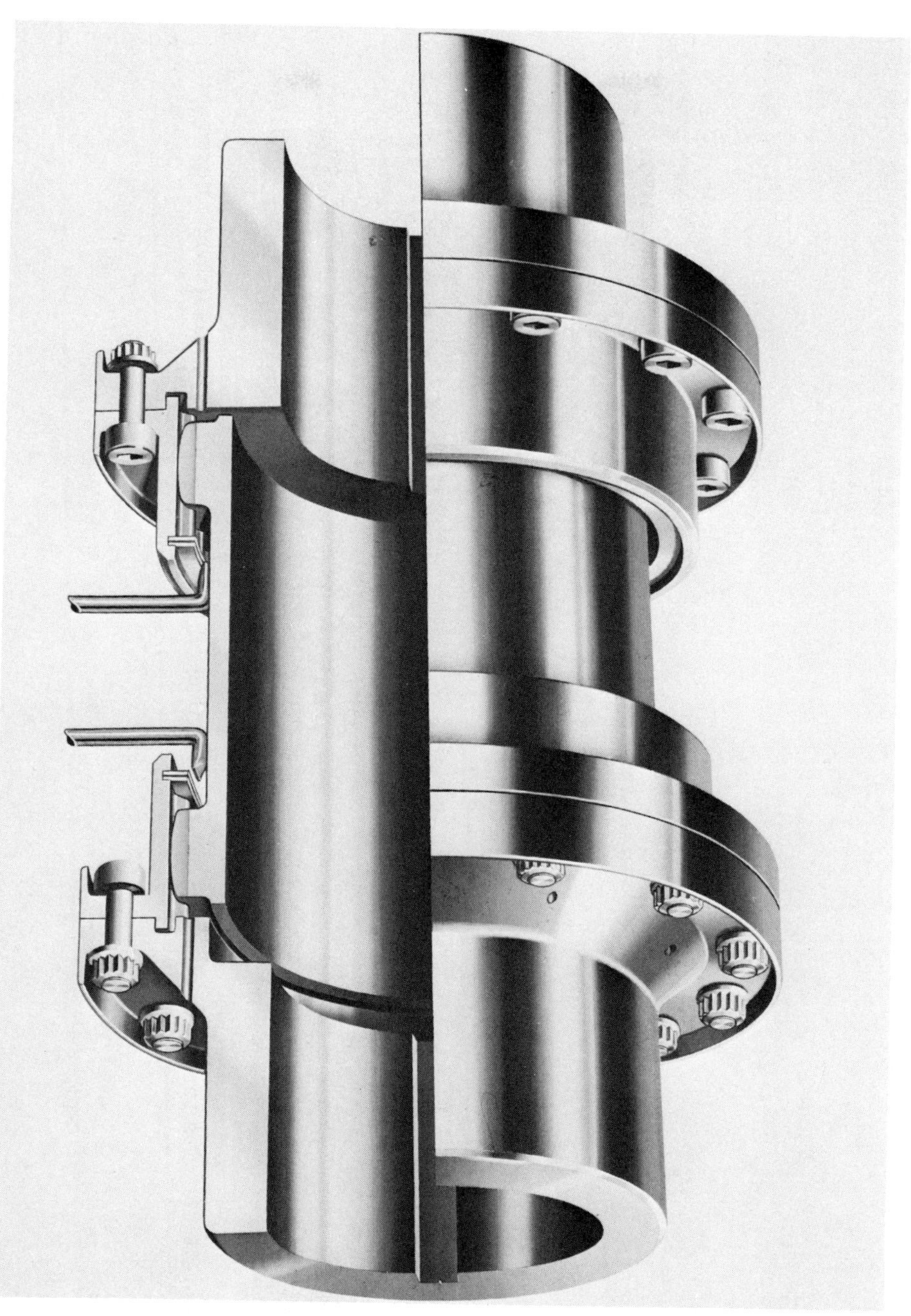

Figure 6.20 Marine-style or flanged-connected gear coupling (courtesy of Koppers Company, Inc., Engineered Metal Products Group).

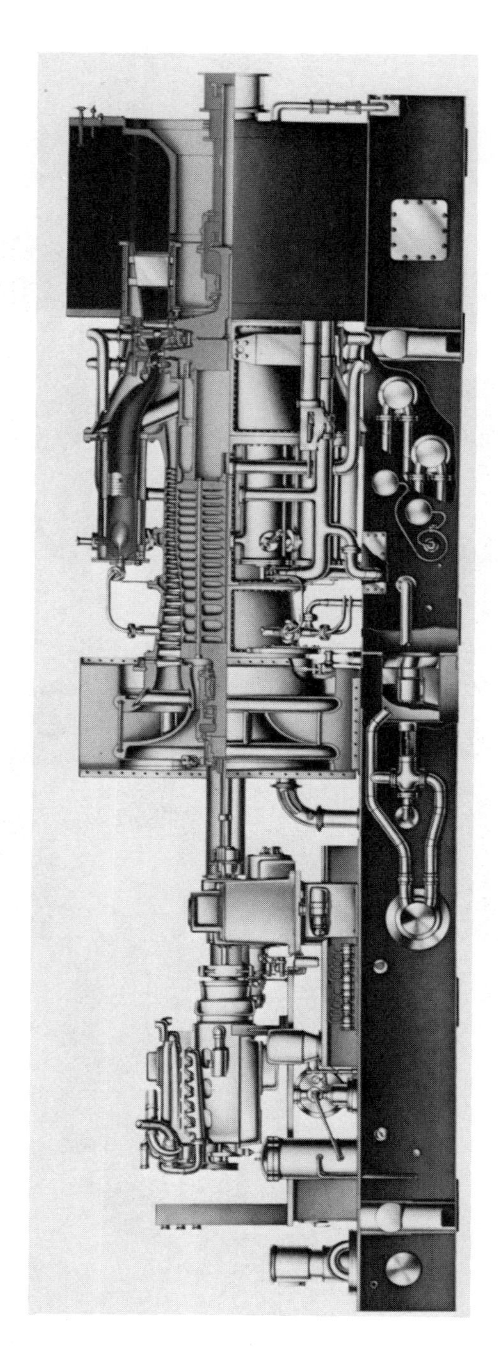

Figure 6.21 Gas turbine application (courtesy of General Electric Schenectady, Gas Turbine Division).

Figure 6.22 Centrifugal compressor application (courtesy of Dresser Clark).

Figure 6.23 Application of a gear spindle (courtesy of Howmet).

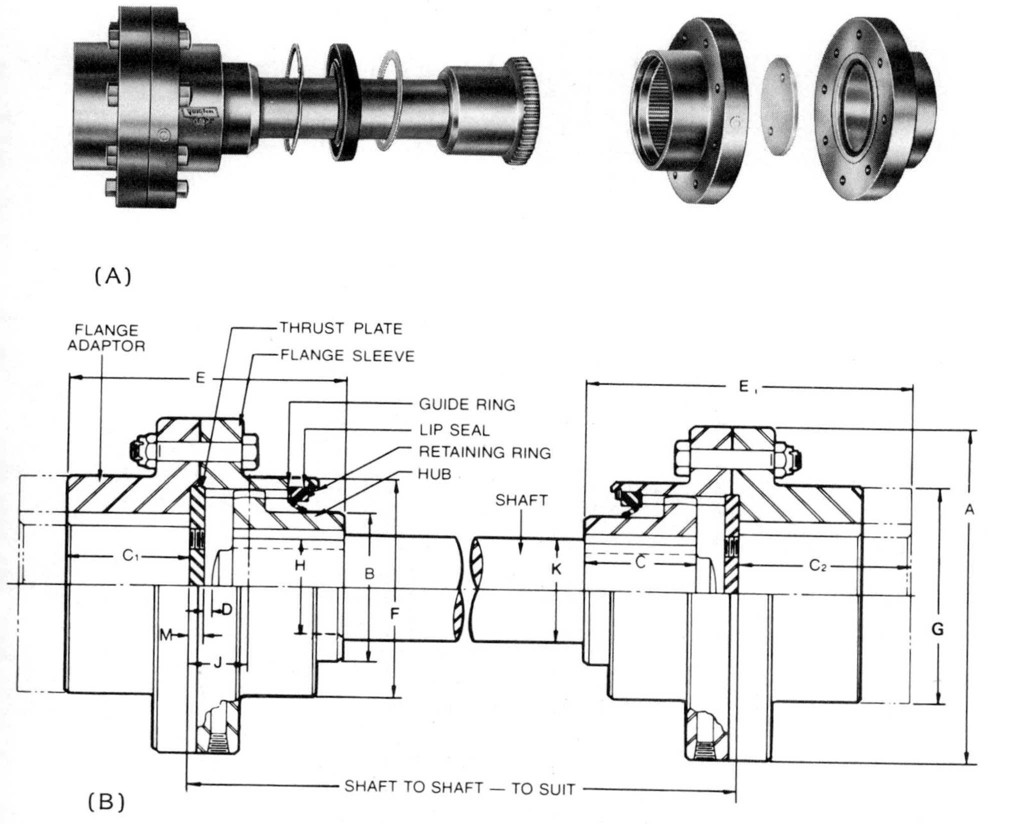

Figure 6.24 High angle/low speed gear couplings (courtesy of Zurn Industries, Inc., Mechanical Drives Division).

SIZE INCHES	ADAPTOR BORE & KEYWAY DATA—INCHES				DIMENSIONS — INCHES													DIMENSIONS WITH MAXIMUM LENGTH ADAPTOR—INCHES	
	SQUARE KEY		REDUCED KEY																
	Maximum Bore	Keyway	Maximum Bore	Keyway	A	B	C	C_1	D	E	F	G	H	J	K	M		C_2	E_1
3 9/16	1 3/4	3/8x3/16	1 7/8	3/8x1/8	3 9/16	1 5/8	1 3/8	1 3/8	1/16	3 9/32	2 5/8	2 9/16	1	11/16	1 1/8	1/4		1 15/16	3 27/32
4	2	1/2x1/4	2 1/8	1/2x3/16	4	1 7/8	1 11/16	1 11/16	1/16	3 15/16	3	3	1 1/4	3/4	1 5/8	1/4		2 1/4	4 1/2
6	2 11/16	5/8x5/16	2 7/8	5/8x7/32	6	2 5/8	1 15/16	1 3/4	1/8	4 7/16	3 7/8	3 7/8	1 3/4	1	1 7/8	1/4		3	5 11/16
7	3 1/4	7/8x7/16	3 1/2	7/8x5/16	7	3 3/8	2 7/16	2 1/4	1/8	5 1/2	5	4 7/8	2 1/8	1 1/8	2 1/4	1/4		3 7/8	7 1/8
8 3/8	4	1x1/2	4 1/4	1x3/8	8 3/8	4 1/8	3 1/32	2 13/16	1/8	6 31/32	6	5 3/4	2 7/8	1 1/2	3	3/8		4 1/2	8 21/32
9 7/16	4 5/8	1 1/4x5/8	5	1 1/4x7/16	9 7/16	5 1/8	3 19/32	3 5/16	1/8	7 31/32	7	6 13/16	3 3/8	1 1/2	3 1/2	3/8		5 1/4	9 29/32
11	5 3/8	1 1/4x5/8	5 3/4	1 1/4x7/16	11	5 5/8	4 3/16	3 7/8	1/8	9 5/16	8	7 3/4	3 7/8	1 3/4	4	3/8		6 1/2	11 15/16
12 1/2	6 1/4	1 1/2x3/4	6 3/4	1 1/2x1/2	12 1/2	6 1/2	4 3/4	4 3/8	1/8	10 7/16	9 5/16	9 1/16	4 7/16	1 7/8	4 5/8	1/2		7 1/8	13 3/16
13 5/8	6 7/8	1 3/4x7/8	7 3/8	1 3/4x5/8	13 5/8	7 1/4	5 5/16	4 15/16	1/8	11 11/16	10 3/8	10 3/16	4 15/16	2 1/16	5 1/8	1/2		8 1/8	14 7/8
15 5/16	7 7/8	1 3/4x7/8	8 3/8	1 3/4x5/8	15 5/16	8 1/2	6 1/32	5 11/16	3/16	13 9/32	11 5/8	11 3/8	5 7/16	2 1/4	5 3/4	1/2		9 3/8	16 31/32
16 9/16	8 3/4	2x1	9 1/4	2x3/4	16 9/16	9	6 5/8	6 1/16	3/16	14 1/2	12 5/8	12 1/2	5 15/16	2 9/16	6 1/4	1/2		10 1/4	18 11/16
18	9 3/8	2x1	9 7/8	2x3/4	18	10	7 13/32	7 5/32	3/16	16 7/16	13 7/8	13 1/2	6 7/16	2 11/16	7	1/2		11 1/4	20 17/32
20 3/4	10 3/4	2 1/2x1 1/4	11 1/2	2 1/2x7/8	20 3/4	12	8 11/16	8 1/2	3/16	19 1/4	15 3/4	15 3/4	6 15/16	2 15/16	7 1/2	1/2		13 1/2	24 1/4

(C)

Figure 6.24 (continued)

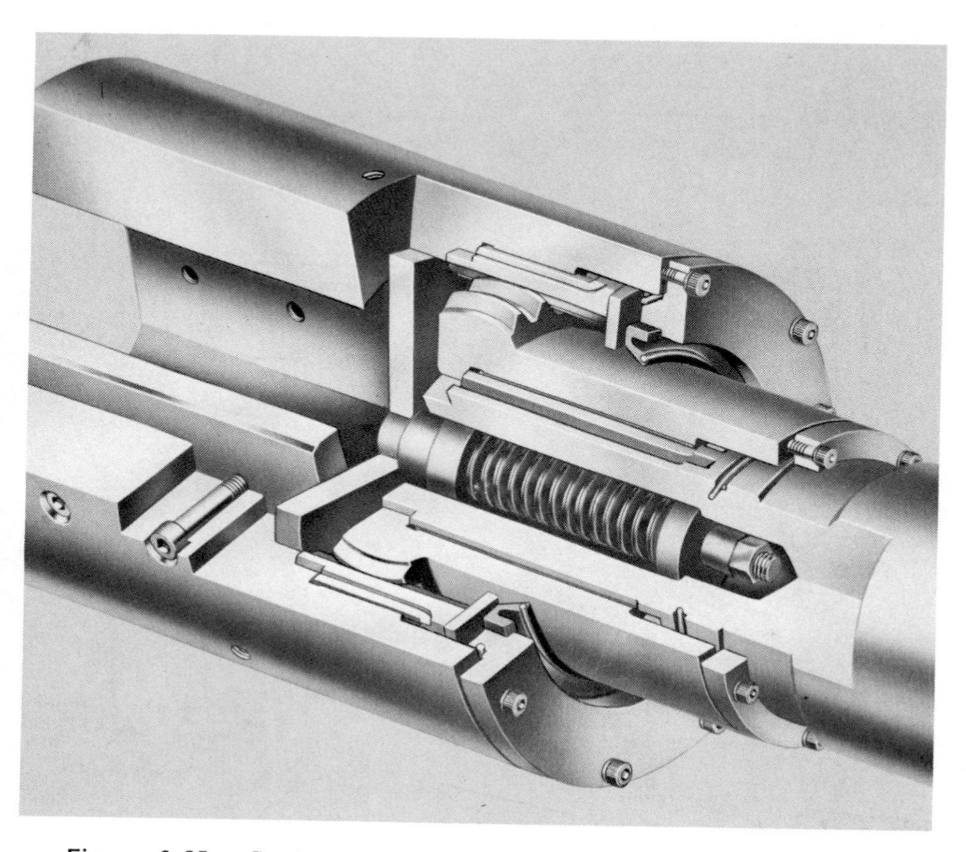

Figure 6.25 Custom-designed gear spindles (courtesy of Zurn Industries, Inc., Mechanical Drives Division).

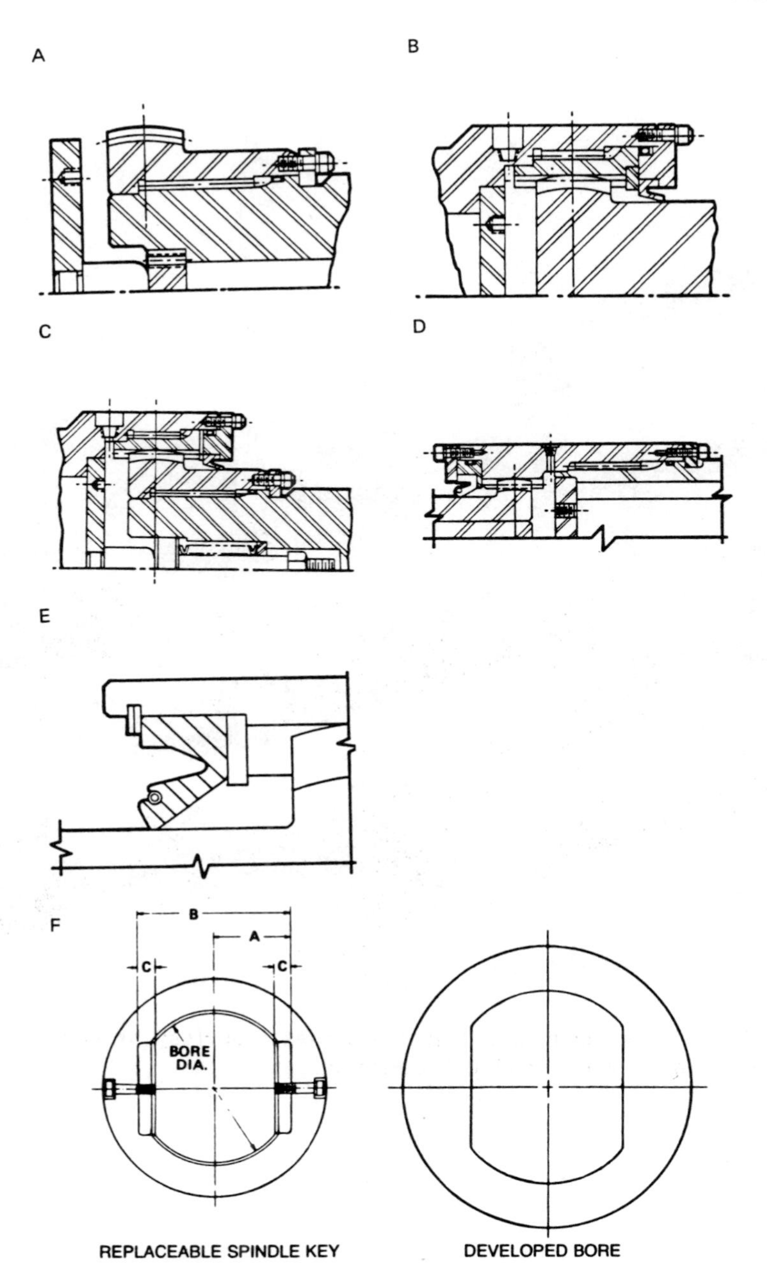
A
B
C
D
E
F
B
A
C
C
BORE
DIA.
REPLACEABLE SPINDLE KEY
DEVELOPED BORE

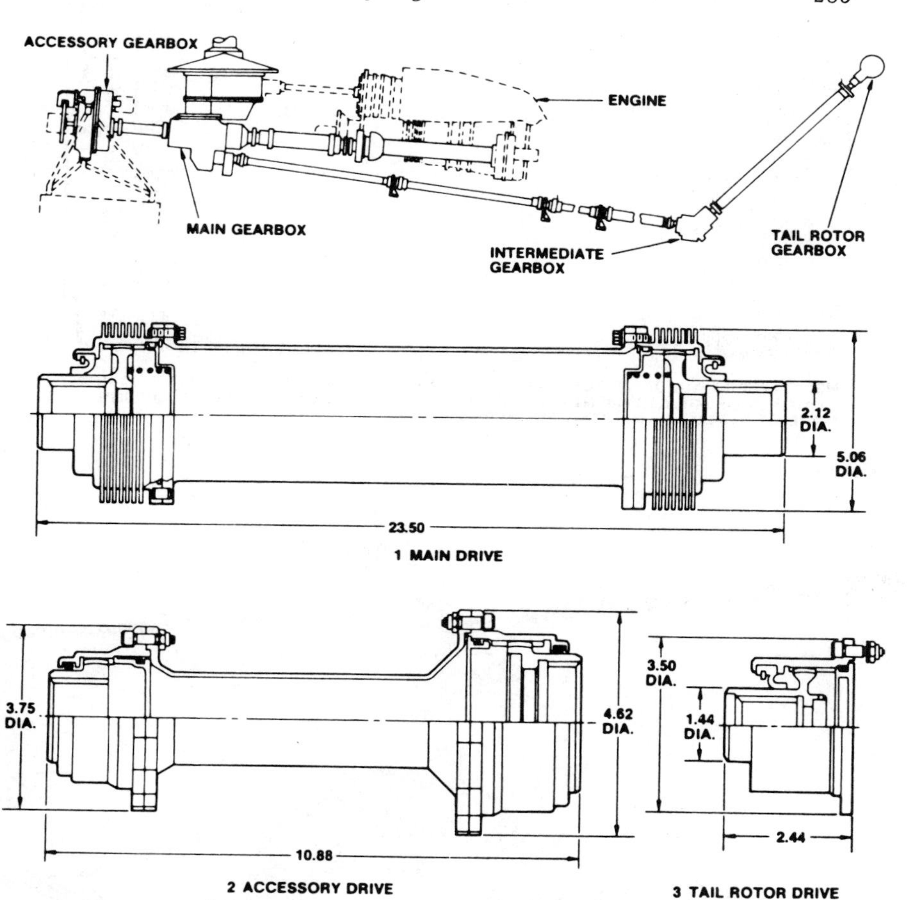

Figure 6.27 Helicopter drives (courtesy of Zurn Industries, Inc., Mechanical Drives Division).

Figure 6.26 Special features of gear spindles: (A) splined hubs; (B) insert gear rings; (C) spring-loaded thrust buttons; (D) splined replaceable sleeves; (E) lipseal; (F) roll end bores (courtesy of Zurn Industries, Inc., Mechanical Drives Division).

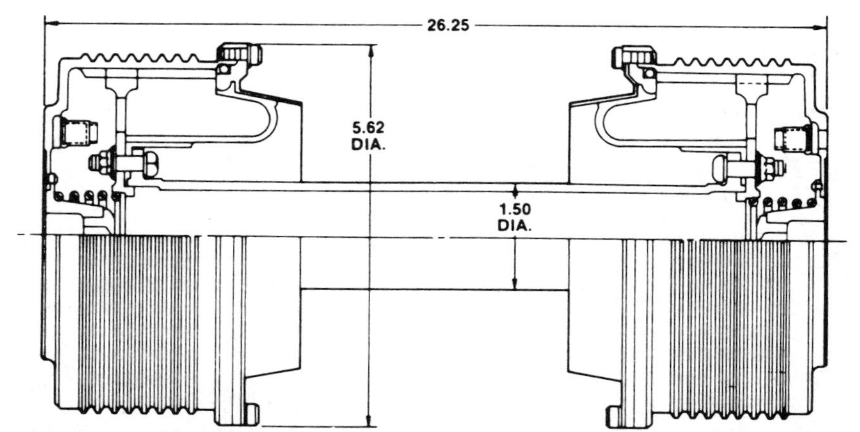

HELICOPTER MAIN DRIVE BETWEEN ENGINE AND TRANSMISSION. 305 HP @ 6,000 RPM. CONTINUOUS MISALIGNMENT 2.5°. INTERMITTENT MISALIGNMENT 4.5°. COUPLING MADE OF NITRALLOY N. NITRIDED LAPPED TEETH.

(A)

Figure 6.28 (A) Helicopter main drive coupling (courtesy of Zurn Industries, Inc., Mechanical Drives Division); (B) helicopter (Courtesy of Soloy Aviation).

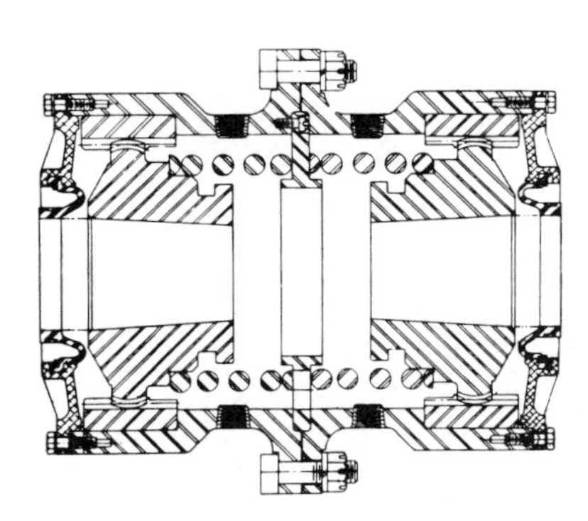

Figure 6.29 Traction drive couplings (courtesy of Zurn Industries, Inc., Mechanical Drives Division).

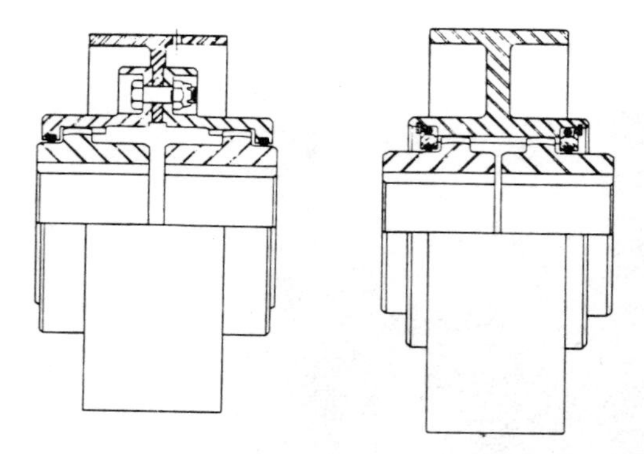

Figure 6.30 Brake drum gear coupling (courtesy of Zurn Industries, Inc., Mechanical Drives Division).

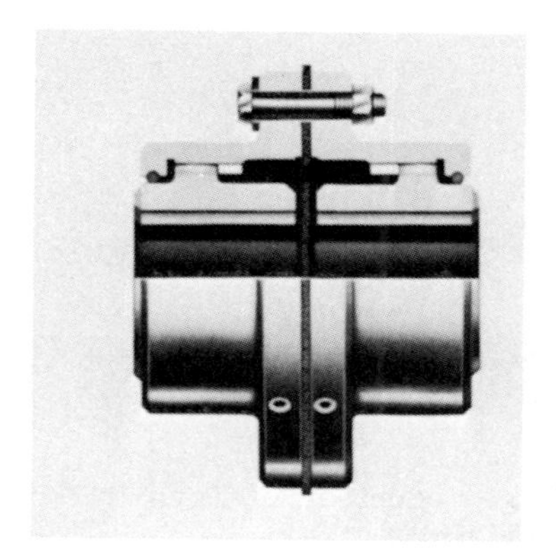

Figure 6.31 Insulated gear coupling (courtesy of Zurn Industries, Inc., Mechanical Drives Division).

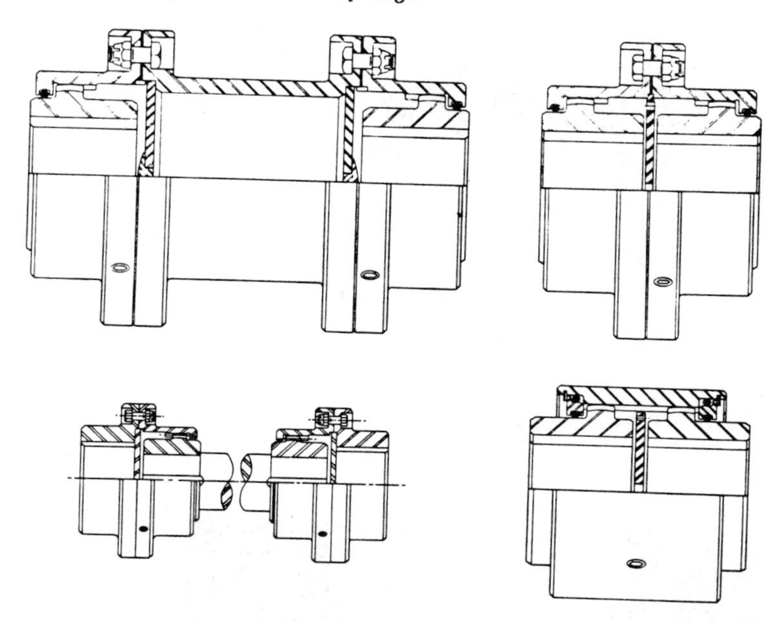

Figure 6.32 Limited end float gear couplings (courtesy of Zurn Industries, Inc., Mechanical Drives Division).

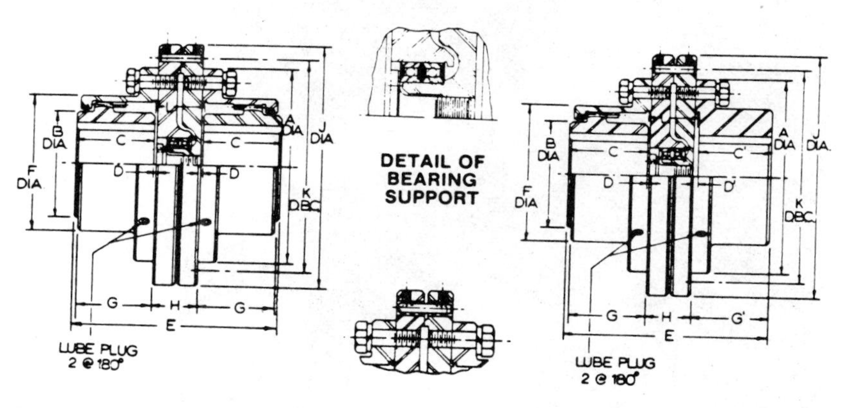

Figure 6.33 Shear pin gear couplings (courtesy of Zurn Industries, Inc., Mechanical Drives Division).

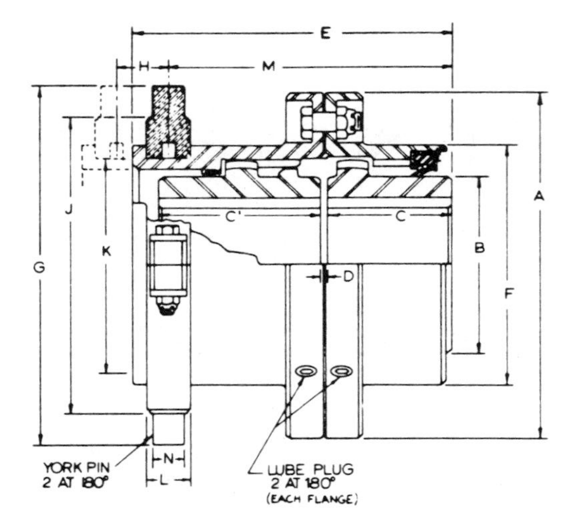

(A)

(B)

Figure 6.34 Disconnect gear couplings: (A) low speed; (B) high speed (courtesy of Zurn Industries, Inc., Mechanical Drives Division).

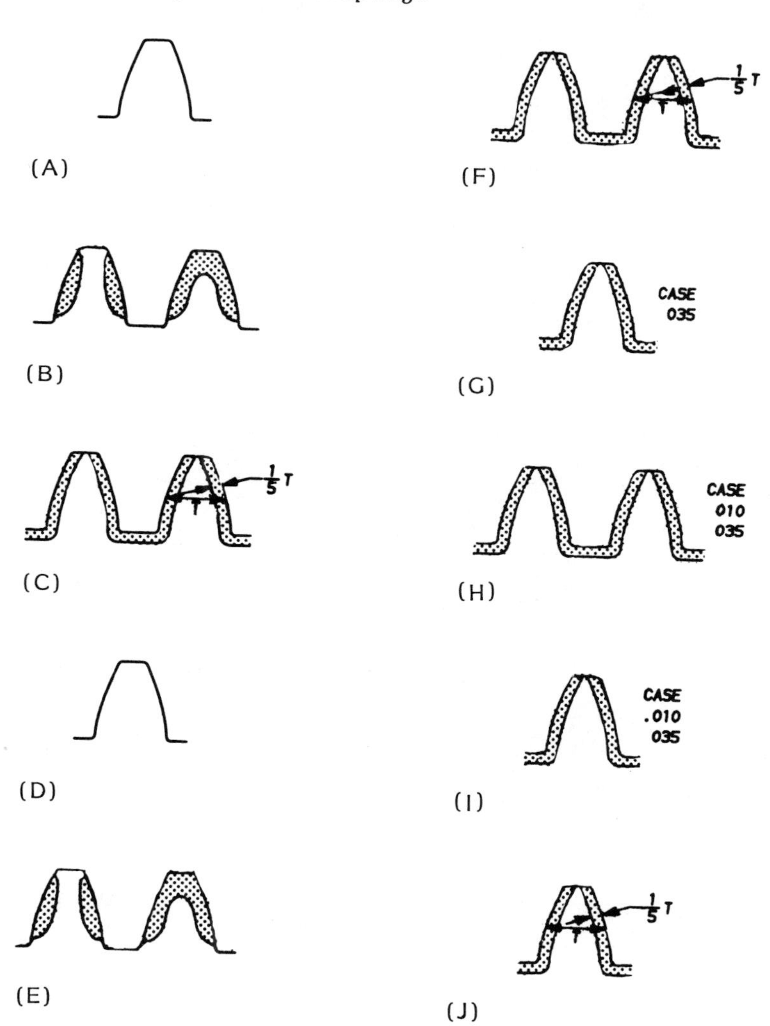

Figure 6.35 Various types of heat-treated gear teeth.

Material		$\frac{1}{4}°$	$1\frac{1}{2}°$	3°
1035—1050		1.25	1	.35
1035—1050 Flank hardened		1.25	1	.5
1035—1050 Fully contoured		2	1.5	.75
4140 & 4340	28 Rc	2	1.5	.6
4140 & 4340 Flank	28 Rc core 50—56 surface	3.75	2.75	.75
4140 & 4340 Fully Contoured	28 Rc core 52—56 surface	3.75	2.75	1.5
4140 & 4340 Nitrided Up to .035	35 Rc core 52—60 surface	3.75	2.75	1.25
4140 & 4340 Nitrided .010—.020 Case	28 Rc core 52—60 surface	2 1.25[a]	1.5 .5[a]	.75 .10[a]
Nitralloy 135 or N	32—35 Rc core 60—65 Rc surface	3 2[a]	2.5 .75[a]	1.25 .15[a]
Carburized 8620, 4320, 3310	32—38 Rc core 59—60 Rc surface	4.5	3.5	1.75

[a]High speed limits.

Figure 6.36 Approximate strength factors of various materials.

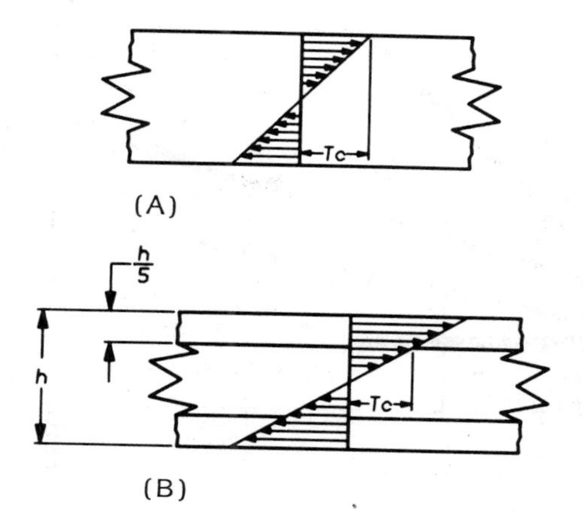

Figure 6.37 Stress distribution: (A) cantilever beam; (B) proportions to tooth depth.

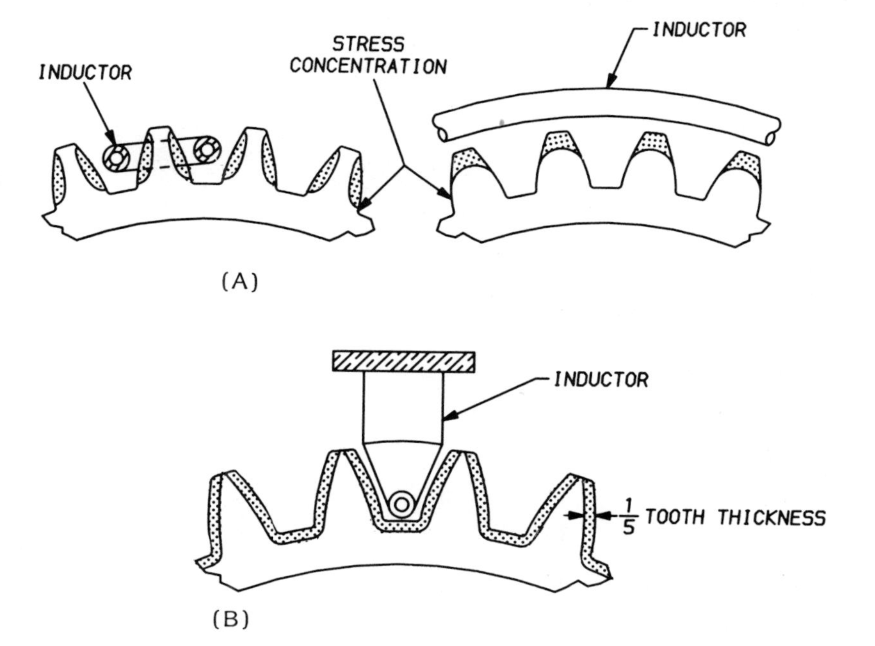

Figure 6.38 (A) Conventional flank hardening; (B) fully contoured hardened patterns.

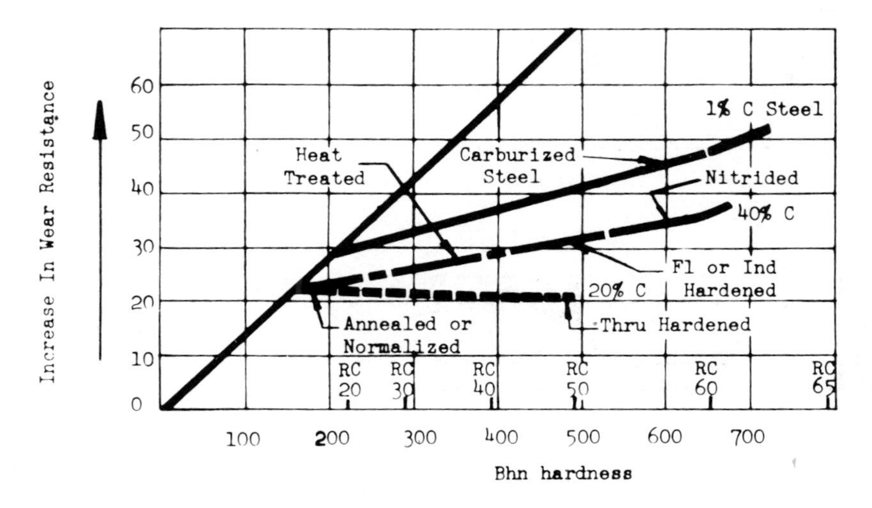

Figure 6.39 Wear resistance versus hardness for various heat treatments.

C = % in Contact	Misalignment
80%	0°
60%	½°
50%	1°
40%	2°
35°	3°
25°	4°
15%	5°
10%	6°

Figure 6.40 Percentage of teeth in contact versus misalignment.

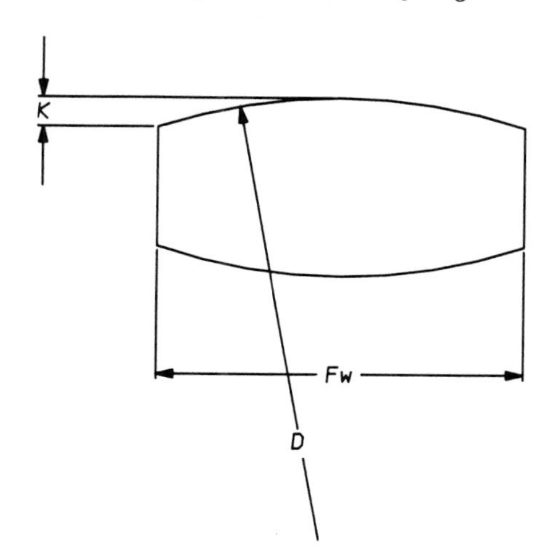

Figure 6.41 Amount of tooth crown.

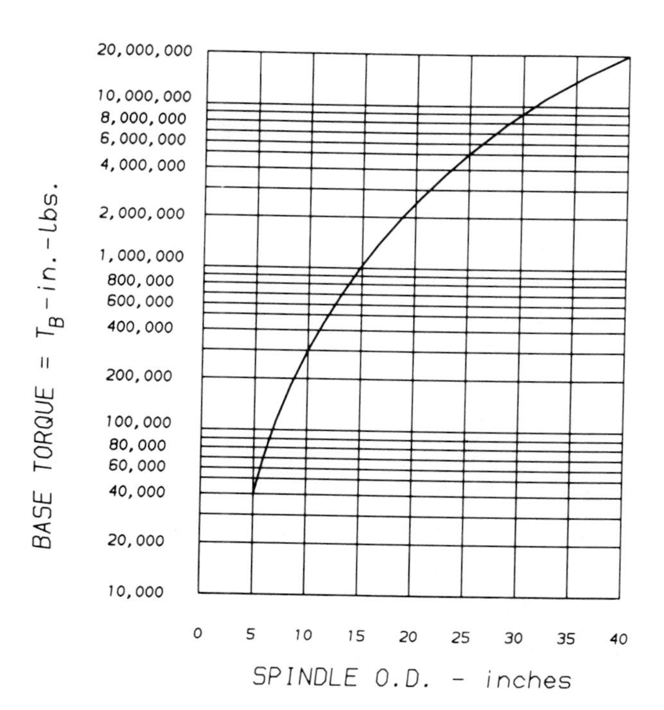

Figure 6.42 Torque capacity versus outside diameter.

Maximum Operating Angle	Misalignment Factor (M)
1°	1.0
2°	0.7
3°	0.5
4°	0.4
5°	0.3
6°	0.2

Figure 6.43 Misalignment factors.

Size Limit	Material and HT	S = Normal Factor at Maximum Operating Angles		
		<4°	5°	6°
All	1045 flank hardened (flame or induction)	1	1	1
12"—28" 4° maximum	1045 full contoured induction hardened	1.5		
All	4140 flank hardened (flame or induction)	1.5	1.5	1.5
12"—28"	4140 full contoured induction hardened	2.75	2.75	2.50
To 13"	4140 & 4340 nitrided	2.75	2.5	2.25
All	8620, 9310, 3310 carburized	3.5	3.5	3.25

Figure 6.44 Material factor.

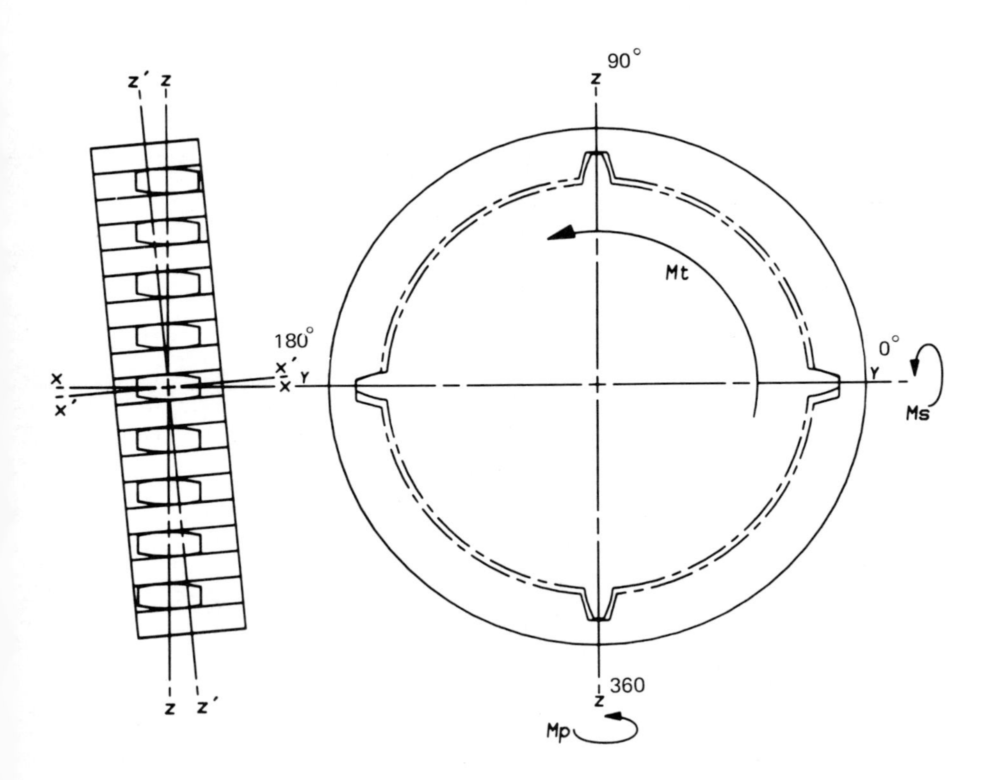

Figure 6.45 Coordinate system for a gear coupling.

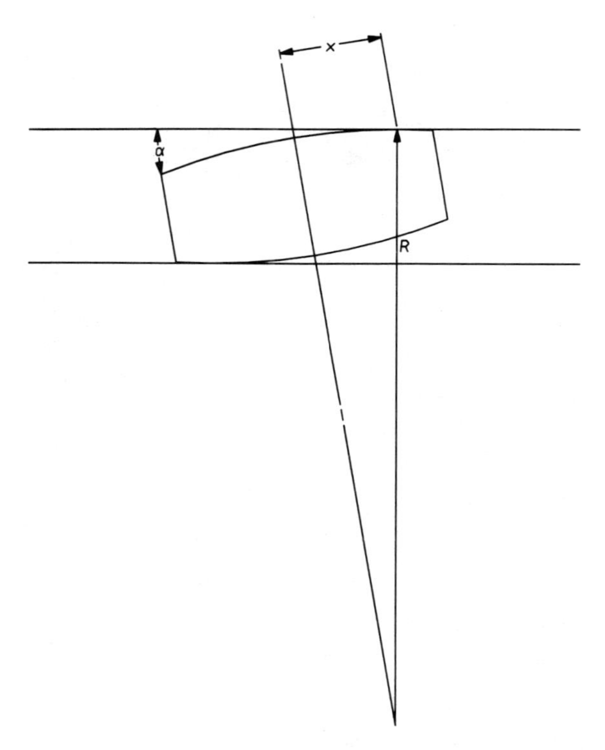

Figure 6.46 Gear tooth sweep.

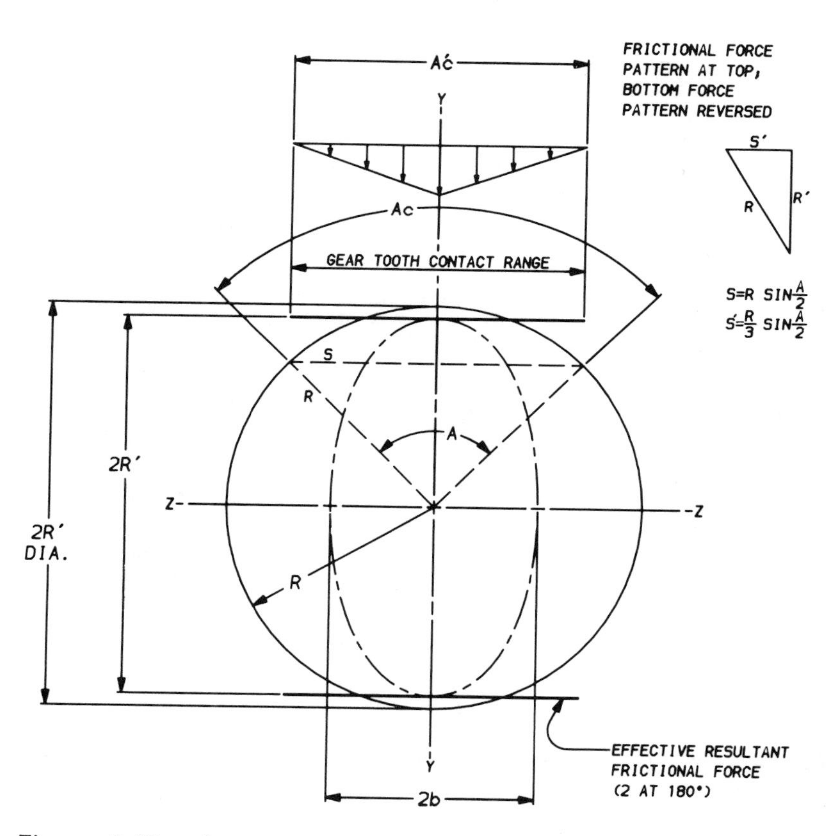

Figure 6.47 Average radius for a misaligned gear coupling.

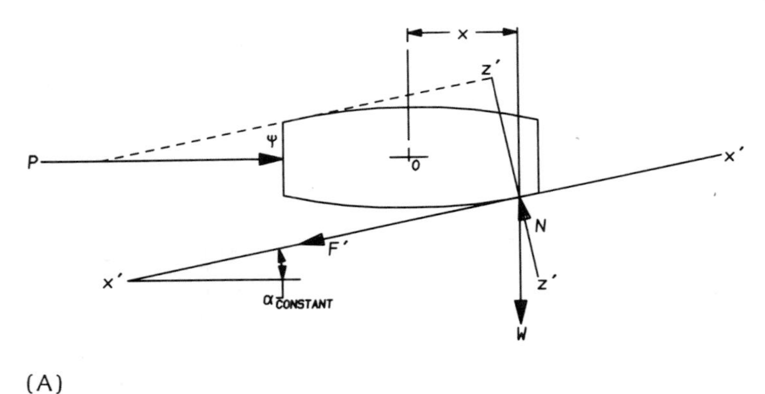

(A)

(MOMENT DUE TO GEAR CONFIGURATION)

Me

HUB FRICTION PATTERN ONE SIDE, OPPOSITE FORCE DIRECTION @ 180° OR OTHER SIDE OF GEAR.

z′ z

Mt

x

x′

SLEEVE STATIONARY

x′

α CONSTANT

HUB STATIONARY

x

F

RESULTANT

z z′

(B)

Figure 6.48 (A) Forces in a gear coupling; (B) moment pattern in a gear coupling.

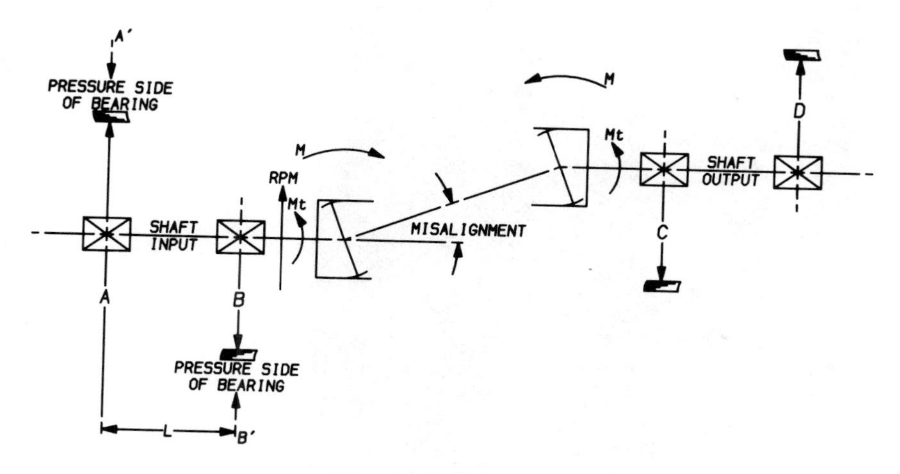

Figure 6.49 Plan view of bearing reactions.

(A)

Figure 6.50 Typical good wear pattern: (A) gear spindle; (B) aircraft gear coupling (courtesy of Zurn Industries, Inc., Mechanical Drives Division).

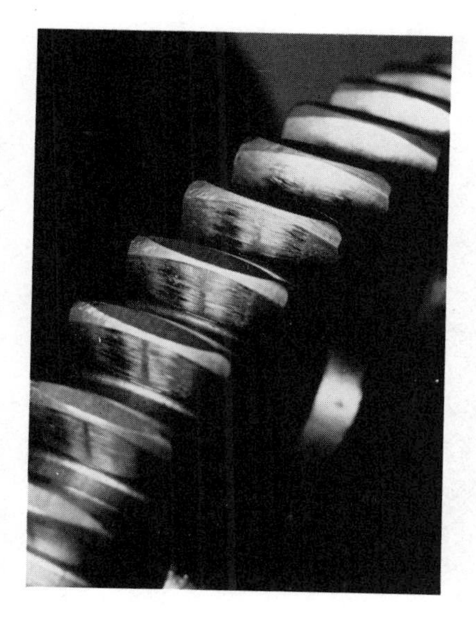

(B)

Figure 6.50 (continued)

Figure 6.51 Failure due to lubricant breakdown.

Figure 6.51 (continued)

Figure 6.52 Improper tooth contact.

Figure 6.53 Failure due to over-misalignment.

Figure 6.54 Failure due to over-misalignment.

Figure 6.55 Failure due to operating at a torsional critical level.

Figure 6.56 Worm tracking failure.

Figure 6.57 Sheared gear teeth.

(A)

(B)

Figure 6.58 Sludge-contaminated gear coupling.

Figure 6.59 Broken spacer from locked-up gear teeth.

(A)

(B)

Figure 6.60 Corroded gear coupling components.

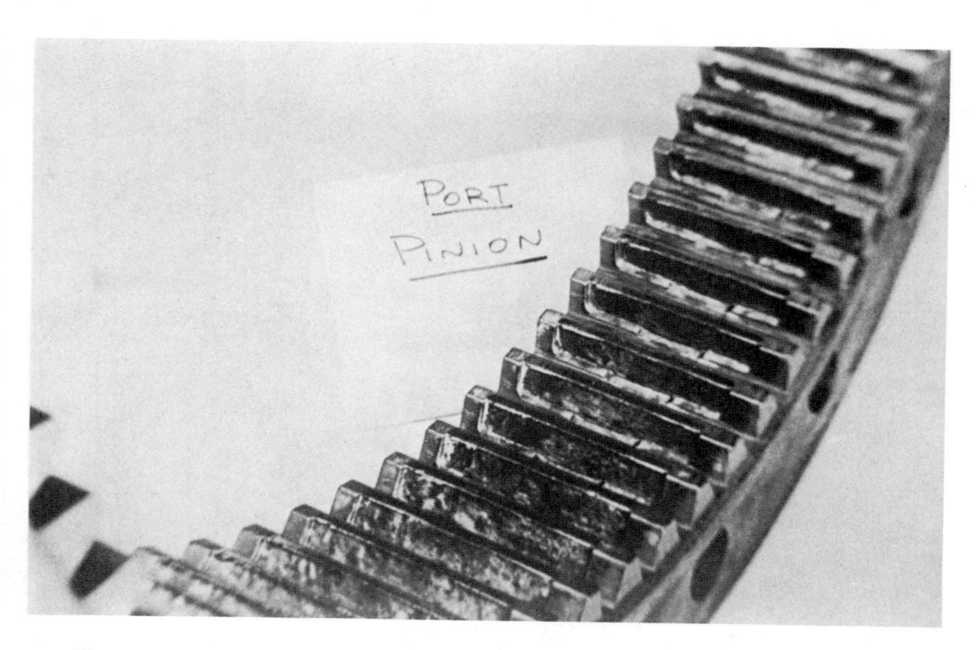

Figure 6.61 Mechanical tooth lock-up from embedding (from Michael Neale et al., *Proceedings of the International Conference on Flexible Couplings for High Power and Speeds*, 1977).

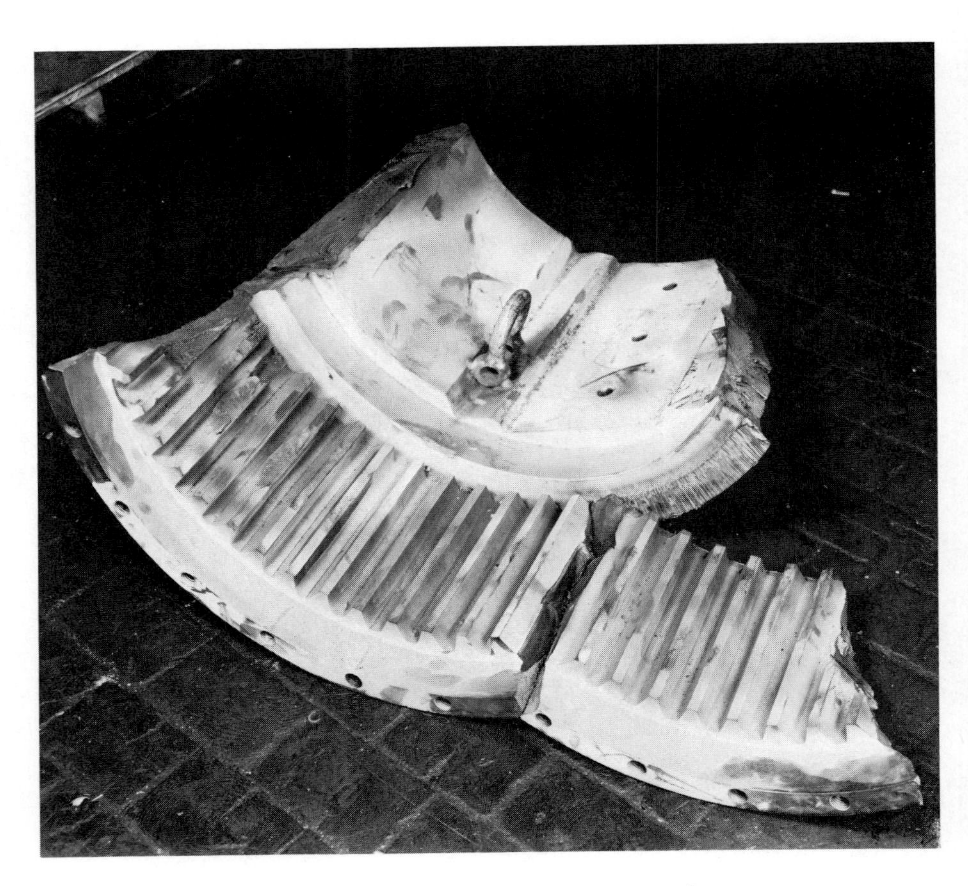

Figure 6.62 Fatigue failure through gear teeth.

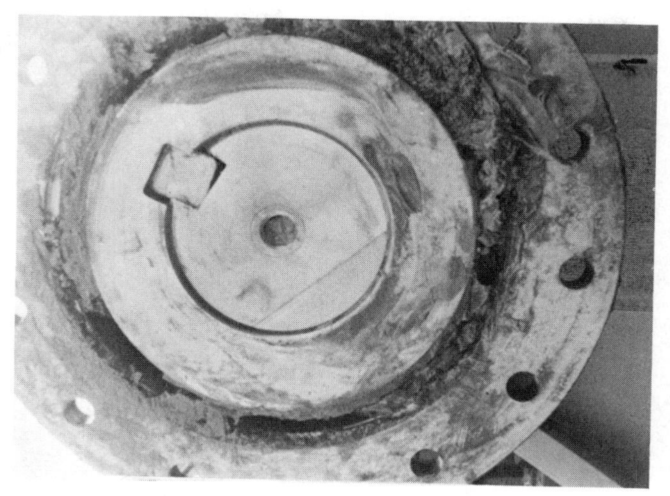

(A)

(B)

Figure 6.63 Key shear rollover.

Figure 6.64 Hub bursting failure.

III. GRID COUPLINGS

The basic metallic grid coupling (see Figure 6.65) consists of two hubs that contain slots, which may be straight, tapered, or contoured. The two hubs are connected by a "serpent"-style metallic grid (spring).

Metallic grid couplings have been used successfully for more than 60 years. Figure 6.66 shows how a grid coupling accommodates for the following three types of misalignment:

1. *Parallel misalignment*: Movement of the gird in the lubricated grooves accommodates for parallel misalignment (Figure 6.66A).
2. *Angular misalignment* (Figure 6.66B): The grid-groove design permits a rocking and sliding action of the lubricated grid and hubs.
3. *Axial movement* (Figure 6.66C): Unrestrained end float for both driving and driven members is permitted because the grid slides in the lubricated grooves. The amount of axial movement can also be limited.

A. Torsional Flexibility of Metallic Grid Couplings

Torsional flexibility is the unique ability of a grid coupling to torsionally deflect when subjected to normal, shock, or vibratory loads, providing flexible accommodation to changing load conditions. Consequently, the coupling is capable of damping vibration and of reducing peak or shock loads.

The following description shows how this is accomplished. Under light loads the grid bears near the outer edges of the hub teeth (Figure 6.67A). The long span between the points of contact remains free to flex under load variations. Under normal loads, as the load increases, the distance between the supports on the hub teeth is shortened, but a free span remains to cushion shock loads (Figure 6.67B). Under shock load, the coupling is flexible within its rated capacity (Figure 6.67C). Under extreme overloads, the grid bears fully on the hub teeth and transmits the full load directly.

B. Where Metallic Grid Couplings Are Used

The metallic grid coupling has been applied to nearly all types of industrial applications, from centrifugal pumps to steel mill applications. Some typical applications are:

Agitators	Machine tools
Blowers	Mixers
Compressors	Paper mills
Cranes	Pumps
Elevators	Rubber industry
Fans	Steel mills and auxiliary equipment
Generators	Textile industry

There are sizes to satisfy a 1-hp motor with a 7/8-in. bore or for a BOF vessel tilting shaft of 40 in. diameter. Several applications of the grid coupling are:

Friction roll drives (Figure 6.68)
Conveyor drive (Figure 6.69)
Centrifugal pump (Figure 6.70)

C. Types of Metallic Grid Couplings

There are two basic types of grid couplings:

1. The horizontal split covered grid coupling (Figure 6.71A)
2. The vertical split covered grid coupling (Figure 6.71B)

A third common type of grid coupling is the spacer type (Figure 6.72). The spacer-type grid coupling is especially suited for pump applications.

The grid coupling is also commonly used with:

1. A brake drum or wheel (Figure 6.73A)
2. A friction disk (Figure 6.73B)
3. A limited end float for use on motors (Figure 6.74)

D. Design and Construction of Metallic Grid Couplings

Metallic grid coupling ratings are usually based on grid fatigue strength. Because of the various grid forms and slots, the design is based primarily on empirical data accumulated over the years. The grids are usually made of spring steel—either high-carbon steel or chrome-vandium steel. In evaluating grid couplings, it is best to ask coupling manufacturers for the following information:

1. Grid material
2. Grid material properties
   a. Yield strength
   b. Endurance limit
3. Stresses in grid for:
   a. Normal torque, no misalignment
   b. No torque, full angular misalignment
   c. No torque, full axial travel

The hubs of a grid coupling are usually made of medium-carbon steel. Without the following information, bore capacities cannot be compared:

1. Hub material
2. Hub properties
   a. Ultimate strength
   b. Yield strength
   c. Endurance limit
3. Hub OD and length dimensions

E. How Metallic Grid Couplings Affect a System

A grid coupling has two very significant effects on a system. It reduces peak shock or impulsive loads by softening (low torsional stiffness) the elastic system, and it absorbs or dampens energy. The torque developed when trying to accelerate a two-mass system is determined using the following formula:

$$T_m = \frac{I_1 I_2}{I_1 + I_2}(\Delta\omega)(\omega_0) \tag{6.30}$$

where

T_m = maximum torque
I_1 = mass moment of inertia-driven equipment
I_2 = mass moment of inertia of the driver
ω_0 = natural frequency of system
$\Delta\omega$ = change of velocity

$$\omega_0 = \frac{K_t(I_1 + I_2)}{I_1 I_2} \tag{6.31}$$

where K_t represents the stiffness of the total system and is determined by the formula

$$K_t = \frac{1}{1/K_s + 1/K_c} \qquad (6.32)$$

where

K_s = stiffness of system
K_c = stiffness of coupling

From the above it can be seen that the softer the coupling, the lower the maximum torque. The resilence or damping of the coupling will tend to reduce the peak torque markedly, as shown in Figure 6.75.

F. Failure Mode for Metallic Grid Couplings

Two of the most common types of failure are wear, usually caused by improper, lost, or deteriorated lubricant, and fatigue of the grids, usually caused by over-misalignment. Other types of failures that have occurred include:

1. *Hub bursting*: results from overtorque or from an improper key/keyway fit
2. *Key shear*: usually caused by overload
3. *Seal failure*: usually caused by loss of lubricant, resulting in excessive heat and deterioration of the seal
4. *Breakage, bolts, spacers, etc.*: typically caused by abuse, overtorqued bolts, over-misalignment, or poor installation and maintenance practices

In general, most coupling failure occurs as a result of misalignment, misuse, abuse, or neglect rather than a design or material deficiency.

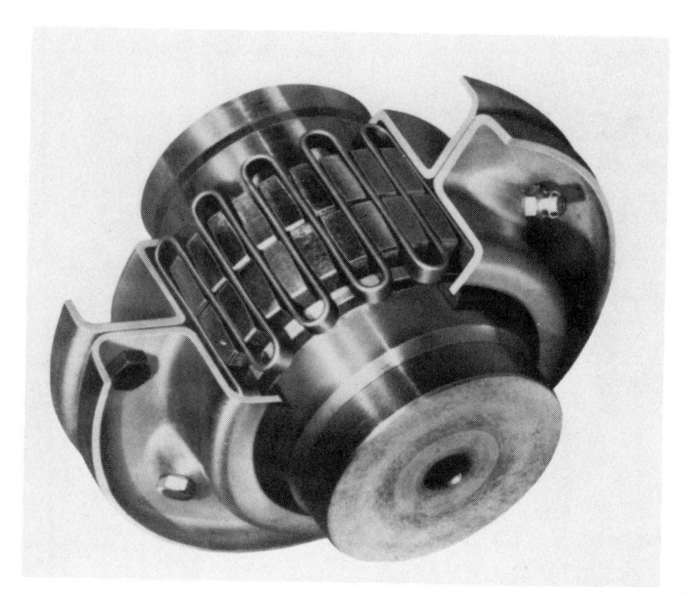

Figure 6.65 Basic metallic grid coupling (courtesy of Browning Manufacturing).

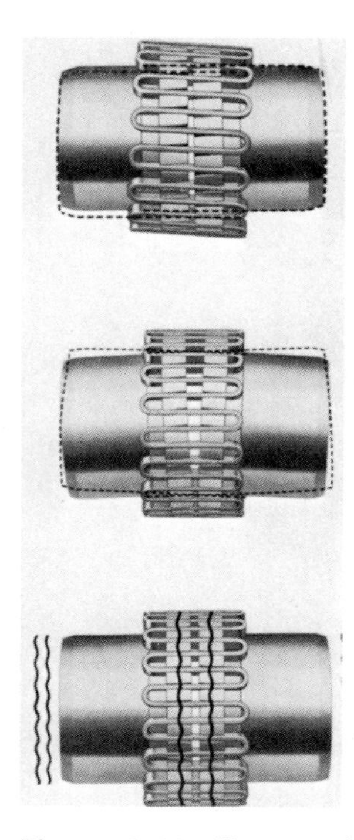

Figure 6.66 How a grid coupling accomodates misalignment (courtesy of Falk Corporation).

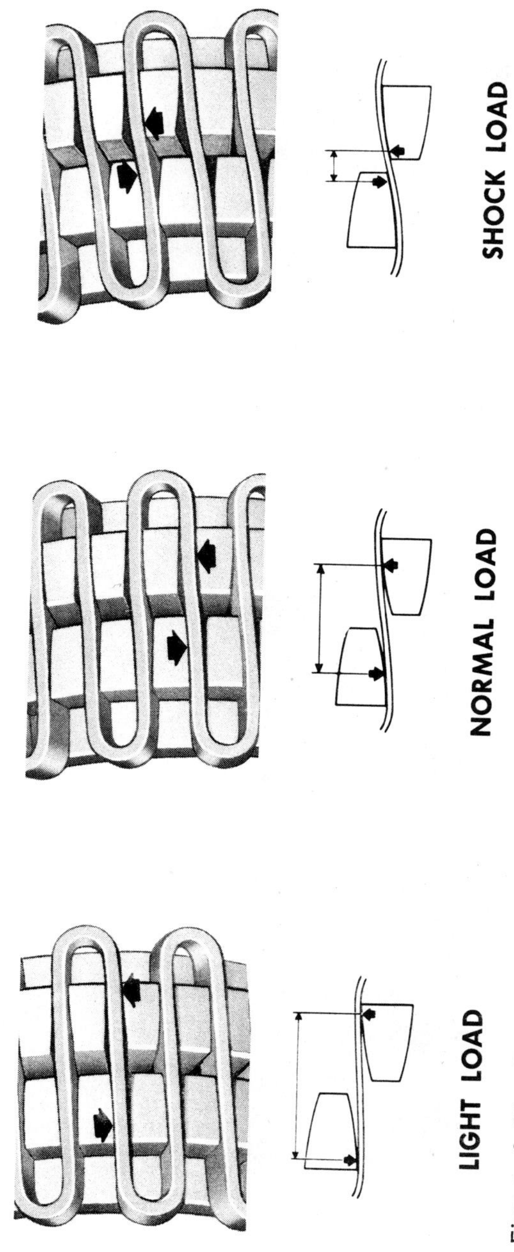

Figure 6.67 How a grid coupling accommodates torsional flexibility (courtesy of Falk Corporation).

Figure 6.68 Grid coupling for friction roll drive (courtesy of Falk Corporation).

Figure 6.69 Grid coupling for conveyer drive (courtesy of Falk Corporation).

Figure 6.70 Grid coupling for centrifugal pump (courtesy of Falk Corporation).

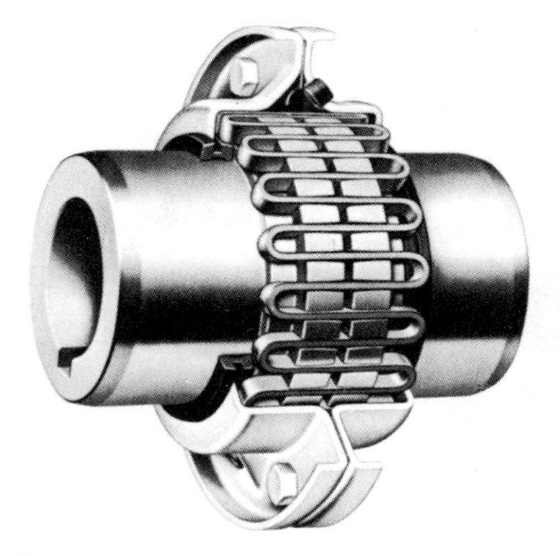

(A)

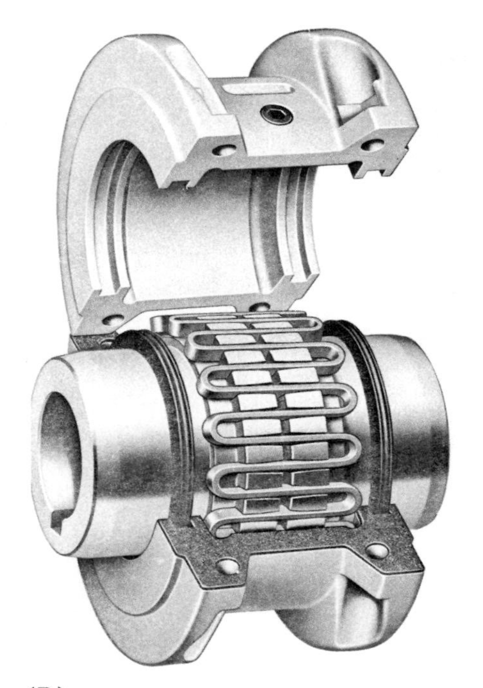

(B)

Figure 6.71 Covered grid couplings: (A) vertical split; (B) horizontal split (courtesy of Falk Corporation).

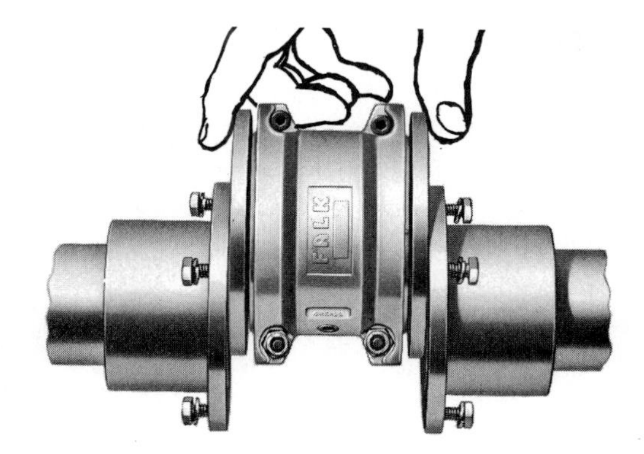

Figure 6.72 Spacer-type grid coupling (courtesy of Falk Corporation).

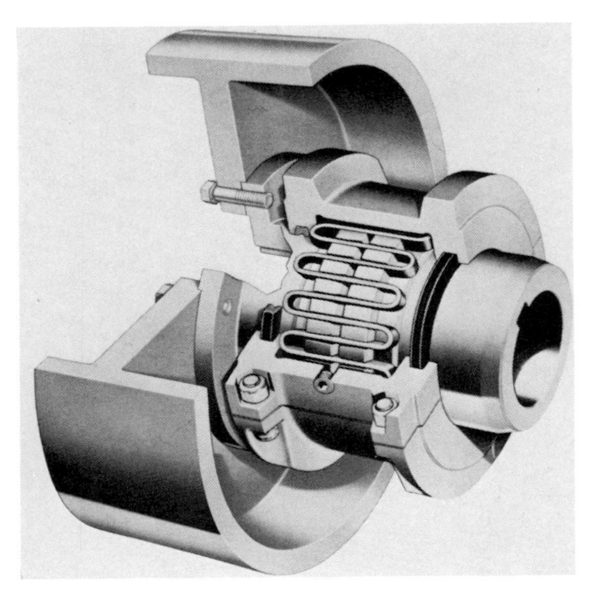

(A)

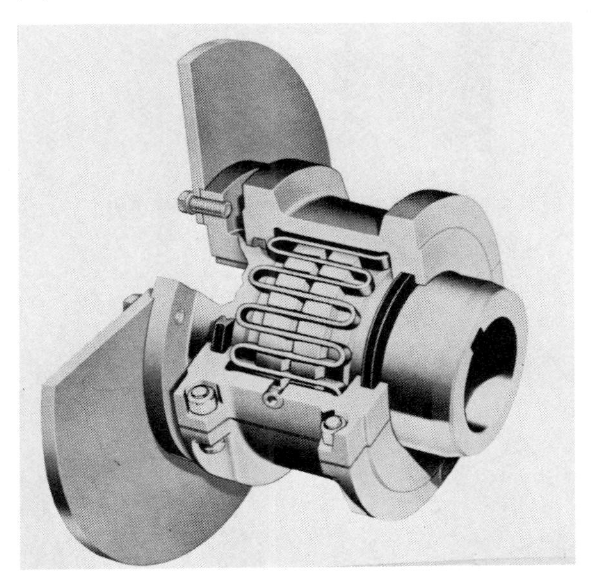

(B)

Figure 6.73 Grid couplings: (A) brake drum; (B) friction disk (courtesy of Falk Corporation).

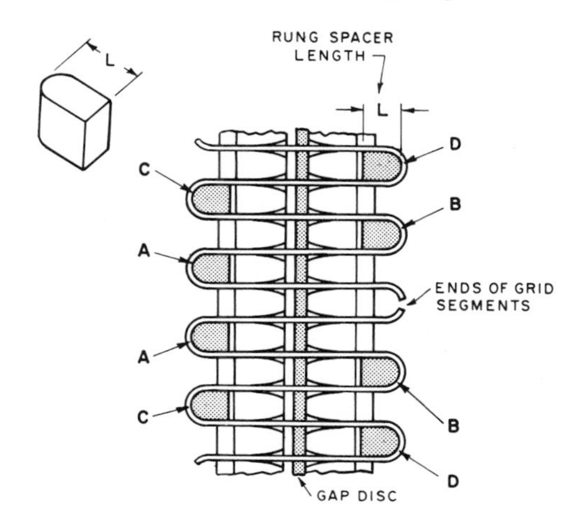

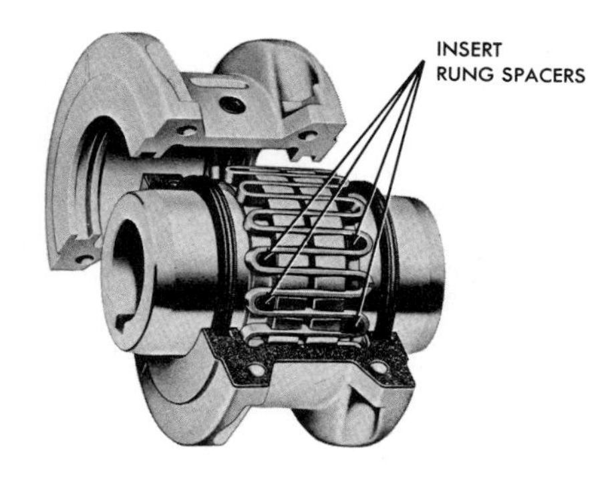

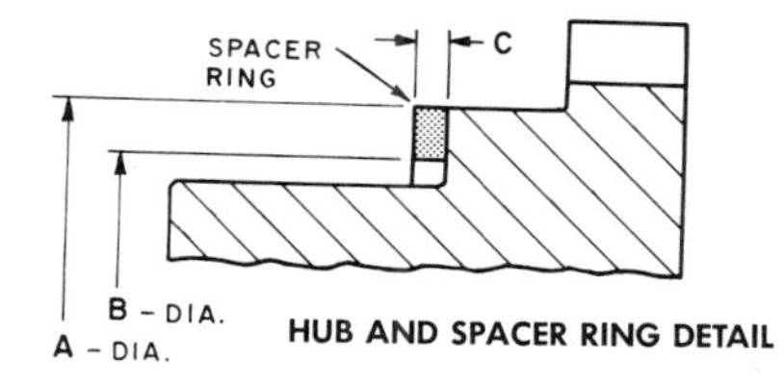

Figure 6.74 Limited end float grid coupling (courtesy of Falk Corporation).

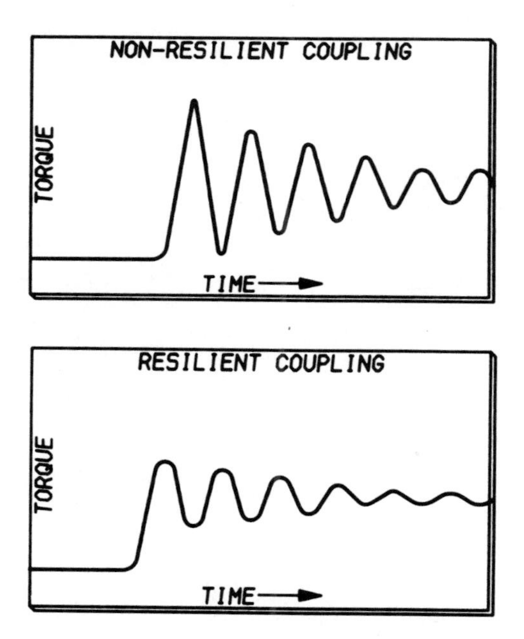

Figure 6.75 Resilient damping of a grid coupling.

IV. CHAIN COUPLINGS

The basic chain coupling consists of two hubs and a length of double-width roller chain (see Figure 6.76). The two hubs have sprockets and are connected by a length of roller chain. Misalignment is compensated by the clearances between the chain and the sprocket (see Figure 6.77).

These couplings have built-in clearances between (1) the chain rollers and the sprocket teeth; (2) the chain rollers and the hardened bushings; (3) the bushings and the pins; and (4) the chain links and the sprocket teeth. All this clearance helps to compensate for angular and parallel offset misalignment as well as shaft end float. Chain couplings can handle up to 2° of angular misalignment and approximately 0.003 in. per inch of coupling outside diameter in offset capacity.

Chain couplings are used on the following types of equipment:

Agitators
Conveyors
Pumps
Feeders
Hoist
Mixers
Machine tools

See Figures 6.78 and 6.79 for typical applications.

A. Types of Chain Couplings

There are three basic types of chain couplings:

1. The double roller chain coupling (Figure 6.80A)
2. The silent chain coupling (Figure 6.80B)
3. The plastic chain coupling (Figure 6.80C)

The double roller chain type is the most common. It is normally used on applications where speeds are moderate, loads are steady, high static or starting loads are present, and ease of connecting driver and driven units is required.

The silent chain coupling is used for heavy duty applications such as on steel mill drives. The high torque capacity is accomplished because this design has a much better load distribution from the chain to the sprocket due to the stacking up of individual links and their angle of engagement with the sprocket teeth.

The plastic chain coupling provides a nonlubricating coupling that can be used in the food-processing and textile industries. It is corrosion resistant, pollution free, and runs more quietly than metal couplings.

B. Design and Construction of Chain Couplings

Chain couplings are usually constructed of high-tensile-strength, heat-treated roller chain and sprockets. The coupling ratings are usually based on the weakest link in the chain, usually the connecting link. As with most mechanical flexible couplings, the most important design consideration is wear and therefore lubrication. It is considered good practice to enclose and lubricate chain couplings because this will help to maintain lubrication between the chain and the sprocket and thus reduce the coefficient of friction, thereby increasing the life of the chain and sprocket and reducing the load on the equipment. See Figure 6.81 for typical coupling covers. As a general rule, chain couplings should incorporate covers under the following conditions:

1. When speeds exceed 500 rpm
2. In dusty atmospheres
3. In moist applications

Figure 6.76 Basic chain coupling (courtesy of Dodge Division of Reliance Electric).

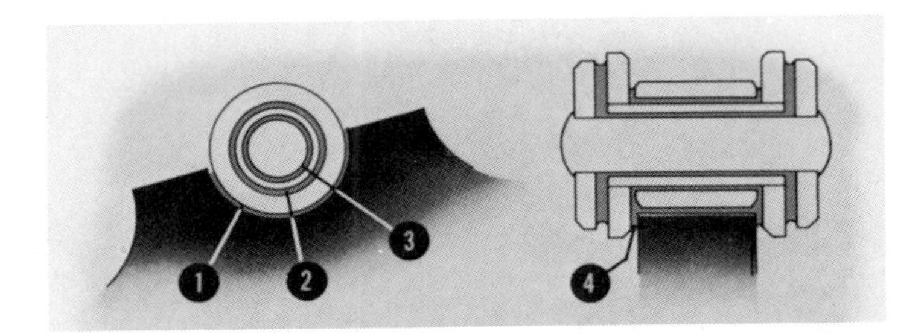

Figure 6.77 Built-in clearances between (1) chain rollers and sprocket teeth, (2) chain rollers and bushings, (3) bushings and pins, and (4) link plates and sprocket teeth.

Figure 6.78 Grain separator drive (courtesy of Dodge Division of Reliance Electric).

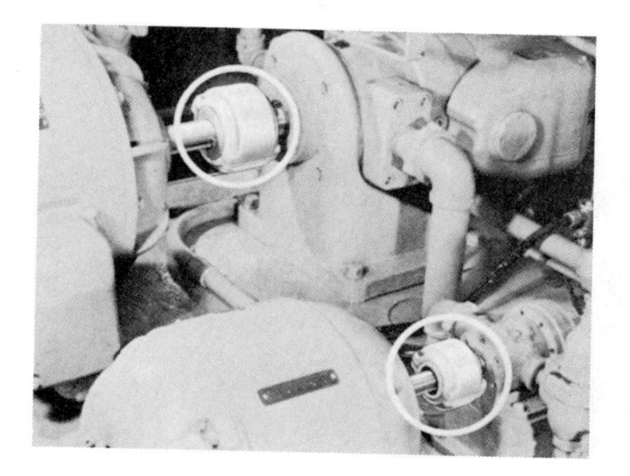

Figure 6.79 Chain coupling on a hydraulic power unit (courtesy of Link-Belt).

(A)

Figure 6.80 Chain couplings: (A) double roller; (B) silent; (C) plastic (courtesy of Morse Industrial Corporation).

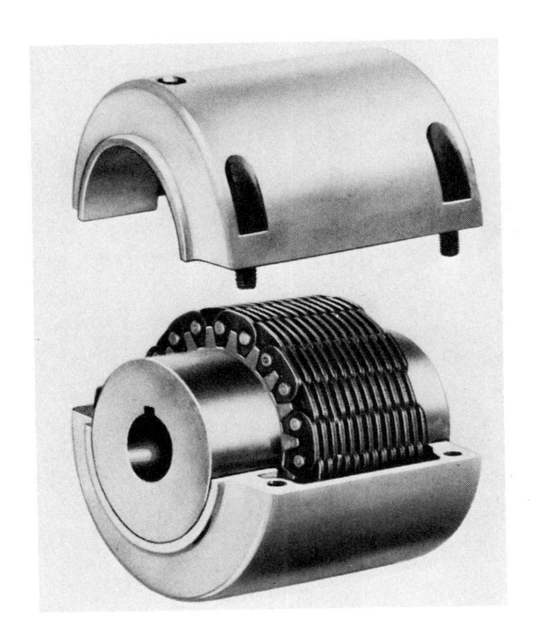

(B)

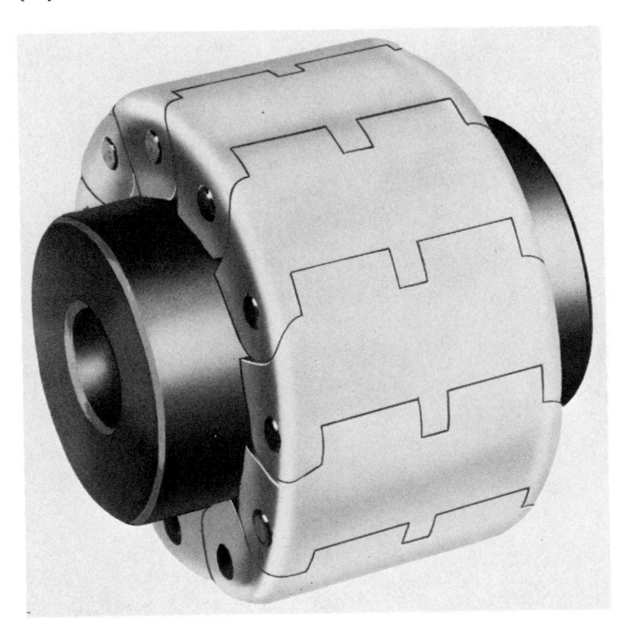

(C)

Figure 6.80 (continued)

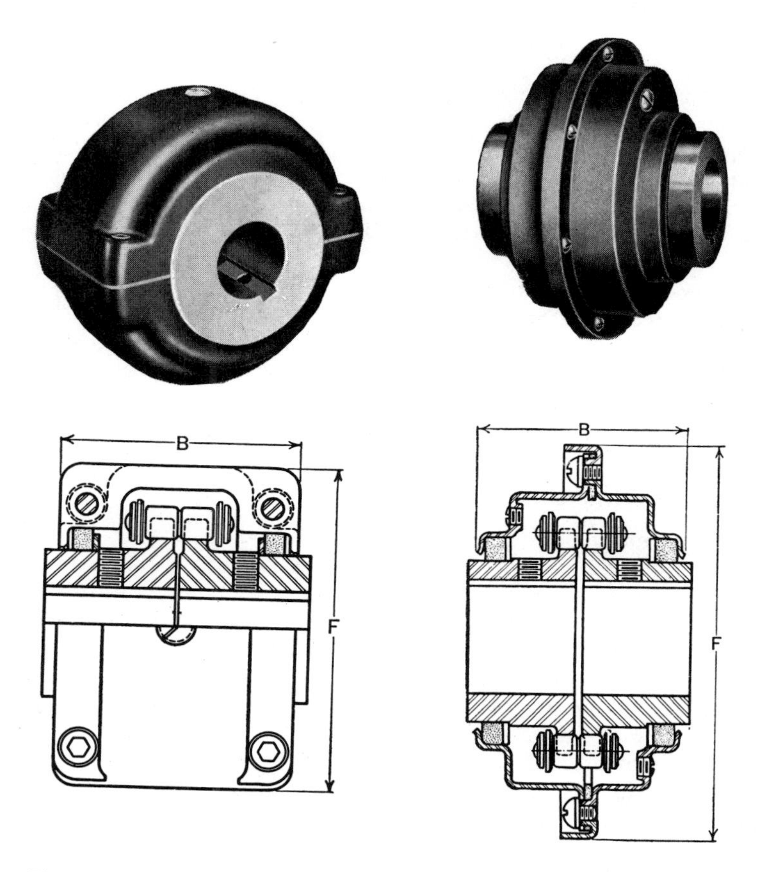

Figure 6.81 Typical chain coupling covers (courtesy of Link-Belt).

V. UNIVERSAL JOINTS

A universal joint is a special type of flexible coupling. A very simple form of it is shown in Figure 6.82. It consists of two shaft shapes drilled at right angles and then connected through a third piece, a cross. Many variations of this joint are in use, some highly complicated. Basically, they are all a form of the Hooke joint. They are able to handle up to 25,000,000 in.-lb of torque.

The universal joint can be used as a single joint or it can be used in pairs. The single universal joint cannot transmit power from one shaft to another shaft that is parallel to it, yet offset; the velocity of the second shaft varies harmonically with each revolution of the first. To analyze and use a universal joint, one must understand its kinematic relationship.

A. Kinematic Relationships of Universal Joints

1. Single Universal Joint. Two shafts inclined at angles to one another are connected by a single universal joint, and the driving shaft is driven at the uniform angular velocity ω_1 (Figure 6.82). With this single universal joint system, the driven shaft rotates in a nonuniform manner with the variable angular velocity ω_2. The amplitude of the sinusoidal variation of angular velocity, or of the angles of lead and lag of the driven shaft, are largely dependent on the angle of deviation.

This characteristic of universal joints, termed Cardan error, must be allowed for in the dimensioning and application of universal joints. It is clear from Figure 6.83 that with one revolution of shaft section 1, four variations of angular velocity ω_2 occur. This means that shaft section 2 passes through the acceleration and deceleration maxima twice in every revolution. With a large amount of angle of deviation and at high rotational speeds, considerable inertial force can become effective. It is thus apparent that a single universal joint can be employed only in applications where low forces are to be transmitted at low speeds and with small angles of deviation and where the nonuniformity of rotary motion in the output shaft is of subordinate significance.

Symbols and their meanings:

G1 = joint between shaft section 1 and shaft section 2
G2 = joint between shaft section 2 and shaft section 3

β_1 = angle between shaft section 1 and shaft section 2 on the plane formed by the two shaft sections (deg)
β_2 = angle between shaft section 2 and shaft section 3 on the plane formed by two shaft sections (deg)
β_s = angle projected onto the vertical plane (deg)
β_w = angle projected onto the horizontal plane (deg)
β = angle of deviation (deg)
α_1 = angle of rotation of shaft section 1 (deg)
α_2 = angle of rotation of shaft section 2 (deg)
α_3 = angle of rotation of shaft section 3 (deg)
ϕ = differential angle (lead and lag angle) (deg)
ω_1 = angular velocity of shaft section 1
ω_2 = angular velocity of shaft section 2
ω_3 = angular velocity of shaft section 3
U = coefficient of cyclic variation

Assuming that the yoke position of joint 1, shown in Figure 6.83 as $\alpha_1 = 0$, is the zero position of angle of rotation α_1, the following relationships apply:

$$\phi = \alpha_1 - \alpha_2$$

$$\frac{\tan \alpha_2}{\tan \alpha_1} = \cos \beta$$

$$\tan \phi = \frac{\tan \alpha_1(\cos\beta - 1)}{1 + \cos\beta \tan^2\alpha_1}$$

This yields for the angular velocities and torque values of shaft sections 1 and 2:

$$\frac{\omega_2}{\omega_1} = \frac{\cos\beta}{1 - \sin^2\beta \sin^2\alpha_1}$$

$$\frac{M_2}{M_1} = \frac{\omega_2}{\omega_1}$$

$$\text{Maximum } \frac{\omega_2}{\omega_1} = \frac{1}{\cos\beta} \quad \text{at } \alpha_1 = 90° \text{ and } 270°$$

$$\text{Minimum } \frac{\omega_2}{\omega_1} = \cos\beta \qquad \text{at } \alpha_1 = 0° \text{ and } 180°$$

$$\text{Maximum } \frac{M_2}{M_1} = \frac{1}{\cos\beta} \qquad \text{at } \alpha_1 = 90° \text{ and } 270°$$

$$\text{Minimum } \frac{M_2}{M_1} = \cos\beta \qquad \text{at } \alpha_1 = 0° \text{ and } 180°$$

For a comparison of nonuniformity, the coefficient of cyclic variation U has been introduced (see Figure 6.84):

$$U = \frac{\omega_{2max} - \omega_{2min}}{\omega_1} = \tan\beta \ \sin\beta$$

$$\tan\phi(\text{max}). = \pm \frac{1 - \cos\beta}{2\sqrt{\cos\beta}}$$

2. Double Universal Joint. The statements made in Section 5.A.1 show that for a single universal joint with a given deviation angle, shaft section 2 always rotates at an irregular angular velocity. If, however, two universal joints are correctly arranged in a "Z" or "W" arrangement as shown in Figure 6.85, a perfect compensation of nonuniform motion of shaft sections 2 and 3—the output shaft—synchronizes with shaft section 1—the input shaft. Synchronous running of shaft sections 1 and 3 is obtained where:

1. All shaft sections of the universal joint lie on one plane (Figure 6.86A).
2. Yokes 1 and 2 of section 2 lie on one plane (Figure 6.86B).
3. The deviation angles β_1 and β_2 of universal joints 1 and 2 are equal (Figure 6.86C).

These conditions ensure that joint 2 works offset in phase by 90° to joint 1 and completely cancels out the influence of universal joint 1. With such an ideal universal joint shaft arrangement, perfect kinematic balance is possible. In practice, we always seek this perfect balance. If any one of the three conditions is not fulfilled, the universal shaft will not work homokinetically. In such cases, it is advisable to consult the universal joint manufacturer, since certain influences can be compensated by specific measures.

3. Determination of Deviation Angle β from Angular Projections. If a universal joint lies spatially on a plane Z as illustrated in Figure 6.87, angles β_W and β_S can be determined from the top and side views of the drawing. For the resultant deviation angle β on plane Z, which is decisive for the dimensioning of the universal joint, the formula

$$\tan \beta = \sqrt{\tan^2 \beta_W + \tan^2 \beta_S}$$

applies. Its graphic representation is shown in Figure 6.88.

Universal joints are used on automobiles, agricultural and locomotive vehicles, horizontal and vertical pumps, winches, paper processing equipment, conveyors, drilling rigs, and crane drive (see Figures 6.89 to 6.93).

B. Types of Universal Joints

There are two basic types of universal joints:

1. Plain cross and bushing
2. Bearing and cross design

The plain cross and bushing is used on low-torque manual controls such as steering columns and on rotating applications that operate at less than 1000 rpm (see Figure 6.94). The bearing and cross design universal joint (see Figure 6.95) is used in applications up to 6000 rpm.

Universal joints are constructed with two basic types of yokes: split eye or solid one-piece constructions (see Figure 6.96). There are many variations in the design of universal joints. Some of these are:

Welded fixed shaft design (Figure 6.97): This joint is used when misalignment angles are fixed.

Welded telescoping shaft (Figure 6.98): This joint is used when the misalignment angle changes in operation.

Flanged fixed shaft (Figure 6.99): This design allows for the replacement of universal joint heads without replacing the shaft.

Flanged universal joint (Figure 6.100): This joint allows for connection of very short shaft-to-shaft distances.

C. Design and Construction of Universal Joints

Of the two types of universal joints, the pin and block design capacity is usually based on the cross. This type of universal joint is constructed of a wide variety of materials. Usually forks, blocks, and pins are hardened either through- or surface hardened.

For universal joints that use bearings, there are many different constructions: needle bearings, roller bearings, split eye yokes, and one-piece yokes. Usually, the bearings are through-hardened. The races and caps on which the bearings are mounted are usually case hardened and ground.

Crosses and yokes are usually made of forged alloy steel. The yokes and/or the crosses on large industrial universal joints are sometimes made of cast steel.

D. Rating and Selection of Universal Joints

There are two types of universal joint ratings:

1. *Strength rating*: This is based on the endurance strength or yield strength of the weak link of the universal joint.
2. *Life rating*: This is based on the B-10 life for the bearing and/or the life rating of the sliding shaft if one is used. A B-10 life is defined as the minimum life expectancy for a 90% probability of survival. The average bearing life would be approximately five times the B-10 life.

Usually, the weakest link of a universal joint is the cross. The strength of the cross is a function of the bending strength of the trunions of the cross.

E. How Universal Joints Affect the System: Loads on Connected Equipment

1. Axial Forces

$$F = \frac{T\mu}{R \cos\beta}$$

where

T = operating torque (in.-lb)
μ = coefficient of friction, 0.15
β = misalignment angle
R = pitch radius of length compensating spline (in.)

2. Radial Forces (see Figure 6.101). A misaligned universal joint produces cyclic forces on connected shafts and bearing. The following equations calculate the maximum amplitude of those forces.

α = angle of rotation (deg)
T = torque (in.-lb)
β_1 = angular misalignment of end A
β_2 = angular misalignment of end B
L = length between joint centers
A, B, C, D = radial bearing forces at 0°
A_1, B_1, C_1, D_1, C_1', D_1' = radial bearing forces at 90°
E, F = distance from joint center to of first bearing
G, H = distance between bearing of equipment at 0°

$$A = T \times \frac{\cos\beta_1 \times E}{L \times G} \times (\tan\beta_1 \pm \tan\beta_2)$$

$$B = T \times \frac{\cos\beta_1(E + G)}{L \times G} \times (\tan\beta_1 \pm \tan\beta_2)$$

$$C = T \times \frac{\cos\beta_1(F + H)}{L \times H}(\tan\beta_1 \pm \tan\beta_2)$$

$$D = T \times \frac{\cos\beta_1 \times F}{L \times H}(\tan\beta_1 \pm \tan\beta_2)$$

Note: Use addition (+) for W arrangement; use subtraction (−) for Z arrangement.

At 90°: For W and Z arrangements:

$$A_1 = B_1 = \frac{T \times \tan\beta_1}{G}$$

$$C_1 = D_1 = \frac{T \times \sin\beta_2}{H\cos\beta_1}$$

F. Failure Mode for Universal Joints

The normal failure mode is wear of the cross or bearing journals. This is usually caused from lose, lack, or deterioration of the lubricant (see Figures 6.102 and 6.103).

Universal joints also experience failure from overload. The failure modes include:

Breakage of crosses (Figure 6.104)
Failure of spline shaft (Figure 6.105)
Failure of spline shaft through the welded area (Figure 6.106)

Figure 6.107 lists some of the most common problems, probable causes, and corrective measures for universal joints.

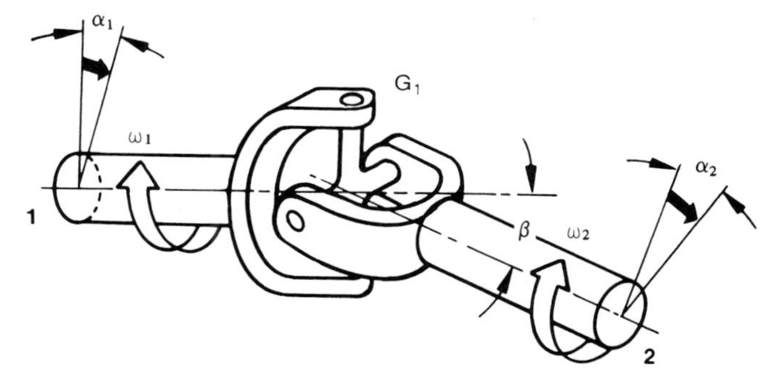

Figure 6.82 Simple universal joint.

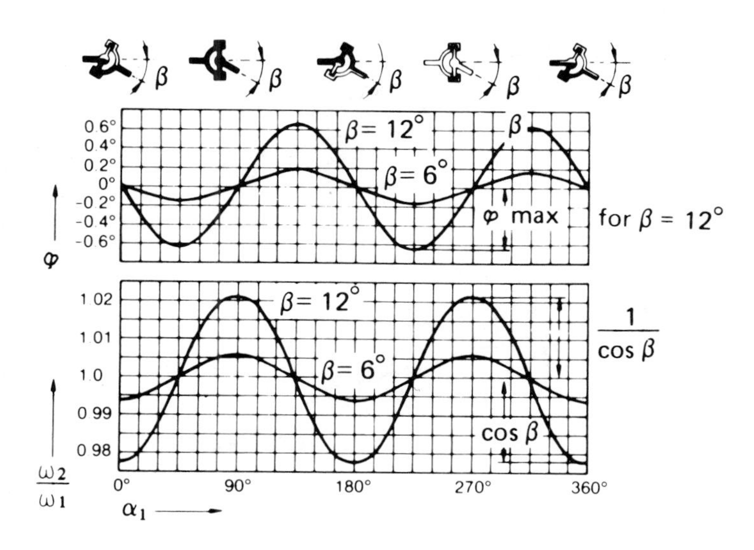

Figure 6.83 Cardan error.

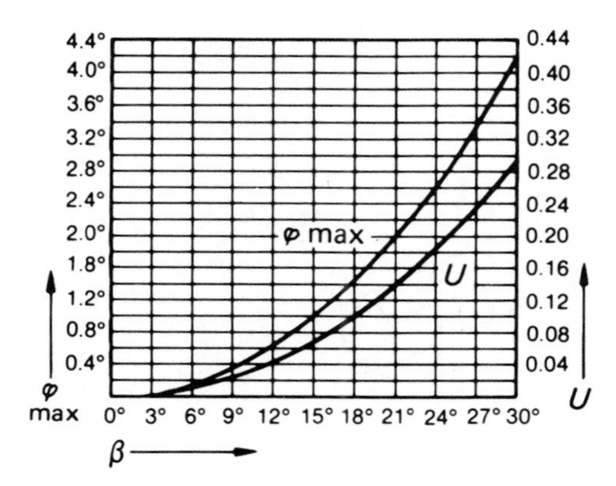

Figure 6.84 Coefficient of cyclic variation.

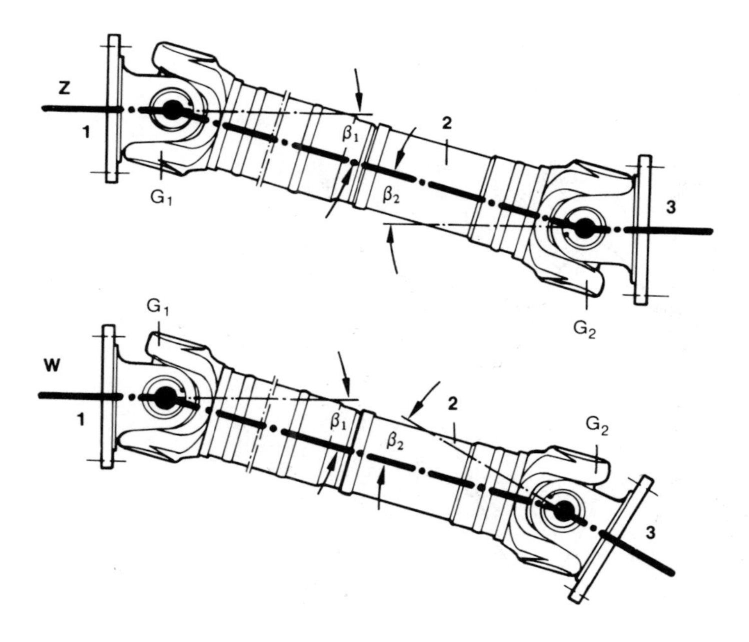

Figure 6.85 Z and W arrangements of a universal joint.

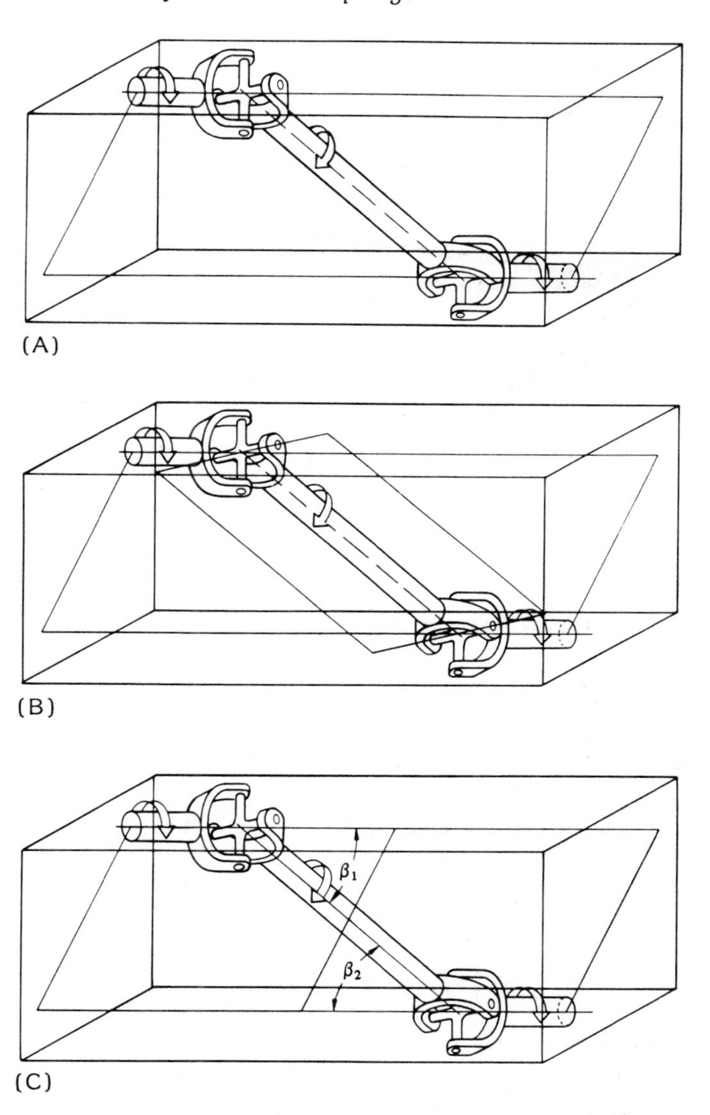

Figure 6.86 Various misalignment phases of a universal joint.

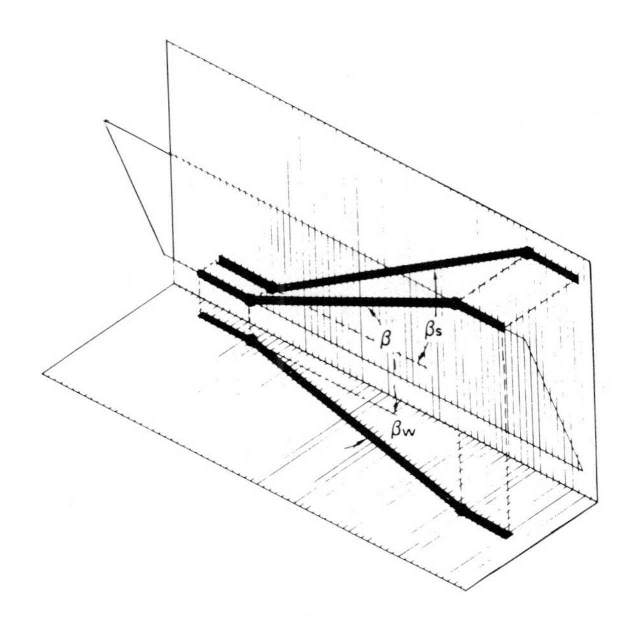

Figure 6.87 Universal joint spatial planes.

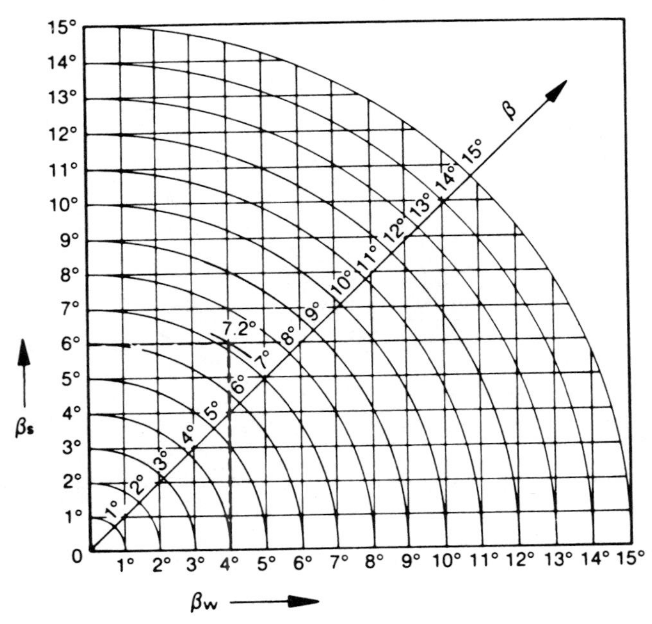

Figure 6.88 Combined misalignment for a universal joint.

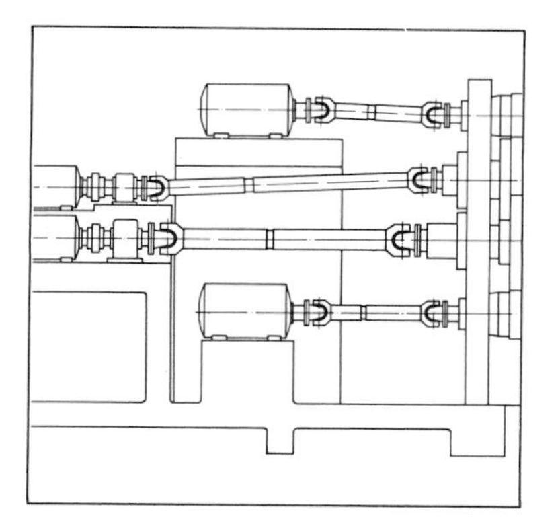

Figure 6.89 Paper machine application (courtesy of Voith Transmit GmbH).

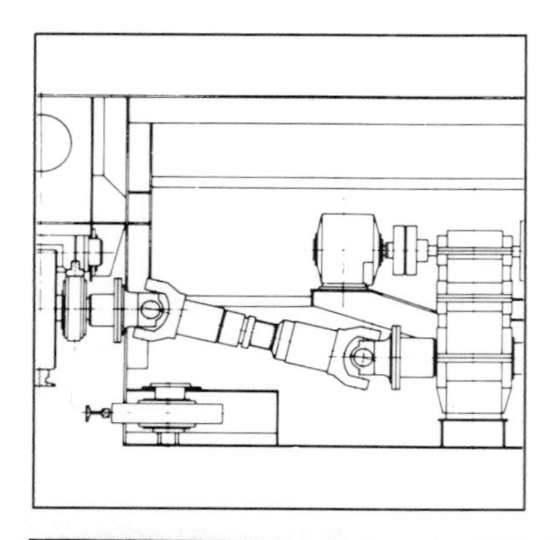

Figure 6.90 Crane application (courtesy of Voith Transmit GmbH).

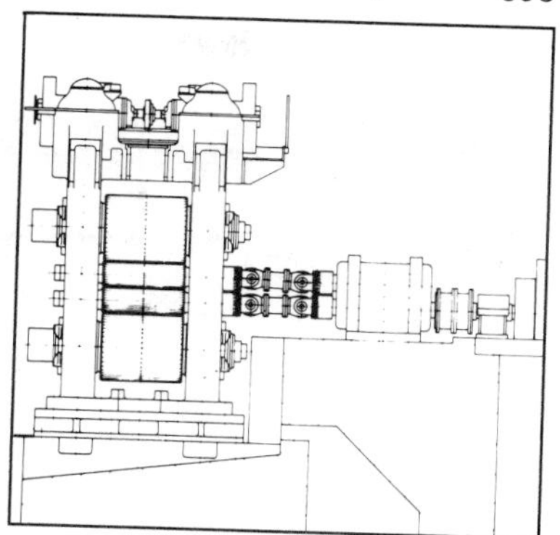

Figure 6.91 Cold-rolling mill application (courtesy of Voith Transmit GmbH).

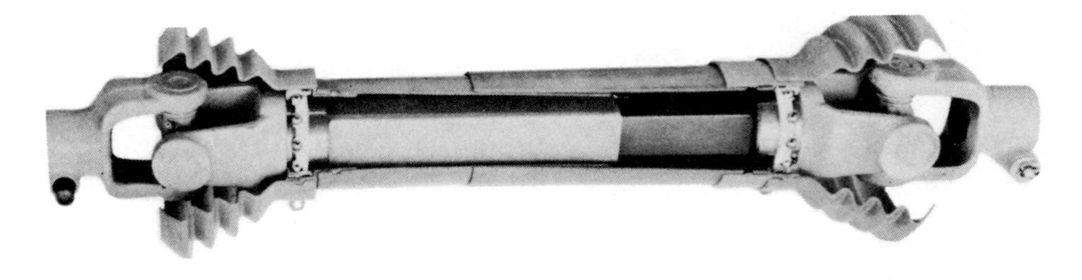

Figure 6.92 Power-takeoff application (courtesy of Koyo International, Inc., of America).

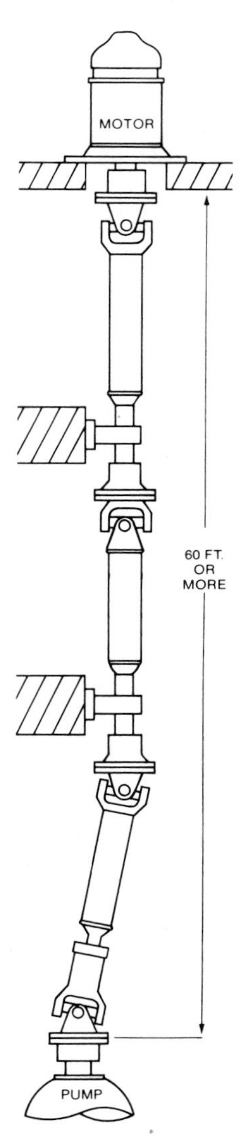

Figure 6.93 Vertical pump application.

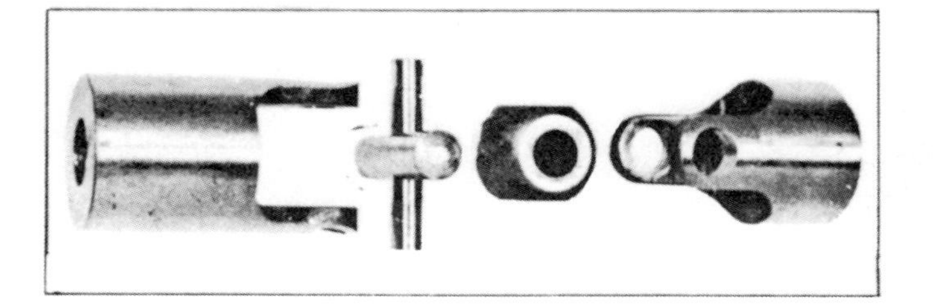

Figure 6.94 Plain cross and bushing universal joint.

Figure 6.95 Bearing and cross design (courtesy of Fenner America Ltd.).

(A)

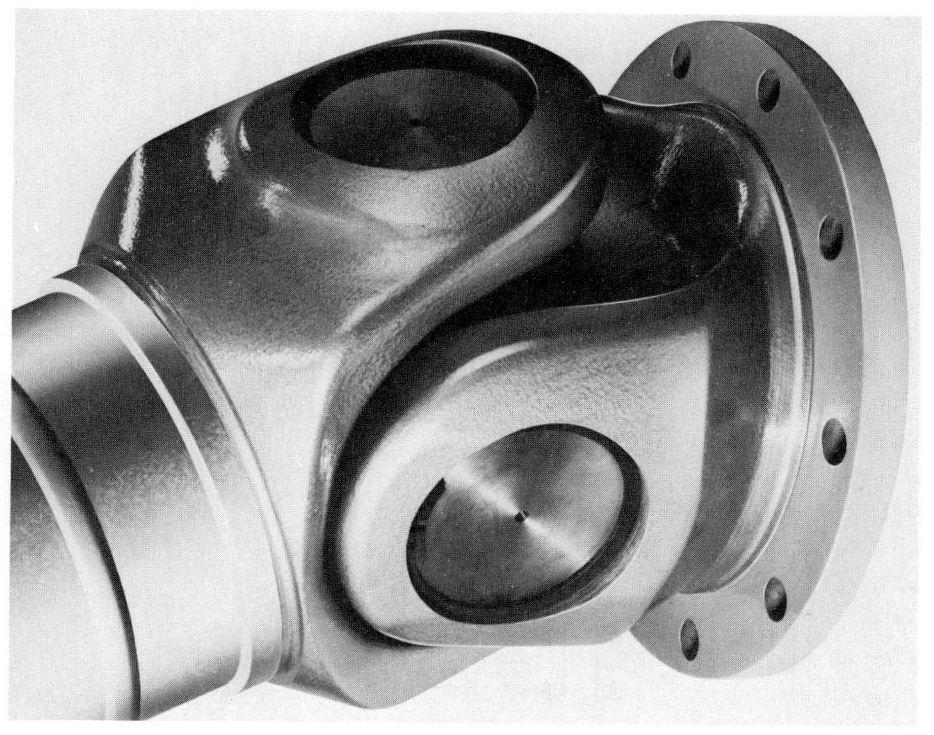

(B)

Figure 6.96 (A) Split eye, and (B) solid yoke design (courtesy of Voith Transmit GmbH).

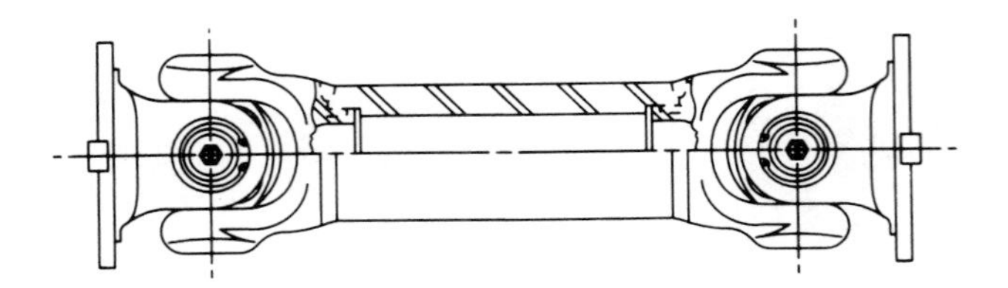

Figure 6.97 Welded fixed shaft universal joint.

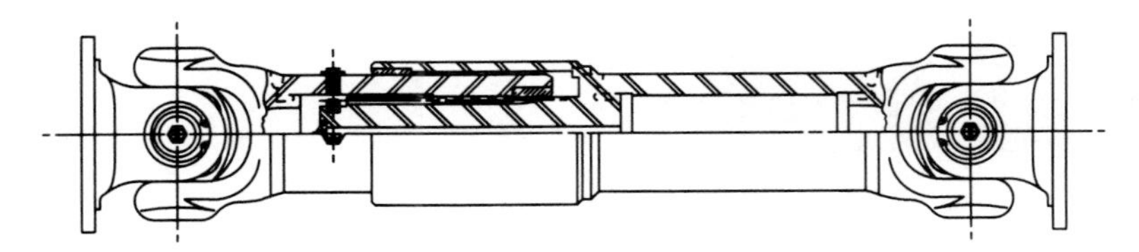

Figure 6.98 Welded telescoping universal joint.

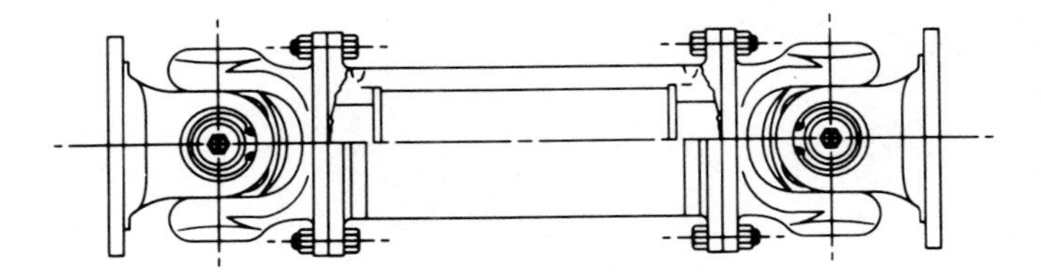

Figure 6.99 Flanged universal joint fixed shaft.

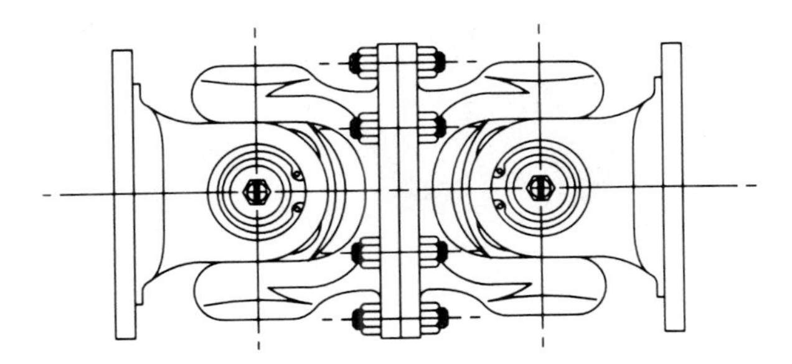

Figure 6.100 Flanged universal joint.

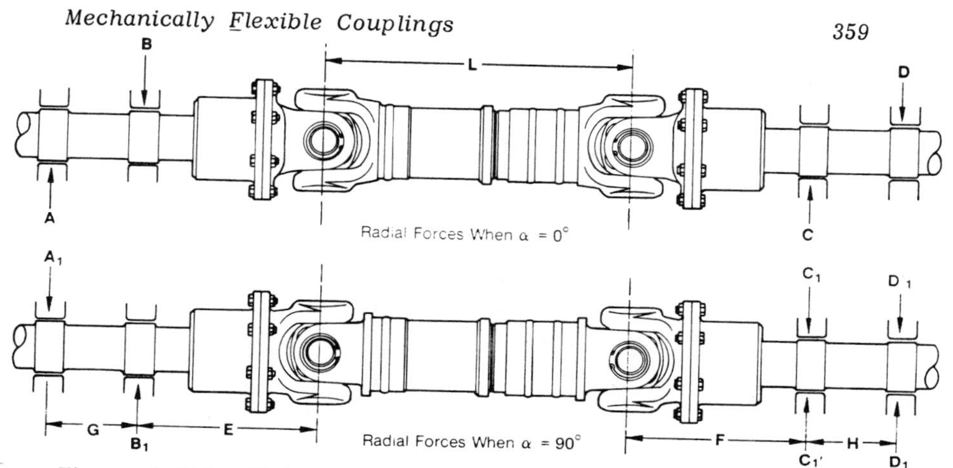

Figure 6.101 Universal joint reaction forces.

Figure 6.102 Wear on crosses.

Figure 6.103 Wear in bearings.

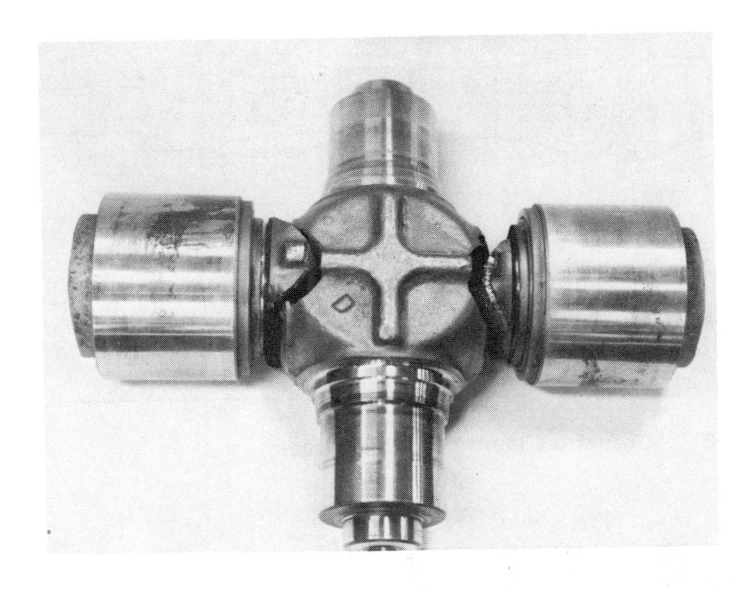

Figure 6.104 Broken crosses.

Figure 6.105 Failure of spline shaft.

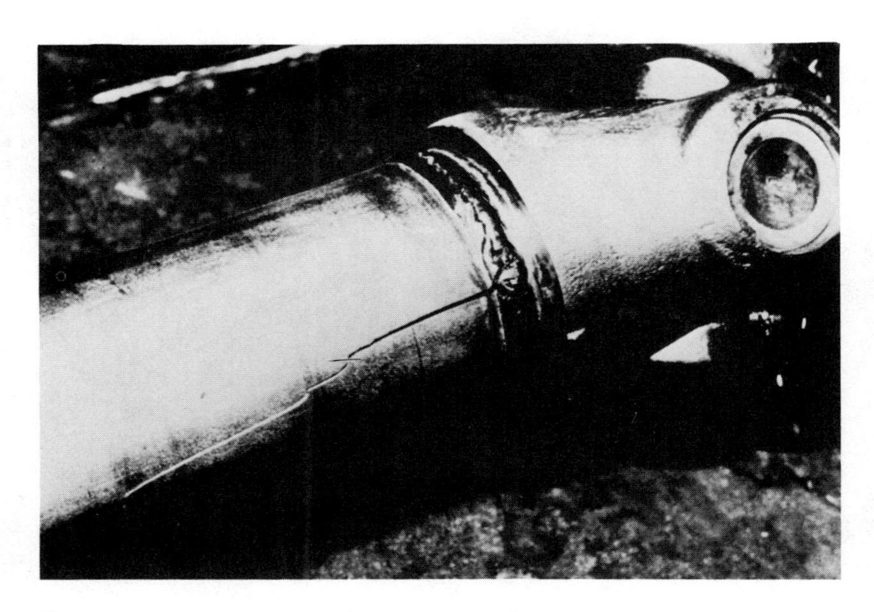

Figure 6.106 Failure of splined shaft through welded area.

Complaint	Probable cause	Correction
	Universal joint center parts	
Fracture of cross or trunnion	Abnormally high loading	Check for abuse of vehicle (stuck, overloading)
	Capacity of universal joint too small	Check for maximum drive-line torque in lowest gear, compare this to capacity of universal joint series; if larger, substitute with larger series
	Operating angles excessive or uneven angles	Measure operating angles, if excessive, decrease to smaller angles
	Defective material	Replace with new parts
Fracture of bushing	Abnormally high loading	Check for abuse of vehicle (stuck, overloading)
	Capacity of universal joint	Check for maximum drive-line torque in lowest gear; compare this to capacity of universal joint series; if larger, substitute with larger series

Figure 6.107 Common problems, causes, and corrective measures for universal joints.

Complaint	Probable cause	Correction
	Universal joint center parts	
	Excessive and/or uneven operating angles	Measure operating angles; if excessive, decrease to smaller angles
	Defective material	Replace with new parts
	Trunnion appearance	
Light brinelling into trunnion	Normal bearing failure	If small area, no need to replace
	Lubrication breakdown	If low hours, use approved lubrication and frequent lubrication cycles
Spalling into trunnions	Lubrication breakdown	Increase frequency of lube cycle and/or use approved grease
	Dirt contamination	Increase frequency of lube cycle, and/or use approved grease
	Normal bearing fatigue	If high hours, replace with new parts
Galling on end of trunnion	Lubrication breakdown	Increase frequency of lube cycle, and/or use approved grease
	Excessive and/or uneven angles	Reduce angles

	Unbalanced assembly	Check runout and balance of assembly
	Yokes: wing-type bushings or split eye	
Loose bolts	Dirt, foreign object, or material (paint, etc.) on mounting pad; look for fretting on mounting pads or drive tang	
Broken bolts	Over- or undertorqued bolts:	
	If there is no indication of fretting on mounting pad or bolt hole but there is fretting on the drive tang, the bolt broke	Surfaces must be free of foreign material and bushings fully seated before bolts are fully torqued
	If there is evidence of fretting, the bolt was loose	Surfaces must be free of foreign material and bushings fully seated before bolts are fully torqued
	Abnormally high angles can also cause bolts to break	Reduce angles to acceptable value

Figure 6.107 (continued)

Complaint	Probable cause	Correction
	Yokes: round bushing type	
Bushing extremely difficult to remove or replace	Hole in yoke distorted or normally close clearances; fretting corrosion and rust buildup	When removing bushing, use a penetrating oil and be careful not to distort ears on the yoke when removing on center cross
		Remove grease, rust, and dirt from mounting surfaces before assembly
New center parts will not flex when installed in yoke	Yoke ears distorted causing center parts to bind	Replace yoke

Figure 6.107 (continued)

7
Elastomeric Couplings

I. INTRODUCTION

As discussed earlier, flexible couplings have three basic functions:

1. To transmit power
2. To accommodate for misalignment
3. To compensate for end movement

Elastomeric couplings not only handle these three functions but also protect equipment from the damaging effects of shock loading and vibration through resilence.

There are two basic types of elastomeric couplings:

1. Elastomeric compression
2. Elastomeric shear

The elastomeric material is the resilient member, made of rubber or synthetic rubberlike material. Figures 7.1 and 7.2 compare the two types of elastomeric couplings: elastomeric compression coupling and elastomeric shear coupling, respectively. Rubber can carry from 5 to 10 times more load in compression than in shear; however, it deflects more under shear than in compression. Thus couplings using rubber in compression can transmit high loads and absorb minimal torsional vibration; those using rubber in shear provide high torsional damping but cannot transmit high loads.

Figure 7.3 shows a typical torque versus angular deflection curve or torsional stiffness curve for various types of flexible couplings. Shown are curves of torque per degree of angular displacement. Curve A is a typical torsional stiffness curve for a metallic membrane coupling (a disk coupling). Curve B is a typical torsional stiffness curve for an elastomeric compression coupling. Curve C is for elastomeric shear coupling.

A. Use of Elastomeric Couplings

Torsional vibration must be controlled to prevent equipment failure. An improper coupling can cause component failure. A proper coupling, on the other hand, can prevent broken shafts, fretted splines, bearing wear or failure, damage to seals, and so on (see Figure 7.4).

When power transmission equipment fails, users often blame the coupling rather than looking for the real cause. If more time and analysis were spent selecting the proper coupling for a system, fewer failures would occur. Equipment users may well assume that they have the best coupling for an application, never suspecting that it is the coupling that is causing the periodic failure of bearings, shafts, and other equipment components. They have thus come to expect these problems, scheduling premature overhauls and inspections that might well have been put off until major equipment overhaul is really required.

The majority of the problems described are caused by vibration, and the phenomena of vibration are as complex as they are misunderstood. Described below is how vibration occurs and how its ill effects can be either eliminated or minimized in rotational equipment. Misalignment is the principal source of most vibration. In many years of industry experience, vibration due to misalignment has been found to be the largest cause of premature failure in rotating equipment. This type of vibration generates its own rotating force with exciting impulses that are related to the amount of misalignment and the speed of rotation.

Whether stationary or moving, every component and every system has what is called a "natural frequency." In rotational equipment, this frequency is a function of many factors, including mass, unbalance, and inertia moments. If a constant force is applied to a system, it can operate at its natural frequency with no problem. But applied forces are rarely constant, particularly in engines and compressors. Thus a system involving the movement of many masses will have many natural frequencies.

As a system rotates, every cycle produces small impulses of energy, called *exciting impulses*, because they can excite system vibration. Most systems are subject to some type of excitation, and with each impulse, a cyclic force is applied to the system. The frequency of these cyclic forces is a function of the rotational speed of the equipment and when this frequency is above or below the natural frequency of the system, the forces are absorbed by the system itself. When they are in phase with the natural frequency, they create resonance at the natural frequency. When the magnitude of these forces exceeds acceptable limits, the system or a component within the system will vibrate and can cause premature failure of a component in system.

The speed at which resonance occurs is known as the *critical speed*, the speed at which exciting impulses are in phase with the natural frequency. If this critical speed is within or too near the operating range of the system, the resulting vibration can cause problems.

Figure 7.5A is an example of a problem. The vertical scale represents the equipment speed, and the horizontal scale merely converts these speeds into frequencies (or cycles per minute) by means of the diagonal line, known as the *critical order*. Or for any given speed, there is a corresponding frequency. The vertical line represents the natural frequency of this hypothetical system, and resonance will occur where the two lines corss, in this case at about 1700 rpm, which becomes the critical speed. The shaded area represents the operating range for the system, showing an idle speed of 1000 rpm and an operating speed of 1800 rpm. Since the operating speed is very close to the critical speed, the system is vulnerable to vibration, which can cause excessive wear and premature failure of many of the equipment's components.

Obviously, the problem outlined in Figure 7.5A could be corrected by lowering the operating speed of the equipment as shown in Figure 7.5B. This forces the critical speed outside the operating speed range of the system, thus avoiding the problem. However, this may not be practical, in which case the other remedy is to alter the critical speed as shown in Figure 7.5C. This can be done by changing the weights (also the inertias) of the system, but that too may be impractical or impossible. Generally, this would mean altering one of the system components; this option may be too late or too expensive once the system is designed and/or built. We have already determined that if it is impractical to alter the operating speed of the system, the other way to move the critical speed away from the operating speed is to change the natural frequency of the system itself. When it is impractical to do this by changing the system's inertia, it can be done with the addition of an elastomeric coupling.

B. Torsional Vibration

Since the primary function of elastomeric couplings is to control torsional vibration, one must understand torsional vibration to be able to apply an elastomeric coupling properly. The coupling manufacturer and/or the system supplier can usually do a system analysis to assure that the selection is done properly. The following is a short dissertation on torsional vibration analysis.

Torsional vibration is continuous angular oscillatory motion. It can originate in a mechanical power transmission system from many sources: from variations in engine pressure which can create peak torques; from motor pulsations; from unbalance forces; from fluctuating load torque requirements; or from torque reversals. When vibratory forces become large, they can produce damage, broken parts, excessive wear, and possibly noise in the rotating system.

Understanding and solving of torsional problems sometimes appears to be complex and very difficult to understand. There are many reasons: torsional problems usually occur less frequently, and the torsional equations contain radii and other factors that make the equations seem more complex.

One way to help understand torsional vibration is to recognize that there is a relationship between "translational" and "torsional" vibration. Both problems can be solved using the same basic formulas. Figure 7.6 shows the relationship between translational and torsional units. The only difference between the two types of vibration is the r or r^2 factors.

1. Terms

Mass (M) is a measure of resistance to rate of change or velocity. A body weighing 1 lb on the earth will require a force of 1/386 lb to accelerate that body at a rate of 1 in. per second per second.

Rotational inertia (I) is a measure of resistance to change in rotation of a body about its center of rotation

r is the radius at which the mass is concentrated with respect to its center of rotation. A torque of 1/386 in.-lb will cause a body with a weight of 1 lb with a radius of gyration r = 1 in. to accelerate at a rate of 1 rad per second per second.

K is a measure of the resistance to displacement. A spring that requires a force of 1 lb to displace it 1 in. has a spring rate of 1 lb per inch. A torsional spring is measured in inch-pounds per radian. If K_R is 1 in.-lb per radian, a torque of 1 in.-lb will cause a rotation of 1 rad.

A radian is a length of arc on the circumference of a circle equal to the radius of the circle. It is approximately 57.3 degrees of arc (see Figure 7.7A).

The relationship between translation and torsional springs may perhaps be better understood by locating one end of a translation spring at point A (in Figure 7.7B). Moving the end of the spring from A to B, a distance r, will require a force K_R (see Figure 7.7C). Since the force K_R is at a distance (r) from the center of rotation (O), the torque caused by the spring is $Kr(r) = Kr^2$. This torque Kr^2 was developed by moving the spring a distance r, or 1 rad along the circumference, so that the result is in terms of inch-pounds per radian, torsional stiffness.

Damping C is a measure of resistance to velocity. A damper with a value of 1 lb per inch per second will have a resisting force of 1 lb when motion across it is at the rate of 1 in. per second. Similarly, a torsional damper C_R will have a resisting torque of 1 in.-lb when motion across it is 1 rad per second.

Remember that damping in elastomers is considered primarily when a system is vibrating at resonance. Usually, rotational systems are operating at a frequency away from resonance. Therefore, the frequency of a rotational system can usually be calculated without consideration of damping, and except at resonance, damping has a negligible effect on transmissibility.

Figure 7.8 shows the same mass, spring, and damper in both translational and torsional systems. The elements have merely been rearranged in the torsional systems to rotate about a point O. This shows the similarity of the systems and the effects of the radius.

For the translational system, assume that the mass, spring, and damper are constrained to vertical motion. In torsional systems, motion about point O is limited to small values, so it is essentially vertical.

The formulas for inertia, rotational damping, and natural frequency in torsional systems indicate the effect of the radii—the distance from the center of the mass, spring, and damper—on the natural frequency of the system. The formula for the natural frequency of the torsional system in Figure 7.8 is identical to that of the translational system except that it is multiplied by the r_k/r_m factor. A torsional system rated in torsional values rather than in modified translational units is shown in Figure 7.8.

2. System Torsional Design. Many designers have little difficulty in understanding single-degree-of-freedom torsional problems. In this case, the base or foundation can be assumed to have an infinite mass or inertia. Usually, however, vibratory systems are force excited and consist of two or more inertias coupled with one or more springs. Solutions to problems in such a system are actually not much more difficult than those in single-degree-of-freedom systems.

The first fact to consider is that most multi-degree-of-freedom torsional systems have two inertias and one spring, or can readily by reduced to an equivalent two-inertia, one-spring system with little error. If a more complex situation is presented, a complete analysis can be utilized by referring to the Holzer analysis, given in many books on torsional vibration.

To compare a single-degree-of-freedom system to two-inertia, one-spring systems, refer to Figures 7.9 and 7.10. Figure 7.9 is a single-degree-of-freedom system with an inertia I_1 connected to a spring K_R attached to a rigid support. Figure 7.10 represents a two-degree-of-freedom system with a force imposed on inertia I_1, through a spring K_R, to inertia I_2, which is also free to rotate. The formulas below the figures give the calculations of natural frequency and transmissibility for each of these systems. Absolute transmissibility, T_{abs}, represents the ratio of torque transmitted through the spring, K_R, to the support (Figure 7.9) or to the driven inertia (Figure 7.10), I_2, to the torque applied to the driving inertia, I_1.

3. Transmissibility. To understand the relationship between single-degree-of-freedom and two-inertia, one-spring systems, examine Figure 7.11. Curve A represents a single-degree-of-freedom system. It shows how transmissibility increases from 1:1 at very low frequencies to infinity when the disturbing frequency equals the natural frequency. Then it moves into the isolation rnage, where transmissibility is less than 1 when the ration of the disturbing frequency to the natural frequency, f/f_n, is more than $\sqrt{2}$ or 1.41.

Curve B shows the torque transmissibility in an application where the inertia of the dirven part of the system is approximately 1/10 that of the driving part. In this case, the transmissibility is 0.1 at very low frequencies. It then increases in the resonance range until the disturbing frequency is equal to the natural frequency, and rapidly decreases in the isolation range.

Refer to the transmissibility formulas in Figures 7.9 and 7.10 and to curves A and B in Figure 7.11. It is obvious that the transmissibility of the two-inertia system is approximately 1/10 of the transmissibility of the single-degree-of-freedom system throughout the frequency range, except at resonance.

Curve C shows how to two-inertia, one-spring system is approximately equal to a single-degree-of-freedom system when the driven inertia is 10 or more times the magnitude of the driving inertia. When I_1 is small compared to I_2, the fraction $I_2/I_1 + I_2$ of a two-inertia, one-spring transmissibility formula approaches that of a single-degree-of-freedom system. Transmissibility based on a single-degree-of-freedom formula will result in a conservative value.

The next factor to be considered in the calculation of torsional systems is gear or belt drives with different speeds between various parts of the system. Where such differences exist, inertias must be corrected for their effective inertia at the coupling speed before frequencies and transmissibilities can be calculated. This correction is made by multiplying the inertia of an element rotating at a speed different from the coupling speed by the factor $rpm_I{}^2/rpm_c{}^2$.

C. Failure Modes for Elastomeric Couplings

When an elastomeric coupling melts or breaks apart and spins off rubber, it is obvious that something went wrong. Failures of elastomeric couplings may be the symptoms of a severe mechanical problem within the system. No matter how an elastomeric coupling fails, it can usually tell us why it failed. There are five common causes of failure:

1. Misalignment
2. High thrust loads
3. High torsional loads
4. Improper selection, operation, or installation and/or maintenance
5. Hostile environment

Figure 7.12 is a list of the ways in which elastomeric couplings typically fail.

1. Misalignment or Excessive Misalignment. Elastomeric couplings are designed to accept certain amounts of misalignment; when this is exceeded, identifiable symptoms are usually recognizable. When, for example, a jaw coupling generates an audible noise, the cause is usually wearing of the rubber spider. Bonded-tire elastomeric couplings often suffer bond failure between the tire and the metallic hub because of the flexing required by excessive misalignment.

In general, when the flexing member shows worn lugs, teeth, spiders, pins, bushings, and so on, excessive angular and/or parallel misalignment is the usual cause. Excessive misalignment can also cause failure of shafts, bearings, and other equipment parts due to excessive loads trying to flex the coupling.

2. High Thrust Loads. Flexible couplings used on equipment such as agitators, mixers, and propeller shafts often must carry axial thrust loads. Excessive "inward" thrust can badly distort shear couplings and place high stresses on them. Reverse or outward thrust cannot be carried by couplings that can be pulled apart. Such loads pull the coupling out of the mesh and heavy loads on the ends of the coupling members (see Figure 7.13).

3. High Torsional Loads. Although flexible couplings are built to carry torsional loads, a sudden impact can cause them to fail. Shear couplings, of course, are designed to rupture under torsional impact to protect the coupled machinery. In failure, these couplings develop a 45° tear or series of tears over the insert.

Torsional peaking loads from reciprocating machinery also fatigue capscrews or fastners that secure pins, bushings, compression members, or hubs. In some instances, these loads can build up to as much as 700% of the normal operating load. To prevent premature failures on rotating equipment, a torsional vibration analysis should be run as part of the coupling selection process. The rotational inertia of the two coupled machines, operating speed, and torsional stiffness of the coupling are the main inputs to the analysis. A torsional analysis should always be checked if any components are changed or replaced.

Manufacturers of reciprocating compressors commonly use torsionally stiff couplings to connect the compressor to the driver. In effect, the driver rotor becomes a flywheel that smooths out the torsional pulses inherent in these drives. A torsionally soft coupling isolates the driver and the compressor, forcing the coupling to absorb both high peak torques and any torque reversals. Elastomeric tire or insert-type couplings fatigue when subjected to high peak torsional loads, while jaw and pin and bushing couplings emit a high-frequency noise. A flywheel, then is a necessity when torsionally soft couplings are used on reciprocating machines.

If torsional overloads are the cause of premature coupling failure, one of the following system changes should be made:

Eliminate the cause of the overload by changing a system component.
Reduce the magnitude of the torsional vibrations by detuning the system; that is, use a coupling that allows the system to operate at noncritical frequencies.
Increase the coupling size.

4. Improper Selection, Operation, Installation, and Maintenance. The three major causes of coupling failure are improper sizing, operation, and installation and maintenance. There are no industry standard service factors for elastomeric couplings. Thus, where one manufacturer uses a 1.5 service factor for centrifugal pumps, another uses 1.25. Also, some coupling ratings are based on constant torque (horsepower varies directly with speed), whereas others are based on an exponential relationship between power and speed. Comparisons among competitive couplings should be made as discussed in Chapter 3.

Frequently, an oversized elastomeric coupling is specified in the hope that it will last longer. However, this may not be true, and it increases the radial loads on the shafts if misalignment exists. Also, if the load transmitted is too low, the oversized elastomeric coupling may actually wear out much faster. Occasionally, a coupling must be oversized to fit the shafts. In these cases, care must be taken to align and install the coupling more accurately than usual.

Speed limits should be strictly adhered to on elastomeric couplings used in high-speed operations. This type of operation generates centrifugal forces that can place high stresses on the flexible coupling components, particularly at the hub bore. Stress concentrates around drilled and tapped holes used to secure the hubs to the shaft. Tire-type elastomeric couplings can "balloon" and eventually rupture at high speeds. Compression-type elastomeric type couplings may whirl or vibrate laterally.

Stated limits can sometimes be increased if high-strength metals hubs and special inserts or flexing members are used. However, hubs and other components may have to be balanced to help prevent vibration. Coupling manufacturers should be consulted regarding when they require that their couplings be balanced. As always, the couplings should be handled carefully during installation and removal. Rough handling can cause internal fractures in the hub casting, which may then fly apart on startup. Extreme care should also be taken when installing elastomer flex elements to prevent tearing or breaking them. Always follow the coupling manufacturer's installation and maintenance instructions.

5. Hostile Environments. Elastomeric couplings are often used in corrosive and abrasive environments and at widely varying temperatures (see Figure 7.14). On food-processing machinery, for example, many couplings must withstand daily washing with hot water as well as resisting corrosion and deterioration from salt spray. These environments can cause corrosion of metallic elements, and/or deterioration of elastomeric elements. In addition, sunlight quickly ages natural rubber flexing members, resulting in the dry rot failure so common to rubber components. Temperatures over 200°F deteriorate neoprene flexing elements, and boiling water turns urethane to a blob. Temperatures below 0°F can cause neoprene flexing elements to shatter if subjected to high torsional impact loads. Low temperatures increase the torsional stiffness of elastomeric flexible couplings. This increase can make the machine operate at a critical torsional frequency, leading to early failure. Fortunately, heat generated when the coupling operates at the critical frequency often warms the flexible member enough to detune the system.

II. ELASTOMERIC COMPRESSION COUPLINGS

A. Types and Uses of Compression Couplings

There are four basic types of elastomeric compression couplings:

1. The jaw coupling
2. The block coupling
3. The pin and bushing coupling
4. The donut or ring (lug/clamped or bushed) coupling

1. Jaw Couplings. The jaw-type elastomeric coupling has an elastomer in compression. One-piece flex members (see Figure 7.15A) are generally used on low-horsepower applications. On high-horsepower applications, individual elastomeric elements can be used (see Figure 7.15B).

Elastomeric elements can be supplied in many forms of rubber (resilent material) and varying degrees of hardness, to suit the load-carrying capacities or for system torsional characteristics. The torsional stiffness, torque capacity, and overall dimensions can be altered by increasing or decreasing the number of jaws, the jaw width, and/or the shape of the jaw.

2. Block Design Couplings (Figure 7.16A). Since block elastomers are almost incompressible, the cavities are constructed so as to allow proper deformation of the blocks, and are completely filled only under conditions of extreme overload, combining high load-carrying capacity with maximum resiliency.

The blocks are usually precompressed into their respective cavities. As torque is applied, alternate driving blocks flex under the increased compressive load and tend to assume the shape of the enclosing cavities. The trailing blocks are relieved, but never completely, and tend (but not quite) to resume their original molded shape (see Figure 7.16B).

The alternating loaded or driving blocks (left) tend to fill cavities as greater deformation occurs. The alternate unloaded or trailing blocks (right), being precompressed, effect a smooth transition upon reversal of torque with complete absence of backlash.

3. Pin and Bushing Couplings. The elastomeric bushing (Figure 7.17A) has been proved in countless applications in power transmissions over many years. It is a universal coupling without working parts in the accepted sense. All relative movement is accommodated by the controlled internal displacement of compressed cyclindrical rubber blocks having specially shaped end profiles. The bore of the bushing is steel bonded into the rubber during vulcanization.

Preloading of the bushing (Figure 7.17B):

A. The diameter of the biscuit in the free state.
B. The diameter of the biscuit after insertion into the housing, showing the biscuit in a preloaded condition. This preloaded condition and the special shape of the biscuit accommodates any movement through controlled internal displacement of the elastomeric material.

Axial displacement resulting from thrust loads (Figure 7.17C):

A. The position of the bushing prior to imposition of thrust load.

B. The position of the bushing after a thrust load has been imposed. The flow of elastomeric material permits free end float without creating undue thrust loads on driveshaft bearings.

Angular deflection (Figure 7.17D):

A. The centerline of the bushing before angular deflection.
B. The displacement of the elastomeric material, as indicated by arrows, compensates for angular misalignment of the connected shafts.

Torsional deflection resulting from torque loads and torsional vibration (Figure 7.17E):

A. The centerline of the bushing before the application of a horizontal load.
B. The imposition of a torque load increases pressure in the direction of the load and reduces pressure in the opposite direction. Because of the initial preloaded condition, the elastomeric bushing is still under compression throughout its volume, even at maximum torque load.

4. Donut or Ring Couplings

a. Elastomeric Donut Lug, Clamped or Restrained. The coupling shown in Figure 7.18 has an elastomeric donut in a precompressed state because of the dimensional tolerances of its mating components. The donut is forced to a diameter smaller than its natural diameter when the capscrews are drawn up. This concept places all legs of the donut in compression before a load is applied. In operation, the trailing legs stretch, but not into tension. Absence of tension benefits the rubber element by minimizing the ozone attack that usually accompanies rubber in tension.

b. Elastomeric Donut (Lug or Bushing) Type (Figures 7.19 and 7.20). This coupling was specially developed for connecting high-inertia drives to low-inertia driven members, such as diesel engines to hydrostatic pumps. The material of the elastomeric element provides a torsionally stiff coupling. High permissible rotational speeds are attained and minor misalignments accommodated. The hardware design allows this coupling to connect directly to flywheels or to connect using simple flywheel adapter plates.

III. ELASTOMERIC SHEAR COUPLINGS

A. Types and Uses of Shear Couplings

There are three basic types of elastomeric shear couplings:

1. The tire coupling
2. The bonded coupling
3. The donut coupling

1. Elastomeric Tire Couplings (Figure 7.21). This coupling has the reinforced flexing element at the outermost radius of the configuration, thereby permitting a small ratio of overall length to torque capacity. The flexing element is subjected to shear. Internal reinforcement and external clamping of the tire increases the torque capacity and overall stiffness over that of a comparable unclamped and unreinforced shear unit of the same package size. Similar designs come with plastic tires to further increase torque capacity, again with increased torsional and lateral stiffness.

a. Three-Bar Linkage Principle. The shape factor or cross-sectional profile of the elastic member connecting the hubs is very important if the coupling is to maintain maximum flexibility and cushioning effect under misaligned conditions. Figure 7.22 shows a simple pinned three-bar linkage that will readily accept movement or misalignment. Figure 7.22A shows the normal position of the linkage. Figure 7.22B shows the action of the bars with the right-hand bar displaced horizontally along the reference baseline. This motion simulates the end float or axial misalignment in a flexible coupling. Figure 7.22C shows the relative position of the bars when the right-hand bar is displaced in a vertical plane above or below the baseline, simulating the parallel misalignment most commonly found in drive system couplings. Angular misalignments in the plane of the paper are illustrated in Figure 7.22D.

If we apply this principle to a coupling, replacing the three pinned bars with a continuous elastic material such as rubber, the motions of misalignment will be approximately the same. Figure 7.22E,F, and G show the translation of the pinned three-bar linkage principle to an arched elastomeric material connected to the hubs of a coupling. The similarity of misalignment capabilities is evident. The elastic material will bend or deform to allow displacement of the coupling hubs for misalignment.

b. Two Types of Elastomeric Couplings. Figure 7.21 is a cutway view of a flexible coupling with an arched profile. A positive method of securing each vertical leg of the connecting member between the shaft hub or flange and an internal clamping ring makes it impossible for the soft resilient member to rub or slide against metal to cause early failure during misaligned operation. There are no projections from either shaft flange to interfere with or bind the coupling during severe misalignment. The flexible element, shaped much like an automobile tire, is split radially at one location through the cross section. This allows it to be installed or replaced without moving or distributing the drive and driven equipment. Since the torsionally soft resilient connecting member carries the full load of torque and compensates for misalignment, it is the only part of the coupling that ever needs replacement. Because there are no rubbing surfaces, dirt or other abrasive material will not shorten the life of the coupling. Problems and high lubrication costs are minimized.

Torsionally soft couplings of this type have been used successfully on many drives where jamming of equipment is unavoidable. In most cases the resilient connecting member absorbs the shock load by virtue of its torsional softness or ability to assume a large angular displacement. When the shock loads reach certain levels, the resilient connecting member will slip slightly between the clamping surfaces of the hub and the clamp ring. Under extreme shock-overload conditions, the connecting member may be torn apart, disconnecting the drive and driven equipment and guarding it from damage. Normally, the resilient connecting member is the softest, least costly, and most easily replaced part of the system. The coupling element can be replaced in 10 to 20 minutes, depending on the size and location of the unit.

The heart of the tire-shaped coupling is the resilient connecting member shown in Figure 7.23, usually a fabrication of vulcanized natural rubber and plies of synthetic cord. The plies of cord are crossed alternately so that regardless of the direction of rotation, half the cord plies will be working in tension to carry the load. This torsionally soft coupling readily accommodates angular misalignment to 4°, parallel misalignment to 1/8 in., or end float up to 5/16 in. It follows that size, misalignments, loads, and speed are significant in determining the service life of the coupling selected for a given application. The design can be adapted to accommodate greater misalignments by using two elements with a shaft (Figure 7.24).

2. Bonded Shear Couplings (Figure 7.25). The basic shear type is elastomeric material bonded between two hubs, used for

low-frequency isolation and accommodation of moderate parallel misalignment.

3. Donut-Type Couplings. The elastomeric shear unclamped donut-type coupling (Figure 7.26) transmits torque by means of shear loading of the element. With a rubber element, it provides low torsional stiffness and low lateral force due to misalignments.

Several materials of donut configuration can increase the load-carrying capacity of a given coupling size. Torsional and lateral stiffness for a given coupling size generally increase as the load-carrying capacity increases. A rubber donut permits accurate theoretical prediction of the dynamic characteristics of this coupling.

This type of coupling is generally used on pumps and motors, and for standard spacing requirements in the pump industry. Donut couplings allow suitable space between shafts so that pump packing can be replaced without disturbing shaft alignment. The coupling consists of two flanges, a sleeve, and two hubs (see Figure 7.27).

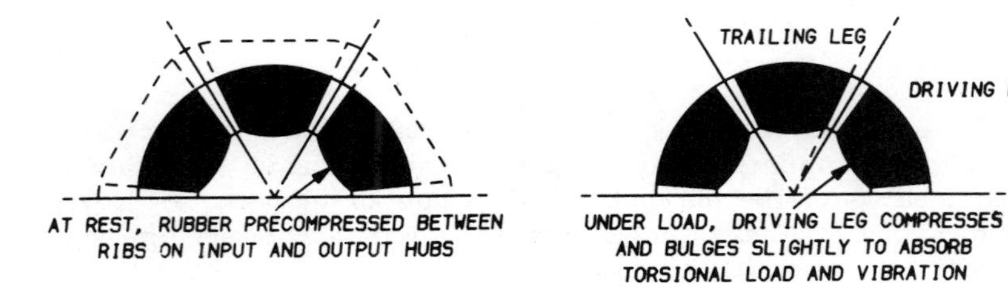

Figure 7.1 Elastomeric compression coupling.

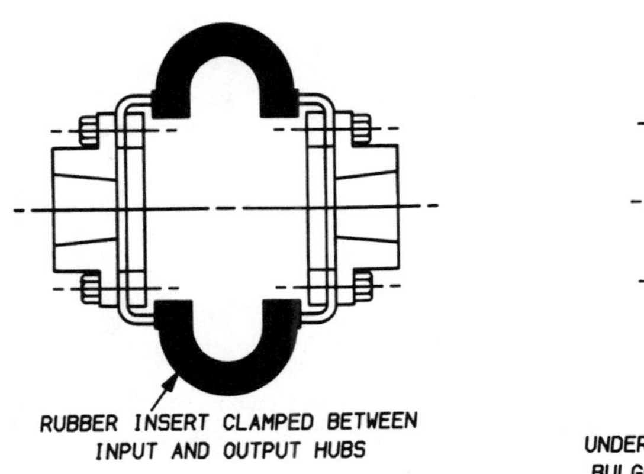

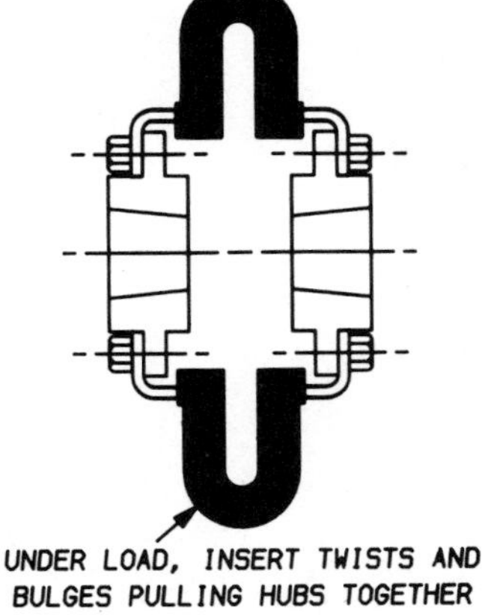

Figure 7.2 Elastomeric shear coupling.

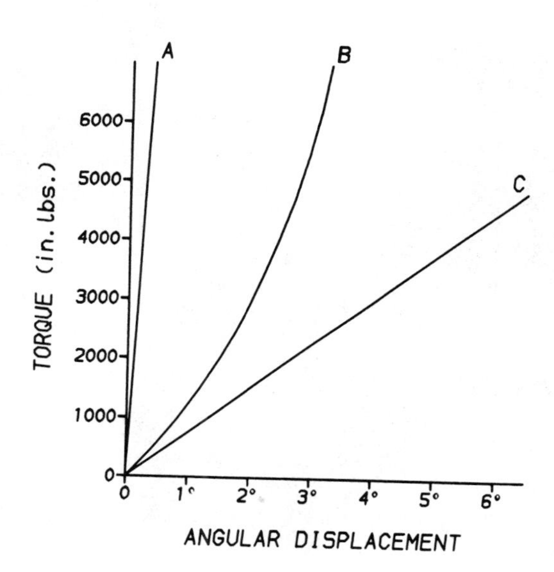

Figure 7.3 Torque versus angular windup.

Figure 7.4 Failed and broken parts (courtesy of Lovejoy, Inc.).

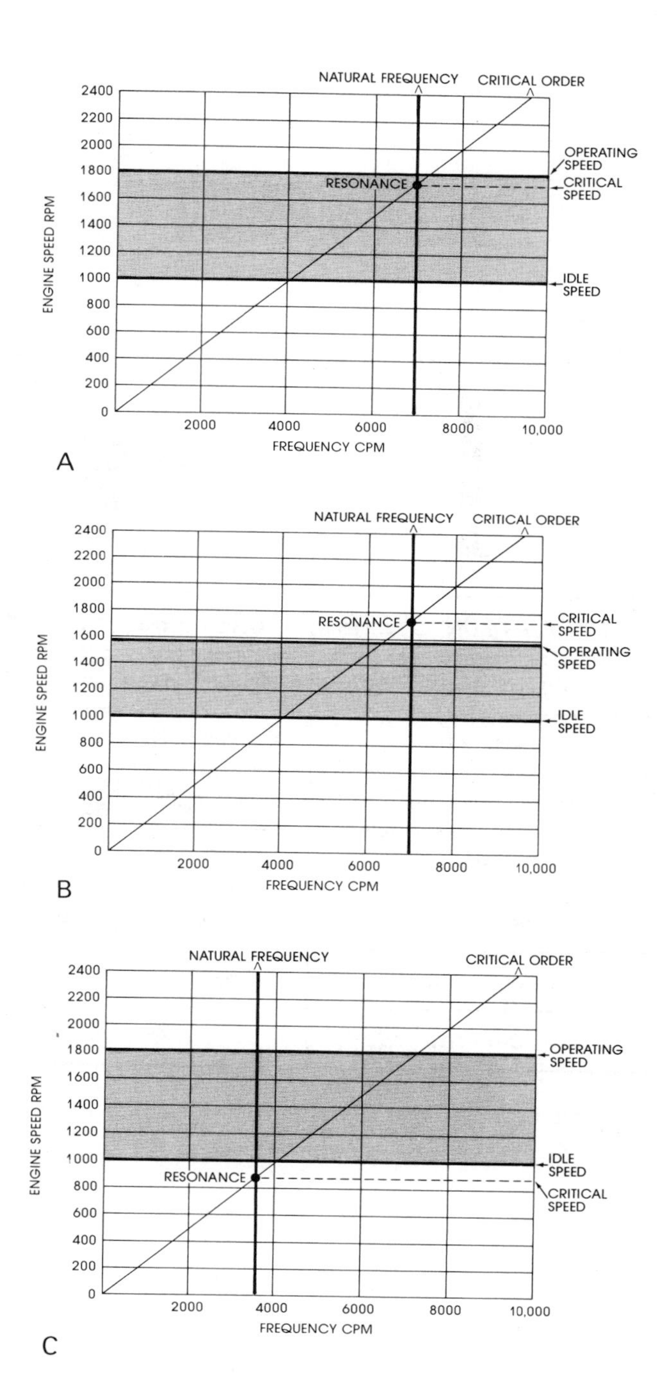

Figure 7.5 (A) Resonance too close to operating speed; (B) lowering the operating speed; (C) changing the inertia.

ELEMENT	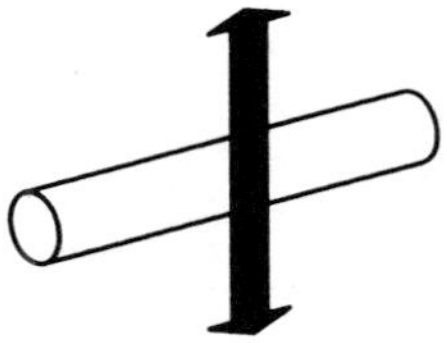 TRANSLATION SYSTEM	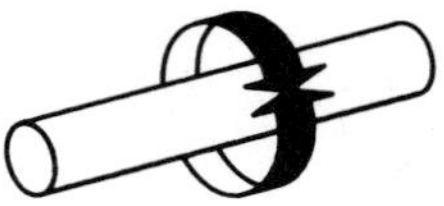 TORSIONAL SYSTEM
MASS/INERTIA	$M=w/g$ (lb sec^2/in.)	$I=Mr_m{}^2=\frac{wr^2}{g}$ (lb in. sec^2)
SPRING	$K=$ lb /in.	$K=Kr_k{}^2$ (lb in./rad.)
DAMPER	$C=$ lb /in./sec	$C_R=Cr_c{}^2 \frac{\text{in. lb sec}}{\text{rad.}}$
FORCE TORQUE	$F=$ lb	$T=Fr_t=$ lb in.
ACCELERATION	$A=\ddot{x}=$ in./sec^2	$\ddot{\theta}=\alpha=a/r=$ rad./sec^2
VELOCITY	$V=\dot{x}=$ in./sec	$\dot{\theta}=\omega=v/r=$ rad./sec
DISPLACEMENT	$D=x=$ in.	$\theta=$ in./$r=$ radians

Figure 7.6 Relation between translational and torsional units.

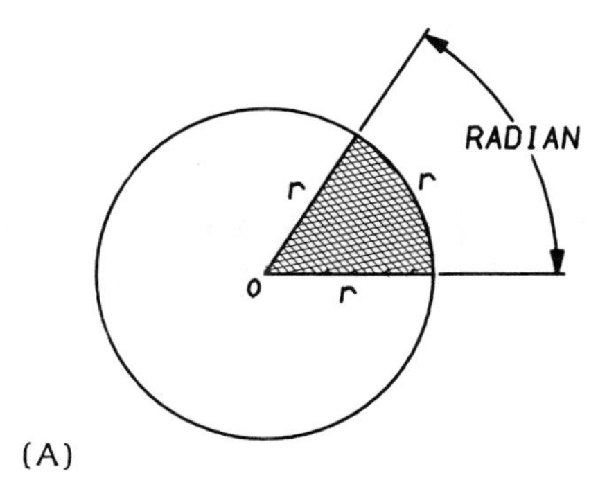

(A)

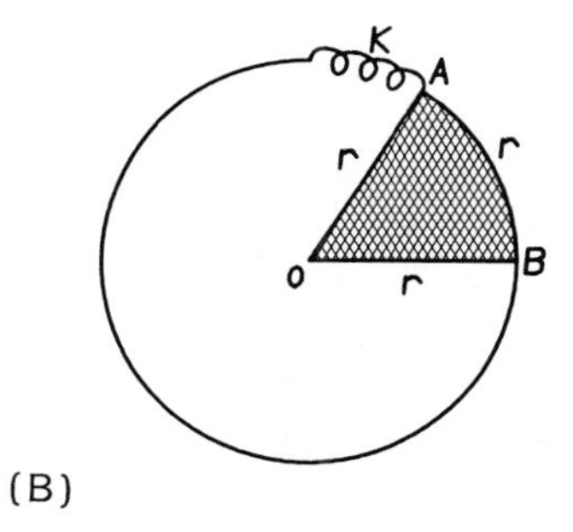

(B)

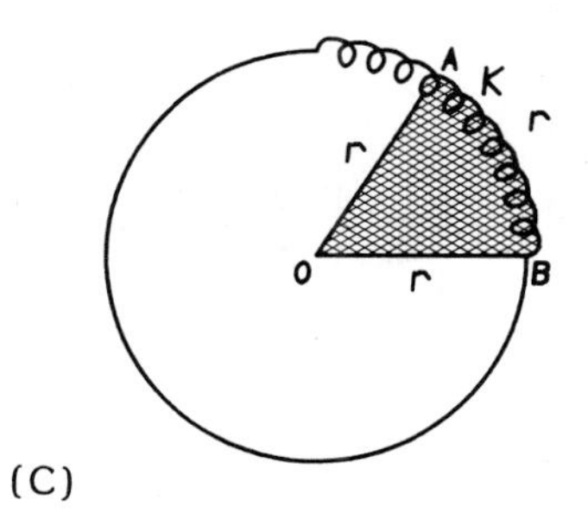

(C)

Figure 7.7 (A) A radian; (B), (C) spring movement.

ELEMENT	TRANSLATIONAL SYSTEM (USING TRANSLATIONAL UNITS)	TORSIONAL SYSTEM (USING TRANSLATIONAL UNITS)	TORSIONAL SYSTEM (USING TORSIONAL UNITS)
	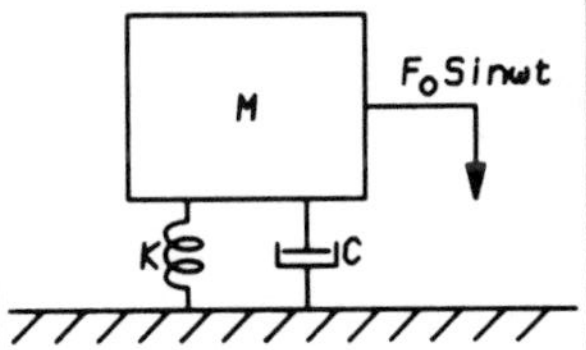	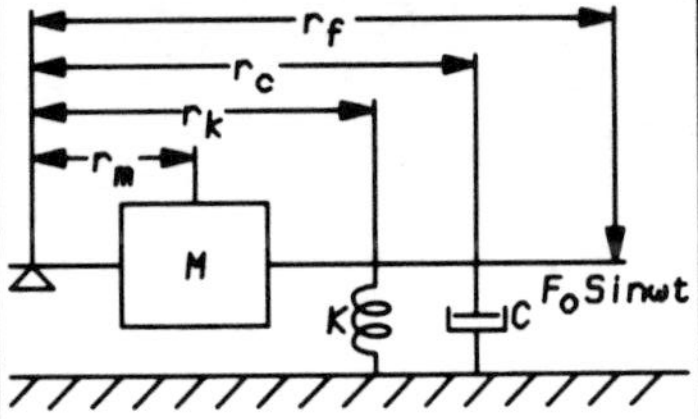	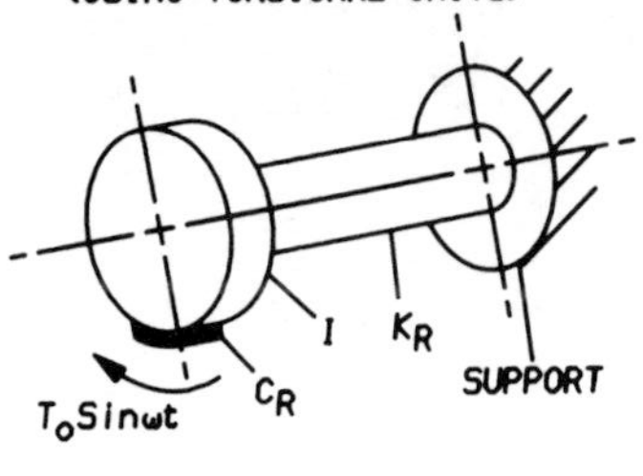
INERTIA:	M	$I = Mr_m^2$	I
SPRING RATE:	K	$K = Kr_k^2$	K_R
DAMPING:	C	$C = Cr_c^2$	C_R
UNDAMPED NATURAL FREQUENCY:	$f_n = \frac{1}{2\pi}\sqrt{\frac{K}{M}}$	$f_n = \frac{1}{2\pi}\sqrt{\frac{Kr_k^2}{Mr_m^2}} = \left(\frac{1}{2\pi}\sqrt{\frac{K}{M}}\right)\frac{r_k}{r_m}$	$f_n = \frac{1}{2\pi}\sqrt{\frac{K_R}{I}}$
EQUATION OF MOTION:	$M\ddot{x} + C\dot{x} + Kx + F_o \sin\omega t$	$Mr_m^2\ddot{\theta} + Cr_c^2\dot{\theta} + Kr_k^2\theta = F_o r_f \sin\omega t$	$I\ddot{\theta} + C_R\dot{\theta} + K_R\theta = T_o \sin\omega t$

Figure 7.8 Comparison of translational and torsional systems.

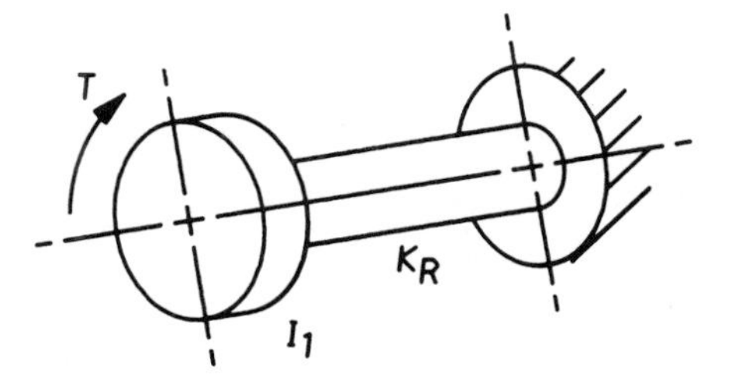

$f_n = \frac{60}{2\pi}\sqrt{\frac{K_R}{I_1}}$ cycles/min
$T_{abs} = \frac{1}{1-\left(\frac{f}{f_n}\right)^2}$

Figure 7.9 Single-degree-of-freedom system.

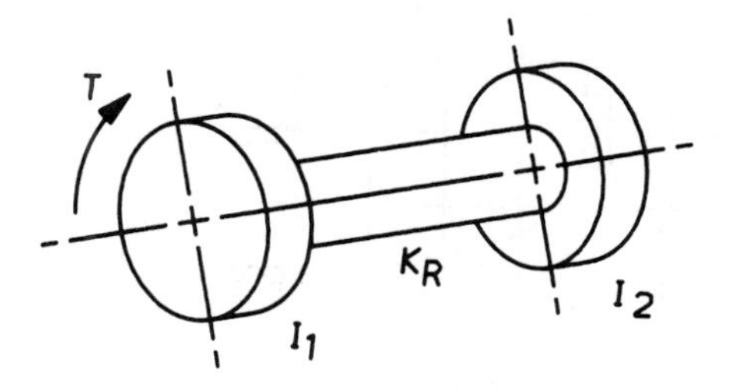

$f_n = \frac{60}{2\pi}\sqrt{\frac{K_R}{I_{Eff.}}}$ cycles/min
$T_{abs} = \frac{I_2}{I_1 + I_2} \times \frac{1}{1-\left(\frac{f}{f_n}\right)^2}$ WHERE $I_{Eff} = \frac{I_1 I_2}{I_1 + I_2}$

Figure 7.10 Two-inertia, one-spring system.

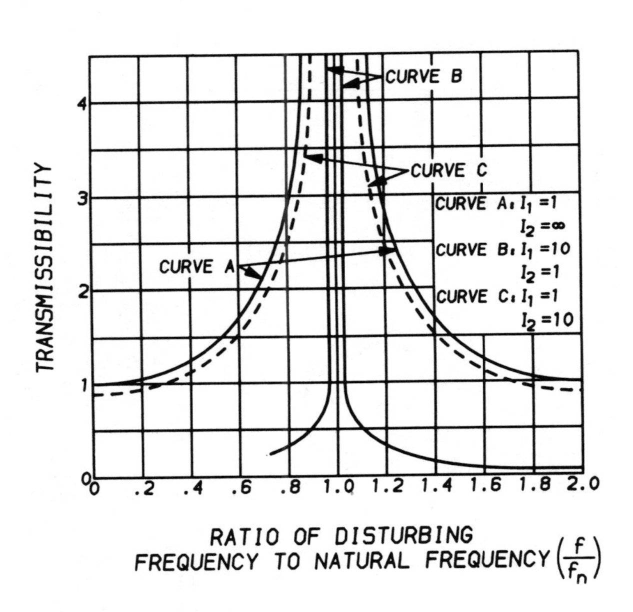

Figure 7.11 Comparative transmissibility curves.

Failure mode	Probable cause	Corrective action
Worn flexing element or shaft bushings Shaft bearing failure High-pitched or staccato noise	Excessive shaft misalignment	Realign coupling and shafts to meet specified tolerances
Ruptured elastomeric flexing element Sheared hub pins or teeth Loose hubs on shaft, sheared keys	Torsional shock overload	Find and eliminate cause of overload Use larger coupling
Fatigue of flexing element Overheated elastomeric tire or sleeve Fatigue of hub pins or discs Worn gear teeth Staccato or clacking noise Loose hubs on shaft, keyseat wallow	Torsional vibration Excessive starts and stops High peak-to-peak torsional overload	Use larger coupling Use larger coupling Add flywheel to hub
Shaft bearing failures High-pitched whine Motor thrust bearing failure	Lubricant failure	Replace or rebuild coupling
Swollen or cracked elastomeric flexing member Lubricant failure Severe hub corrosion	Chemical attack	Use more chemically resistant flexing member or hub Coat hubs Coat hubs
Distorted or deteriorated elastomeric flexing member Lubricant failure	Excessive heat	Use more heat-resistant flexing member or lubricant
Shattered flexing member Lubricant failure	Low temperature (below 0°F)	Use special low-temperature rubber compounds and lubricants

Figure 7.12 Typical elastomeric coupling failures.

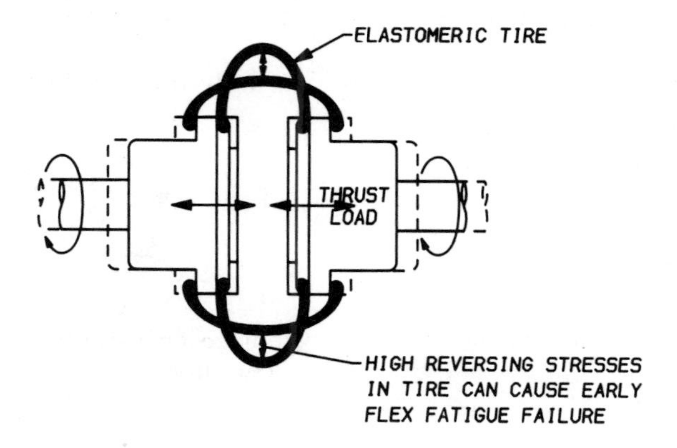

Figure 7.13 Thrust-load distortion of coupling members.

Resistance to:	Neoprene[a]	EPDM[a]	Urethane[a]	Hytrel[a]
Acetone	2	1	3	2
Ammonium hydroxide solutions	1(158°F)	1	1	—
Benzene	3	3	3(158°F)	2
Butane	1	2	1	1
Carbon tetrachloride	3	3	3(122°F)	3
Chloroform	3	3	3	3
Ethyl alcohol	1(158°F)	1	3	1
Ethylene glycol	1(158°F)	1	2	1
Fuel oil	1	—	2	1
Gasoline	2	2–3	2	1
Glycerin	1(158°F)	—	1	1
Hydraulic oils	1	—	2	1
Hydrochloric acid (20%)	1	—	2	2
Isopropyl	1	—	3	1
Kerosene	2	—	3	—
Lubricating oils	2(158°F)	—	2	1
Methyl alcohol	1(158°F)	—	3	1
Mineral oil	1	—	1	1
Naphtha	3	3	2	1
Nitric acid (10%)	2	—	3	3
Nitrobenzene	3	1	3	3
Phenol	2	—	3	3
Soap solutions	1(158°F)	—	1	1
Sodium hydroxide (20%)	1	1	—	1
Sulfuric acid (50–80%)	2–3	—	3	3
Toluene	3	3	3(122°F)	2
Trichlorethylene	3	—	3	3
Turpentine	3	3	3	—
Water	1(212°F)	1(158°F)	1(122°F)	1(158°F)
Xylene	—	3	3	2

[a]1, little or no effect; 2, minor to moderate effect; 3, severe effect. Where temperatures are noted, resistance drops off quickly above the stated temperature.

Figure 7.14 Chemical resistance of flexible coupling materials.

(A)

(B)

Figure 7.15 (A) One-piece flex element; (B) individual elastomeric elements (courtesy of Vulkan Corporation).

(A)

(B)

Figure 7.16 (A) block elastomeric coupling; (B) loading elastomeric blocks (courtesy of Koppers Company, Inc., Engineered Metal Products Group).

(A)

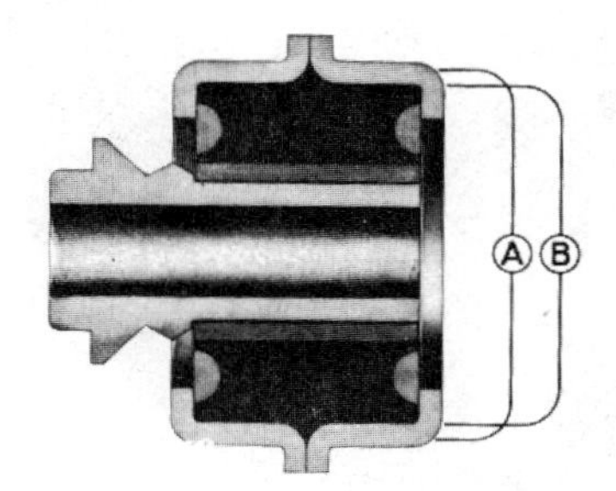

(B)

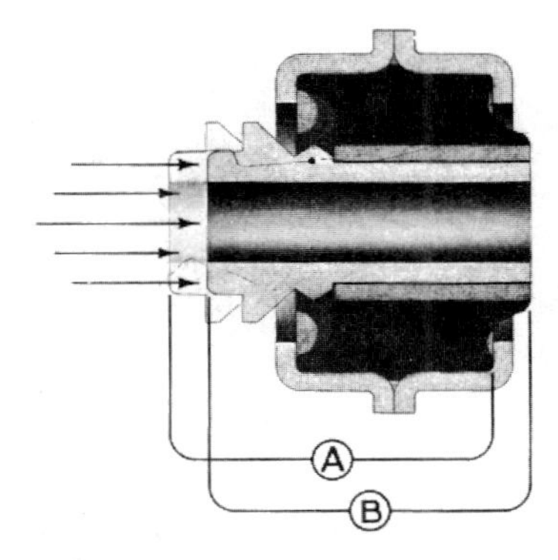

(C)

Figure 7.17 (A) Typical pin and bushing coupling; (B)–(E) loading bushings (courtesy of Morse Industrial Corporation).

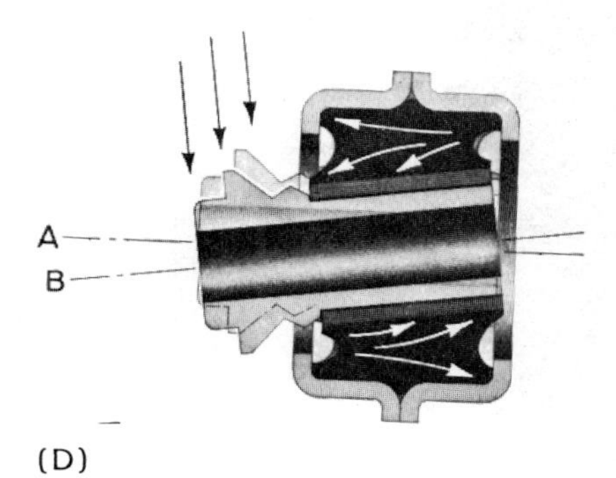

(D)

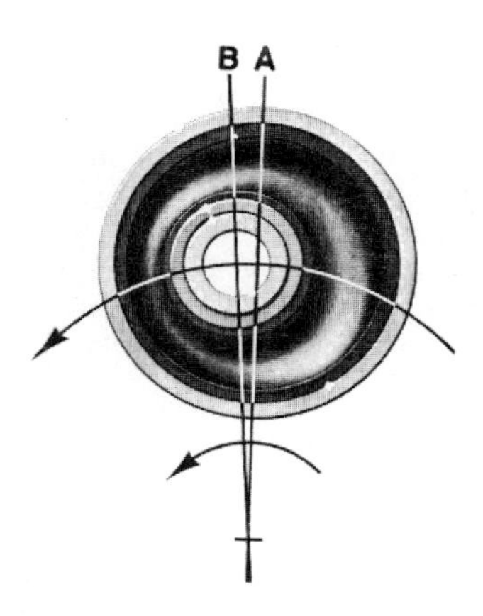

(E)

Figure 7.17 (continued)

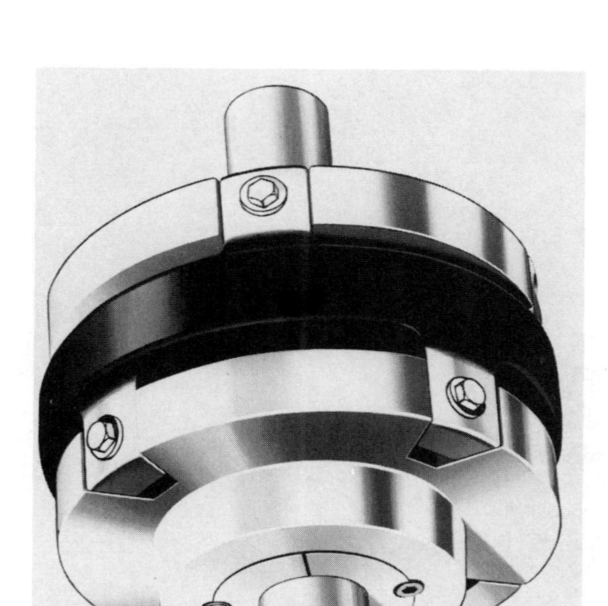

Figure 7.18 Elastomeric donut (courtesy of Koppers Company, Inc., Engineered Metal Products Group).

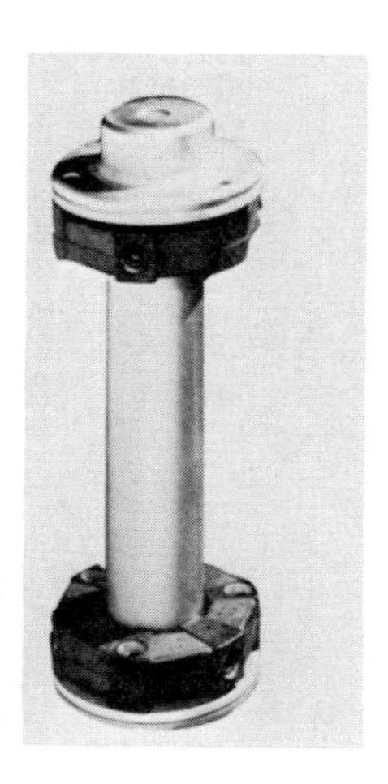

Figure 7.19 Elastomeric donut with lugs (courtesy of Lovejoy, Inc.)

Figure 7.20 Elastomeric donut with bushing (courtesy of the Lord Corporation).

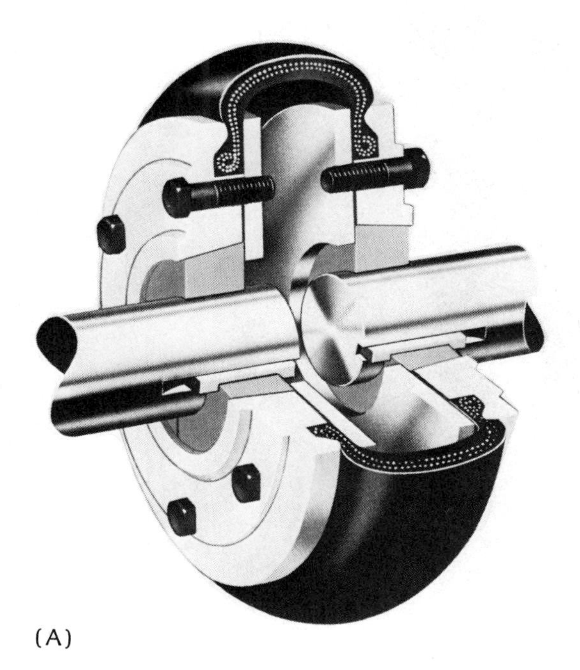

(A)

Figure 7.21 (A) Elastomeric tire coupling; (B) tire coupling for runout table (courtesy of Dayco Corporation).

(B)

Figure 7.21 (continued)

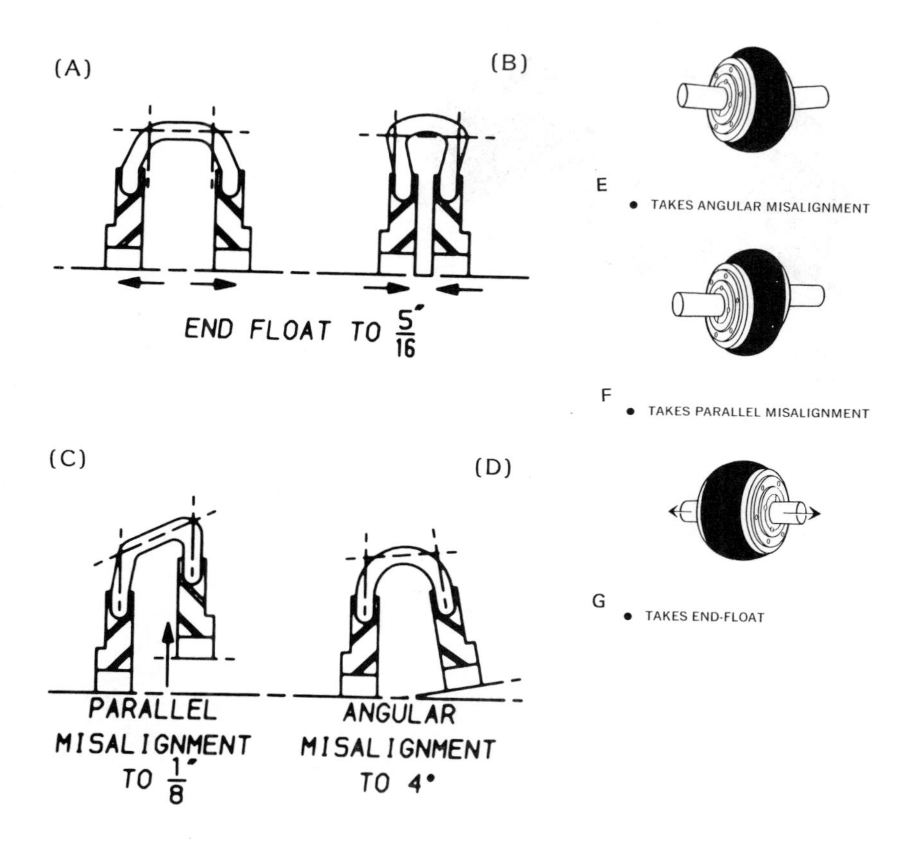

Figure 7.22 (A)–(D) Three-bar linkage principle; (E)–(G) misalignment motion of tire coupling (courtesy of Dayco Corporation).

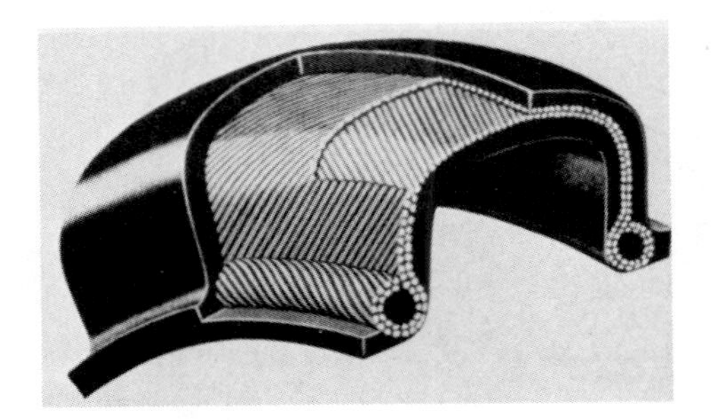

Figure 7.23 Cutaway of a rubber tire coupling (courtesy of Dodge Division of Reliance Electric).

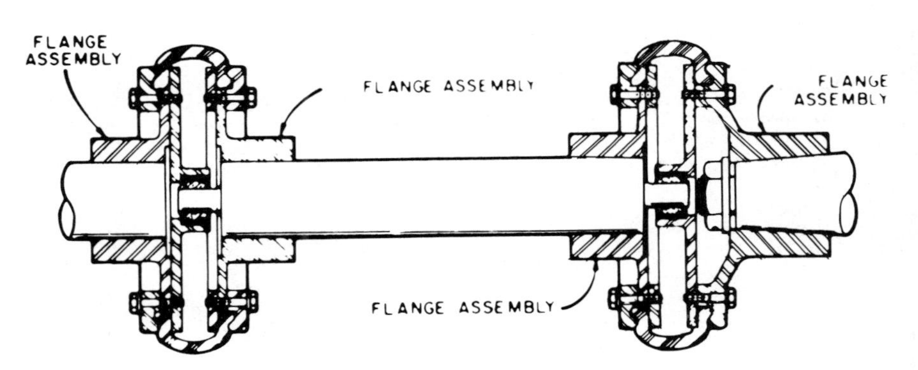

Figure 7.24 Tandem rubber tire coupling (courtesy of Dodge Division of Reliance Electric).

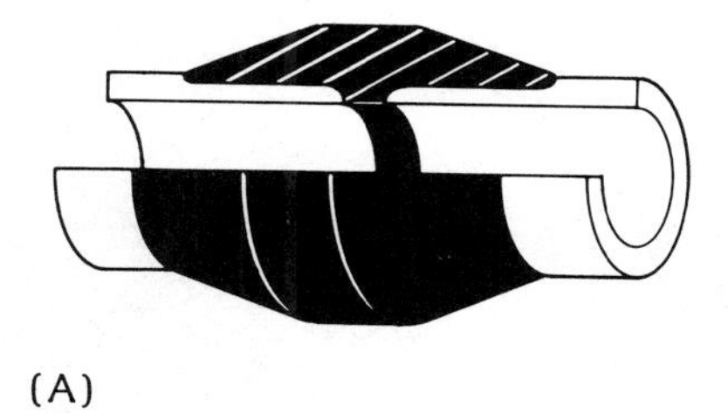

(A)

(B)

Figure 7.25 (A) Bonded shear coupling; (B) bonded shear coupling application (courtesy of Lord Corporation).

(A)

(B)

Figure 7.26 (A) Unclamped donut coupling; (B) unclamped donut coupling application (courtesy of T.B. Wood's Sons Company).

(A)

(B)

Figure 7.27 (A) Donut spacer coupling; (B) donut spacer coupling application (courtesy of T.B. Wood's Sons Company).

8
Metallic Membrane Couplings

I. INTRODUCTION

The development of metallic membrane couplings came about through a desire to eliminate the problems associated with lubricated couplings. The basic designs of most metallic membrane couplings have been around longer than most mechanically flexible or elastomeric couplings. Use of these couplings was very limited prior to the 1950s, and they did not become widely used until the late 1960s. These couplings rely on the flexure of metallic materials to accommodate for misalignment and axial movement.

There are two basic types of metallic membrane couplings. One is the disk coupling, which usually consists of several flexible metallic membranes which are alternately attached with bolts to the opposite flanges (Figure 8.1). The other type is the diaphragm coupling, which consists of one or more metallic membranes which are attached at the outside diameter of a drive flange and transfer torque through the diaphragm to an inside-diameter attachment (see Figure 8.2).

A. Stresses

Metallic membrane couplings are designed for infinite life. In applying flexible membrane couplings to industrial applications the most important design consideration in relation to the disk, contoured diaphragm, flexible frame, and multiple convoluted diaphragm relates the operating stresses in the flexible membrane. These stresses must be designed to be under the endurance limit of the material used.

The types of stresses that must be considered are shown below. A multiple convoluted diaphragm is used as an example. Most flexible membrane couplings have similar stresses and the approach is very similar.

1. Diaphragm Stresses. To understand diaphragm stresses, it is important to understand how the diaphragm reacts to the various types of misalignment. Figure 8.3 is exaggerated to demonstrate more clearly the diaphragm's reaction to various forces. It is also important to realize that a diaphragm on the centerline of the diaphragm pack reacts differently than a diaphragm off the centerline, when angular misalignment is imposed on the entire pack.

Some of the stresses resulting from the diaphragm deflection shown are continuous during the entire period of operation, and these are termed *steady-state*. On the other hand, some of the stresses not only vary, but go through complete reversals during each revolution. These are termed *alternating*.

a. Steady-State Stresses. The steady-state stresses are considered to be the stresses that result from axial displacement of the diaphragm (Figure 8.3A), torque transmission, centrifugal effects, and thermal gradient effects.

Axial stress: S_A is determined by the amount of deflection imposed on the diaphragm. W is axial force and d is distance moved. The stiffness of a pack is equal to $K_a = W/d$. The stiffness of a coupling is equal to $K_A = W/2d$.

Shear stress: τ occurs when torque is transmitted through the diaphragm pack and is dependent on the size, number, and thickness of the diaphragms. Shear stress (τ) is highest at the inside diameter of the diaphragm. *Note:* In cases where torque is cyclic or reversing the stress or part of it must be combined with other dynamic stresses.

Centrifugal stress: S_c always results when the coupling is rotated, and this rotational effect on the diaphragm must be combined with the other steady-state stresses.

Thermal gradient stress: S_t applies only where there is a temperature differential across the surface of the diaphragms and/or where there is a coefficient of expansion difference. If this case exists the combined thermal stress must be calculated.

b. Alternating Stress (S_b). Two cases are considered which contribute to the total alternating stress: angular flexure stress (Figure 8.3B) and stresses due to offset deflection (Figure 8.3C).

1. *Flexure stress* (Figure 8.3B): S_F is the result of the angular misalignment of the coupling. α is misalignment angle in degrees.
2. *Offset stress* (Figure 8.3C): S_o is caused by angular deflection of the inside diameter of the diaphragm with respect to the outside diameter. The diaphragm elements which are spaced axially away from the centerline of flexure experience a stress proportional to the distance that they are removed from the centerline. Figure 8.3C shows a cross section of an exaggerated pack and how the outer diaphragms are compressed or stretched due to the distance(s) removed from the center of misalignment.

c. Combined Diaphragm Stress. All the stresses presented above are calculated at the inside of the diaphragm and for the diaphragm farthest from the centerline, which are then combined in the following manner to give the highest stress point in the diaphragm pack. Axial stress, thermal stress, and centrifugal stress are additive. Total steady-state normal stresses (S) are computed as follows:

$$S = S_A + S_t + S_c \tag{8.1}$$

These are then combined with the shear stress to produce the combined steady-state stress (S_M):

$$S_M = \frac{S}{2} + \sqrt{\left(\frac{S}{2}\right)^2 + \tau^2} \tag{8.2}$$

Total alternating stress (S_B) is conservatively determined by the simple summation of the offset and flexure stresses. Where no cyclic torque is present,

$$S_B = S_o + S_F \tag{8.3}$$

If cyclic torque is present:

$$S_B = \frac{S_o + S_F}{2} + \sqrt{\left(\frac{S_o + S_F}{2}\right)^2 + \tau^2} \tag{8.4}$$

Finally, the mean stress and the alternating stress resulting from bending of the diaphragm can be plotted on a *modified Goodman line* for various misalignments and total coupling axial displacements. Using a typical Goodman equation, the value of the design factor can be calculated. The results are shown in Figure 8.4.

$$\frac{1}{S.F.} = \frac{S_M}{S_{ult}} + \frac{S_B}{S_{end}} \tag{8.5}$$

where

$S.F.$ = safety factor
S_{ult} = ultimate strength of material (psi)
S_{end} = endurance limit of material (psi)

B. System Design Considerations

When a metallic membrane coupling is used to connect rotating equipment, three system questions arise:

1. How to predict the coupling axial natural frequency
2. How to analyze the coupling when calculating the lateral critical speeds of the system
3. How much heat will be generated and what the operating temperature will be in the coupling's enclosed guard

1. Axial Natural Frequency.

The axial natural frequency is a function of the coupling stiffness and the coupling's center weight. Figure 8.5 shows two cases that could represent motion and vibration response of a diaphragm coupling as connected to a system. A response plot for case 1 is given in Figure 8.6. For this plot, C is the viscous damping, w the driving frequency, and w_n the natural frequency, F_n the natural frequency in cycles per minute, and W the weight of center body (lb).

$$F_n = \frac{60}{2\pi}\sqrt{\frac{2K_a g}{W}} = 375\sqrt{\frac{K_A}{W}} \tag{8.6}$$

δ_{st}, the static deflection, equals the driving force divided by 2K. K_a is the stiffness of a diaphragm pack (lb/in.) and K_A is the stiffness of a coupling.

The plot in Figure 8.6 is the response to the forcing function, and the peak of the curve is defined as the magnification ratio Q. The Q for case 2 is one-half of Q for case 1. If Q for a coupling is found to be 24 and if the equipment shown is moved at a frequency equal to the natural frequency at an amplitude of 2 mils peak to peak, the center body will vibrate at 24 mils peak to peak.

Many metallic membrane couplings have linear axial stiffness or exhibit linear stiffness's within certain axial travels. The following couplings exhibit linear stiffness:

1. Multiple convoluted diaphragm couplings
2. Wavy contoured diaphragm couplings
3. Flexible frame couplings
4. Certain disk couplings

The following couplings exhibit nonlinear stiffness:

1. Tapered contoured couplings
2. Multiple straight diaphragms couplings
3. Most disk couplings

Refer to Figure 8.7 for typical linear and nonlinear stiffness's of various couplings. For couplings that have linear stiffness, only one ANF value exists. For nonlinear stiffness many ANF values exist. Because of this, depending on the rate of stiffness change and the magnitude of excitation, the coupling may never become excited.

2. Lateral Critical Speed. Figure 3.4A is a schematic of a coupling as connected to a system. Figure 3.4B is a free-body diagram depicting the mass-spring system that models the coupling as connected to a system.

The critical speed of a flexible membrane coupling is conservatively calculated by modeling the coupling as a distributed mass with pinned supports, with the supports having a stiffness equal to the lateral stiffness of the diaphragm packs.

$$y_1 = \frac{5WL^3}{384EI}$$

$$I = \frac{\pi}{64}(D_o^4 - D_i^4)$$

$$N_1 = 211.4 \sqrt{\frac{1}{y_1}}$$

$$y_2 = \frac{W}{K_{LL} \times N \times 2}$$

$$N_2 = 187.7 \sqrt{\frac{1}{y_2}}$$

$$N_c = \frac{1}{\sqrt{\frac{1}{N_1^2} + \frac{1}{N_2^2}}}$$

where

- y_1 = tube deflection (in.)
- W = weight of floating member (lb)
- L = length between element centerlines (in.)
- E = 30×10^6 (psi)
- I = area moment of inertia ($in.^4$)
- D_o = spacer tube outside diameter (in.)
- D_i = spacer tube inside diameter (in.)
- N_1 = critical speed of tube (cpm)
- y_2 = deflection of diaphragms (in.)
- K_{LL} = lateral stiffness per diaphragm (lb/in.)
- N = number of diaphragms per pack
- N_2 = critical speed of diaphragms (cpm)
- N_c = critical speed of coupling (cpm)

The critical speed of a coupling as connected to a system must also consider the stiffness of the equipment shaft and the equipment bearing stiffness. These stiffness's tend to lower the actual measured critical speed and can be calculated as follows:

$$N_s = \frac{30}{\pi} \sqrt{\frac{K_e g}{W_c}} = \text{system critical speed (cycles per minute)}$$

$$\frac{1}{K_e} = \frac{1}{2(K_c/2)} + \frac{1}{2K_{LL}} + \frac{1}{2K_s} + \frac{1}{2K_B}$$

$$N_c = 211.4 \sqrt{\frac{1}{\Delta}}$$

$$\Delta = \frac{5WL^3}{384EI}$$

$$K_c = \frac{W_c}{g}\left(\frac{\pi N_c}{30}\right)^2$$

where

K_s = equipment shaft stiffness (lb/in.)
K_B = equipment bearing stiffness (lb/in.)
W_c = coupling weight (lb)
K_c = stiffness of coupling shaft (lb/in.)

3. Heat Generation and Windage Loss. The rotation of a coupling within a stationary guard may result in an increase of the temperature in the guard due to frictional resistance within the enclosure. The calculation method represents the coupling as a disk at its maximum diameter and length (see Figure 8.8).

The temperature of the air inside the guard may tend to be higher than the guard temperature due to frictional heating. We have assumed the guard temperature to be the air temperature, which will result in a conservative answer in the windage loss calculation.

a. Disk Windage Loss. The disk windage power accounts for frictional losses in both ends of the guard. The correlation with rpm and diameter is

$$\text{Disk windage power} \propto \text{rpm}^3 \times \text{diameter}^5$$

The equation to find disk windage loss for a coupling is

$$hp_{\text{loss disk}} = \text{rpm}^{2.85} \frac{1}{K_1}\left(\frac{S}{D}\right)^{1/10}$$

where D and K_1 are found on the horsepower loss constant chart (Figure 8.9) for the proper size of coupling. S (in.). This equation is to be applied to each end of the coupling if disk loss is present. If S/D is greater than 1.0, use S/D = 1.0.

b. Cylinder Windage Loss. The cylinder windage power correlation with rpm, diameter, and length is

$$C_{losses} \propto rpm^3 \times \Sigma \text{ diameter}^4 \times \text{length}$$

The equation to find the amount of horsepower loss for each cylindrical section is

$$hp_{loss\ cylinder} = E \times L \times Cf$$

where

$E = rpm^3 \times D^{3.859} \times 5.5 \times 10^{-15}$
D = diameter of section (in.)
L = length of section (in.)
Cf = cylinder friction coefficient

$$Cf = \frac{128B + 2.075}{B(rpm)(D)^2} + 0.0015$$

$$B = \frac{\text{diameter of guard} - \text{diameter of section}}{\text{diameter of section}}$$

This is done for each section and all the values of $hp_{loss\ cylinder}$ are summed together.

c. Total Windage Loss. Add the disk hp loss and cylinder hp loss for the total hp loss.

$$hp_{loss\ total} = hp_{loss\ disk} + \Sigma\ hp_{loss\ cylinder}$$

To find the assumed temperature of the guard, take the area of the guard, A_g (ft^2), and divide into the hp loss total, where

$$A_g = \pi \times D_G \left(L + \frac{D_G}{2}\right)$$

D_G = diameter of guard (ft)
L = length of guard (ft)

From Figure 8.10 for the total hp/A_g and the correct ambient temperature, T_a (°F), find T_{gl} the assumed temperature of the guard (°F). If $T_{gl} > 175°F$, use this scaling technique to find the actual hp_{loss} and operating temperature of the guard.

$$hp_{loss\ total} \times \frac{590}{T_{gl} + 560} = hp_{loss\ actual}$$

Again from Figure 8.10, use $hp_{loss\ actual}/A_g$ and ambient temperature to find the operating temperature of the guard (°F).

II. DISK COUPLINGS

The flexible disk coupling is available in a number of forms, but they all have one thing in common: the driving and driven bolts are on the same bolt circle. The types shown in Figure 8.11 are some of the most commonly used disk shapes. The flexibility that each type can obtain is dependent on the free span of the material between the bolts. Torque is transmitted by the pull between the driving and driven bolts.

The loading of the disk in Figure 8.11A and B will produce a simple tensile stress in the blades. The disk in Figure 8.11C will produce compression stresses in the outer diameter and tension stress at the inner. If the line of action between the driving and the driven bolt should be outside the material of the lamination, the tensile stress could be high. To ensure that this line of action falls in an acceptable position, it is sometimes necessary to use more driving bolts than would otherwise be required, and this reduces flexibility. A membrane of such a design requires the use of more laminate material than is absolutely required, or a larger diameter, to reduce the stress levels. The disk in Figure 8.11B has the inoperative portion of the membrane removed. An attempt has been made to produce a shape giving a uniform tensile stress pattern over the whole of the driving portion of the blade.

Bending stresses in the lamination due to misalignment will occur at the anchor points and these will be very heavily influenced by the tensile pull in the blade. Any reduction in the tensile stress that can be made in this area will therefore be of advantage and leave a greater margin for the fluctuations in bending stress due to angular misalignment. The form shown in Figure 8.11A makes an attempt to do this by allowing more material in the laminations in the area where the largest bending stresses occur, with the result that tensile stresses will be lower in the region of the anchor points for a given stress in the midspan of the blade. The disk transmits torque by a simple tensile force between alternate driving and driven bolts on a common bolt circle diameter and flexibility which is derived from the free span between adjacent bolts and will vary as the cube of this length. When portions of the flexible element between bolts is considered as a beam, the advantages of a thin laminate construction, as opposed to one thick disk, will be obvious in terms of flexibility and forces transmitted due to misalignment. Certain manufacturing considerations make it impractical to use very thin laminations for disk. Typically, thicknesses ranging from 0.005 to 0.025 in. have been found to be satisfactory. The form in Figure 8.11D has some advantages in manufacturing and with larger units also aids with assembly and disassembly of the links into the coupling.

The degree of flexibility required and the limits of acceptable bearing loads determine the number of driving and driven bolts used. This can be an important factor when selecting a coupling, as by choosing a unit with a greater degree of freedom than is actually required, not only will the diameter and weight be larger than necessary, but the cost could also be too high. It is possible for a coupling of this type to have any even number of bolts, but the simplest form, with two driving and two driven bolts, will provide the largest degree of flexibility and generate the lowest forces. At the other extreme, it is practical to have a large number of bolts giving a short flexing length when it is necessary to restrict the movement of a rotor with no axial location while still allowing for some small misalignment.

A. Construction of Disk Coupling

The disk coupling is used from fractional-horsepower drives to very large drives (100,000 hp). In general we can categorize the disk coupling in two groups: one for general-purpose applications, the second for high-speed applications.

The general-purpose disk coupling's torque-transmission components are made from low- to medium-carbon steels. Flexing disks are usually made of spring steel, AISI 1050 to 1080, 300 series stainless steel. The high-speed disk coupling's torque-transmission components are usually made of alloy steels. The flexing disk is usually made of corrosion-resistant steel, AISI 300 series stainless steel, PH stainless steel, or high-strength nickel alloy.

B. Applications of Disk Couplings

The disk coupling is used in a variety of applications. The most common is on medium-horsepower pumps. They are also used on marine drives, cooling tower drives, generators, compressors, mill equipment, fans, and machine tools. See Figures 8.12 to 8.15 for some typical applications.

C. Types of Disk Couplings

There are many types of disk couplings, including:

A closed-coupled double flex disk coupling (Figure 8.16) usually used for low-torque applications.
A tandem disk coupling (Figure 8.17) typically used to connect long spans.
A spacer-type disk coupling (Figure 8.18), typically used on medium-horsepower pumps.
A spacer disk coupling using a square rather than a circular disk (Figure 8.19).
A spacer disk coupling that uses individual links rather than a one-piece disk (Figure 8.20).
A spacer disk coupling similar to the above except that the disks are square and several disks are used between the driving and driven flanges to increase flexibility (Figure 8.21).
A spacer disk coupling that uses a unitized pack to help maintain balance and decrease the number of parts during assembly and disassembly (Figure 8.22).

D. Failure Modes for Disk Couplings*

The most common form of failure is disk fatigue due to excessive flexure. This is usually caused by poor initial alignment of the

---

*This section was supplied through the courtesy of the Coupling Division of Rexnord, Thomas Couplings.

connected machines. It also can be brought about by operational conditions.

Machines connected by flexible couplings should be aligned with the greatest possible accuracy. The better the initial alignment, the more capacity the coupling has to take care of subsequent operational misalignment. Changes from the initial condition can occur through bearing wear, settling of foundations, base distortion due to torque, thermal changes, and vibrations in the connected machines. The alignment of the machines should be checked at regular intervals and corrected as necessary.

In Figure 8.23A the disk is broken adjacent to the washer face. This usually indicates excessive shaft misalignment during operation. This type of disk failure generally starts with outer disk in the pack and progresses through the disk pack. The equipment must be realigned and the disk pack replaced. A hot check of alignment should be made to assure that it is within coupling misalignment capacity.

In Figure 8.23B the disk is broken adjacent to the washer face, with heavy corrosion present along the area of the break. Also, iron oxide will probably be bleeding from the disk pack. This failure is typical of disks that have been in service for several years and/or have been operating in a corrosive atmosphere. Breaks will first appear in the outer disks and progress into the disk pack. The disk pack should be replaced and the equipment realigned. If excessive corrosion exists on other areas of the disk or coupling component parts, they should be replaced with proper corrosion protection: stainless steel, plating, painting, and so on.

The disk coupling is easily inspected. Visual analysis may point to possible drive system problems. Proper evaluation of the disk packs and connecting parts may save considerable maintenance costs and downtime. Following are some of the more evident visual inspection criteria and recommended corrective procedures.

1. The disk is broken through the bolt hole (Figure 8.23C). This indicates loose coupling bolts. Replace the disk pack and tighten bolts to the specified torque value.
2. Disks are embedded in the bolt body. This is usually the result of a loose bolt or a severe torque overload. This may also appear when turning the bolt during installation. Replace the bolt and tighten the locknut to the proper torque. Do not turn the bolt during the locknut tightening process.

3. The disk pack is wavy and the dimension between flange faces is smaller than indicated on the installation instructions or applicable assembly drawing (Figure 8.24A). The coupling has been installed in a compressed condition or equipment has shifted axially during operation. Check for thermal growth problems. If the application is a sleeve-bearing motor, make sure that the operating centerline of the motor rotor is properly positioned. Make necessary adjustments to relieve disk pack compression during operation.
4. The disk pack is wavy and the dimension between flange faces is larger than specified on the installation instructions or applicable assembly drawing (Figure 8.24B). The coupling has been installed in an elongated position or equipment has shifted axially during operation. Realign the axial position of equipment so that the coupling operates with a neutral flat disk pack. If the application is a sleeve-bearing motor, make sure that the operating centerline on the motor rotor is in the proper position.
5. The disk pack has a bulge near the center, or is bowed toward one flange in alternate chord positions, or bolts are bent (Figure 8.25). This condition is a result of a large torque overload induced into the system above the peak overload capacity of the coupling. The remaining disk pack chordal sections will be very straight and tight. Check the coupling selection; apply the proper size and style for the characteristics present.

III. DIAPHRAGM COUPLINGS

The flexible diaphragm coupling is available in three basic forms (Figure 8.26). This coupling obtains its flexibility from the free span between the diaphragm OD and ID. Torque for these couplings is transmitted between the OD and the ID. The diaphragm elements can be of constant or variable thickness, usually with maximum thickness at the smaller diameter. All three shapes commonly used have some shape modification that helps to reduce their size and increase their flexibility.

1. *Tapered contoured diaphragm* (Figure 8.26A): This diaphragm shape is designed for constant shear stress from ID to OD. The tapered shape greatly increases the coupling's flexibility capabilities.

2. *Multiple straight diaphragm (with spokes)* (Figure 8.26B): By using multiple thin plates rather than one thick plate the flexibility of this couple is greatly increased compared to that of a single diaphragm. With thin diaphragms in parallel, the stress are usually lower. The stresses, moments, and forces of a diaphragm increase with the third power of the thickness (T^3). Hence the use of several thinner diaphragms produces lower values (Nt^3), where Nt is equal to T for single diaphragm coupling. The coupling shown also has material removed (forming spokes) to increase flexibility further.
3. *Multiple convoluted diaphragm* (Figure 8.26C): This coupling incorporates all the flexibility of the one shown in Figure 8.26B but also has a convoluted shape, which helps to increase its flexibility in the axial direction and provides linear axial stiffness (refer to Figure 8.7).

A. Construction of Diaphragm Couplings

The diaphragm coupling is usually used on high-performance equipment and therefore must have high reliability. Because of this, the torque-transmission components of these couplings are usually made of high-strength alloys such as AISI 4140 and 4340.

Flexing diaphragms are made from high-strength alloys that exhibit good fatigue properties. Tapered contoured diaphragms are usually made from AISI 4100 or 4300 steels coated for corrosion protection. Multiple diaphragms (straight and convoluted) are typically made from cold/reduced 300 series stainless steel (1/4 to 1/2 hard condition), may use some of the PH stainless steels (15–5PH or 17–7PH), or may be made from some of the high-strength nickel alloys (such as 718).

B. Applications of Diaphragm Couplings

The diaphragm coupling is typically used in high-performance applications where a failure or downtime can be costly (see Figures 8.27 and 8.28).

C. Types of Diaphragm Couplings

There are many types of diaphragm couplings, including:

Tapered contoured diaphragm (Figure 8.29)
Multiple straight spoke diaphragm (Figure 8.30)

Multiple convoluted diaphragm (Figure 8.31)
Double diaphragm pack (diaphragms are in series) (Figure 8.32)
Reduced moment multiple convoluted diaphragm (Figure 8.33)

D. Failure Modes for Diaphragm Couplings

Some of the typical failure modes for the three basic type of diaphragm couplings are as follows:

Typical failure of tapered contoured diaphragm from over-misalignment and excessive axial travel (Figure 8.34)
Multiple straight diaphragm failure from overtorque (Figure 8.35)
Multiple convoluted diaphragm coupling failure: stress corrosion fatigue (Figure 8.36)

Failed outer diaphragm (Figure 8.36A)
Failed outer diaphragm with corrosion pitting (Figure 8.36B)

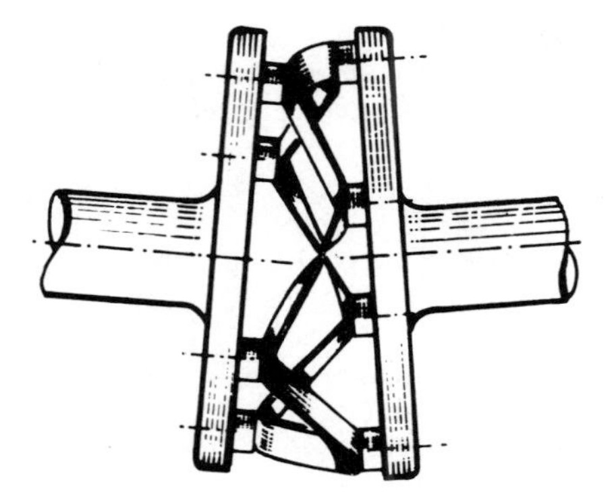

Figure 8.1 Disk coupling.

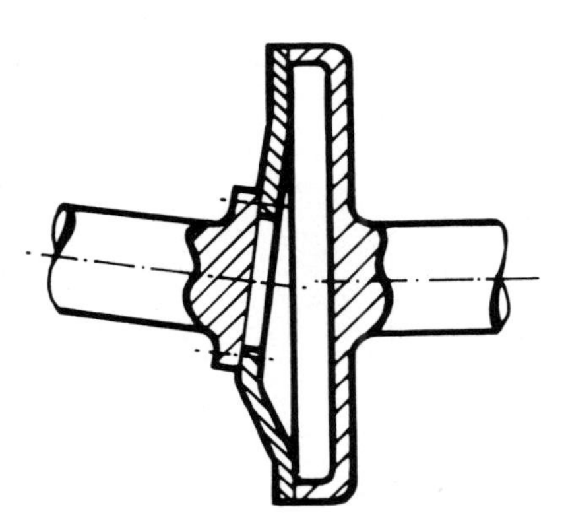

Figure 8.2 Diaphragm coupling.

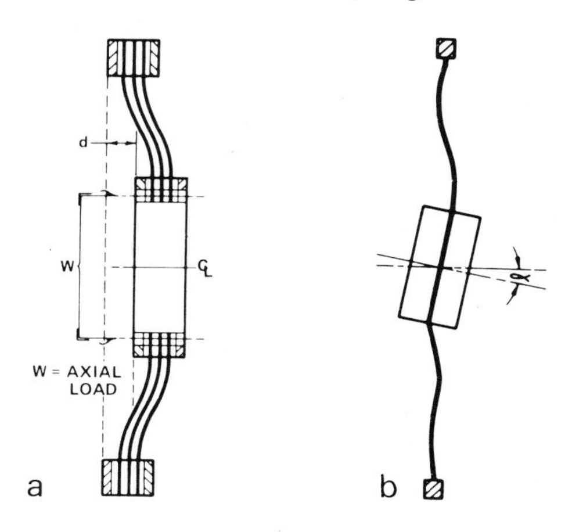

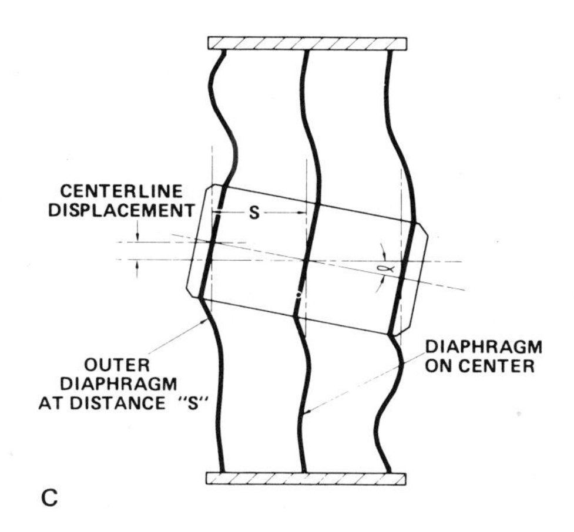

Figure 8.3 (A) Axial displacement of a diagram coupling; (B) flexure of a diaphragm coupling due to misalignment; (C) offset displacement of a multiple diaphragm.

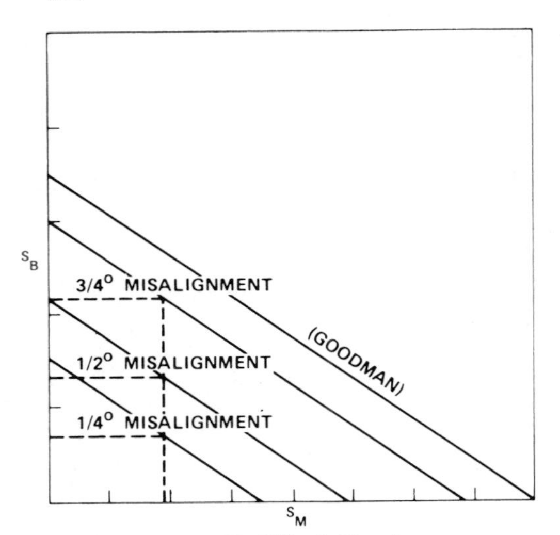

Figure 8.4 Modified Goodman curve.

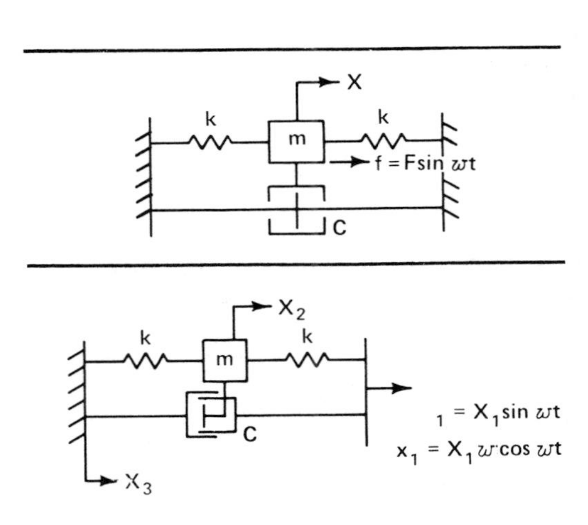

Figure 8.5 Free body of a metallic membrane coupling

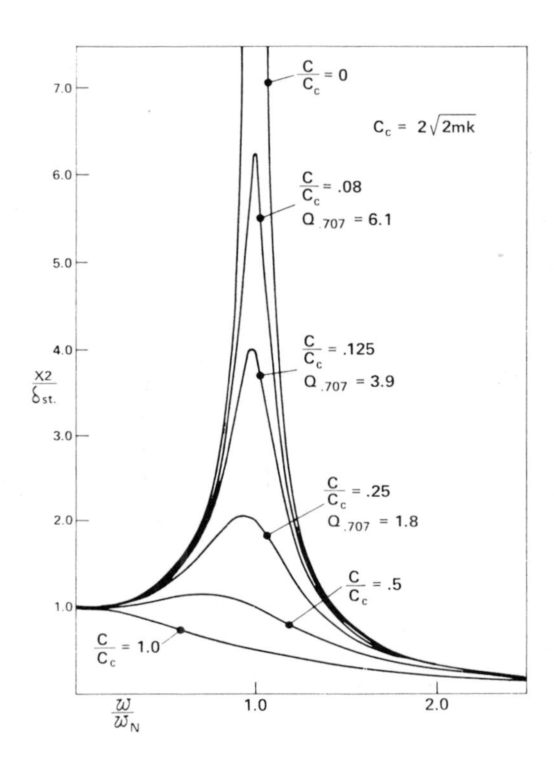

Figure 8.6 Plot of response versus forcing function

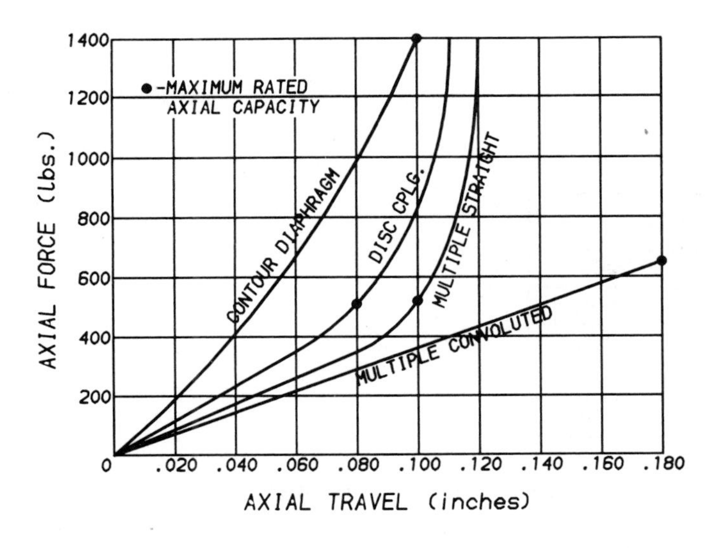

Figure 8.7 Linear and nonlinear axial stiffness

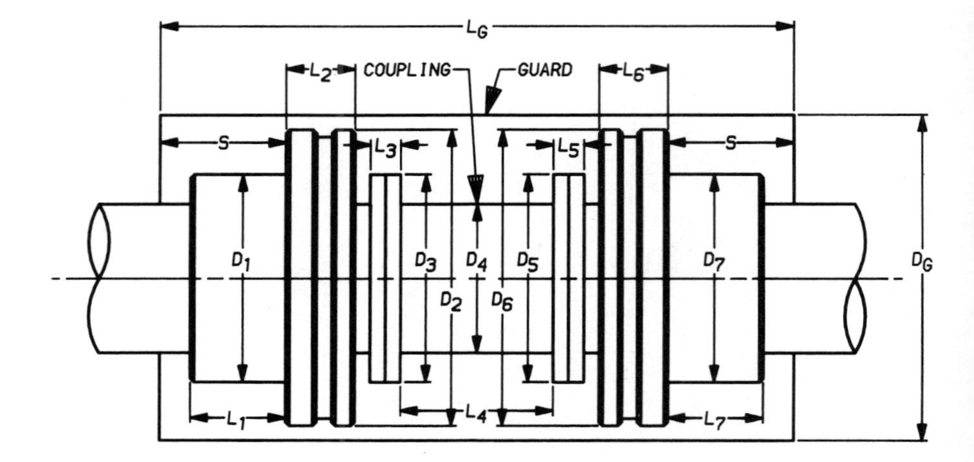

Figure 8.8 Coupling outlined in a coupling guard.

D (in.)	$K_1 \times 10^{10}$
6	2460
9	372
12	67.6
16	19.3
24	4.76

Figure 8.9 Typical horsepower loss constant chart.

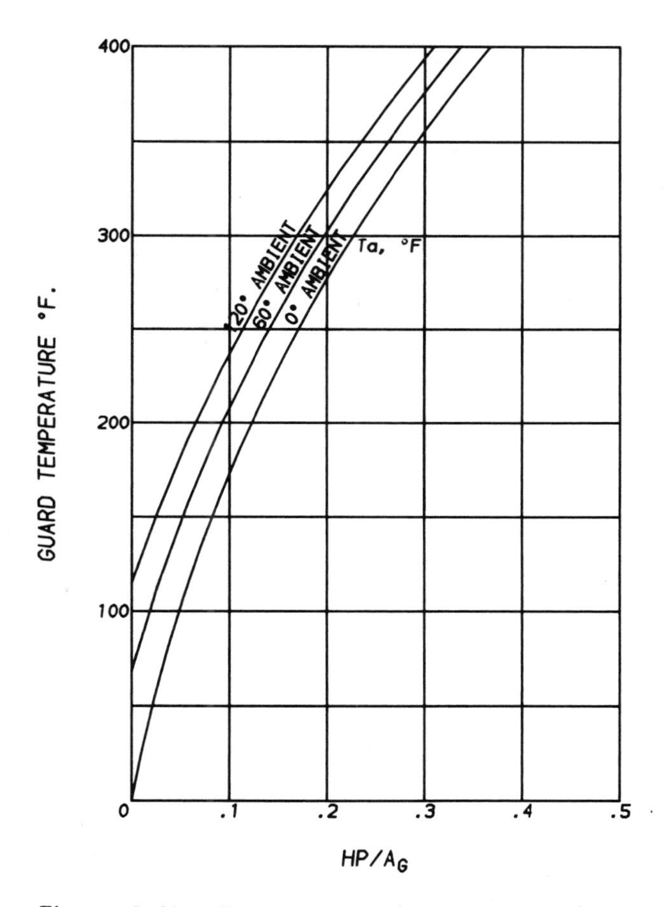

Figure 8.10 Temperature rise versus HP/Ag.

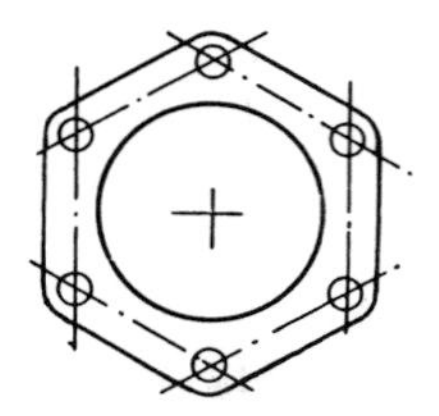

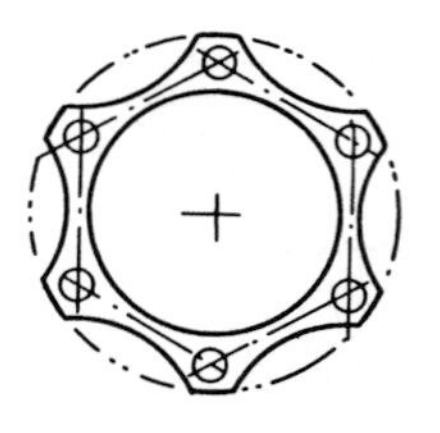

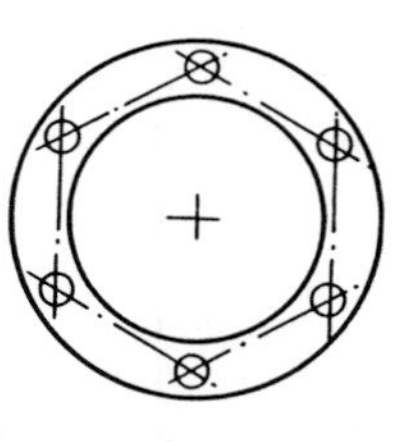

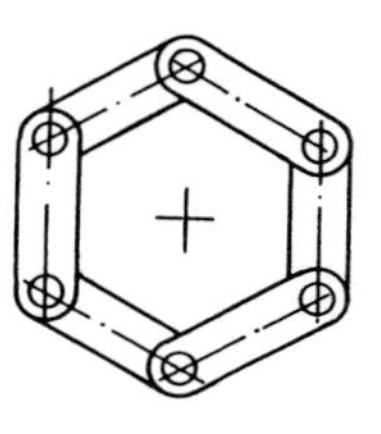

Figure 8.11 Various types of disk shapes.

Figure 8.12 Centrifugal pump application (courtesy of Coupling Division of Rexnord, Thomas Couplings).

Figure 8.13 Marine diesel application (courtesy of Formsprag Division of Dana Corporation).

Figure 8.14 Cooling tower application (courtesy of Formsprag Division of Dana Corporation).

Figure 8.15 Generator application (courtesy of Coupling Division of Rexnord, Thomas Couplings).

Figure 8.16 Closed coupled disk coupling (courtesy of Coupling Division of Rexnord, Thomas Couplings).

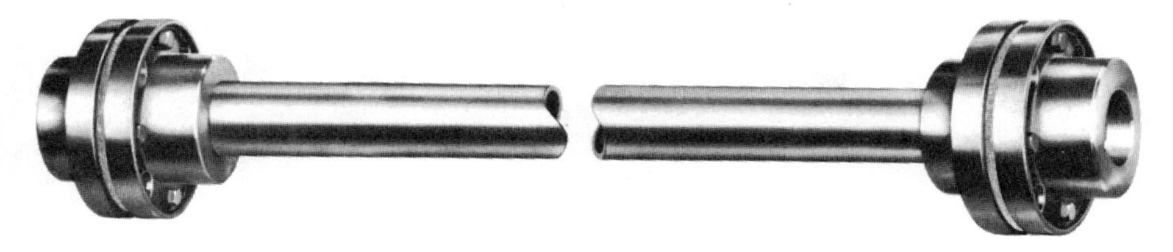

Figure 8.17 Floating shaft disk coupling (courtesy of Coupling Division of Rexnord, Thomas Couplings).

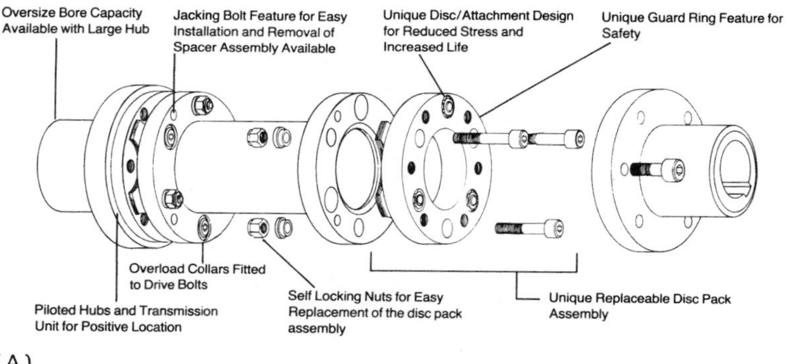

(A)

(B)

Figure 8.18 Replaceable disk pack coupling (courtesy of Flexibox International, Inc., Metastream Coupling).

Figure 8.19 Spacer disk coupling with square disk (courtesy of Formsprag Division of Dana Corporation).

Figure 8.20 Spacer disk coupling with separate links (courtesy of TGW Thyssen Getriebe).

Figure 8.20 (continued)

Figure 8.21 Spacer disk coupling with multiple frames (courtesy of Kamatics Corporation, Kaflex Coupling).

Figure 8.22 Spacer disk coupling with unitized packs (courtesy of Coupling Division of Rexnord, Thomas Couplings).

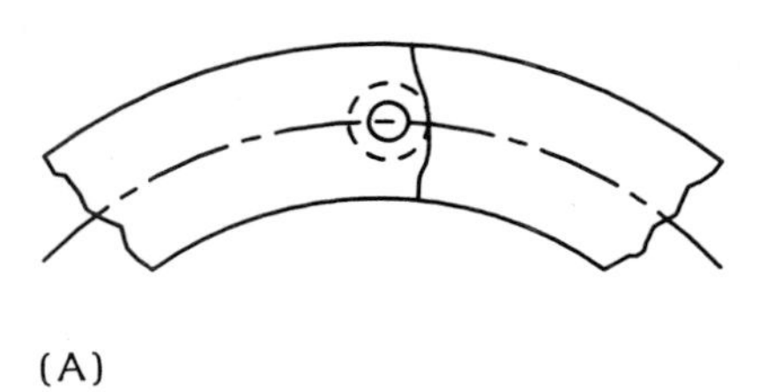

(A)

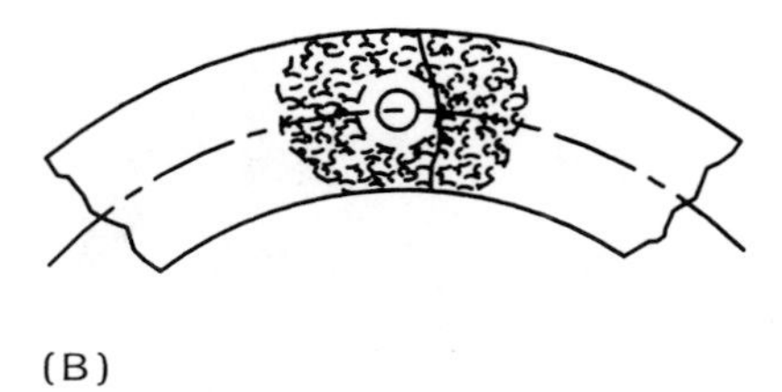

(B)

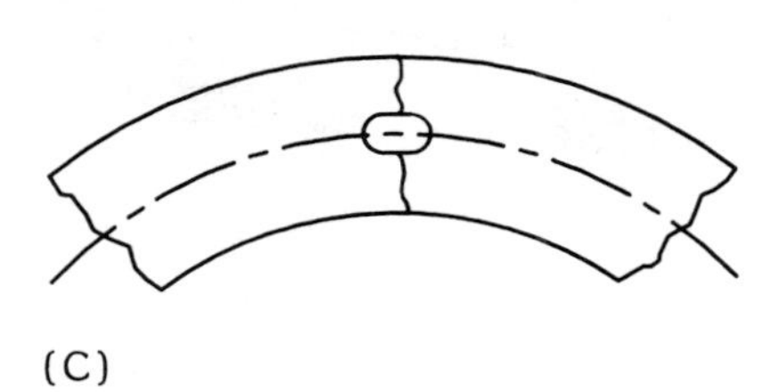

(C)

Figure 8.23 (A) Misalignment failure; (B) fatigue failure; (C) elongated bolt holes.

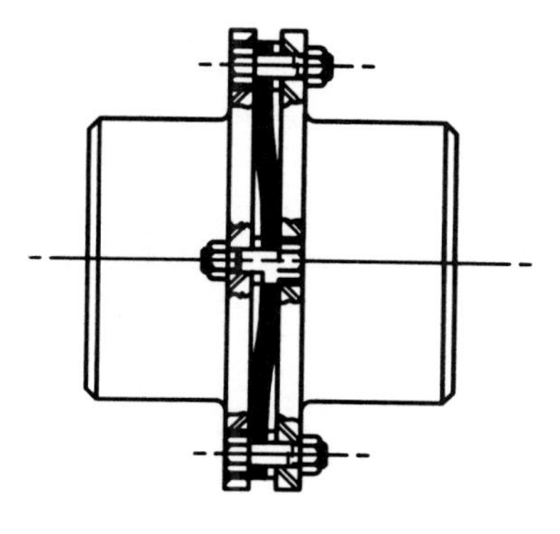

(A)

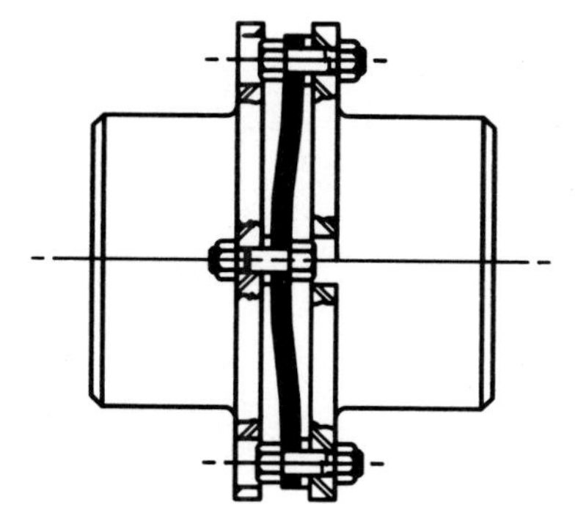

(B)

Figure 8.24 (A) Compression damage and (B) elongation damage of a disk coupling.

Figure 8.25 Torque-overload failure.

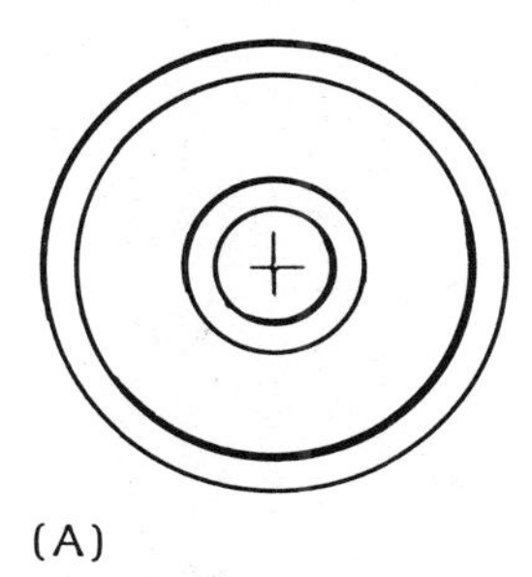

(A)

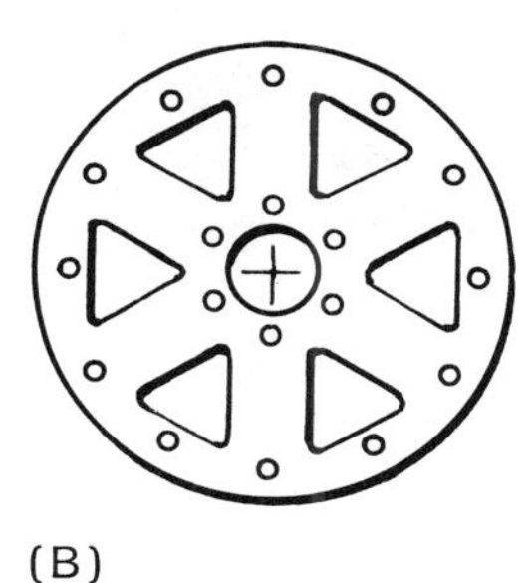

(B)

(C)

Figure 8.26 (A) Tapered contoured diaphragm (courtesy of Bendix, Fluid Power Division); (B) multiple straight diaphragm (courtesy of Flexibox International, Inc., Metastream Coupling); (C) multiple convoluted diaphragm coupling (courtesy of Zurn Industries, Inc., Mechanical Drives Division).

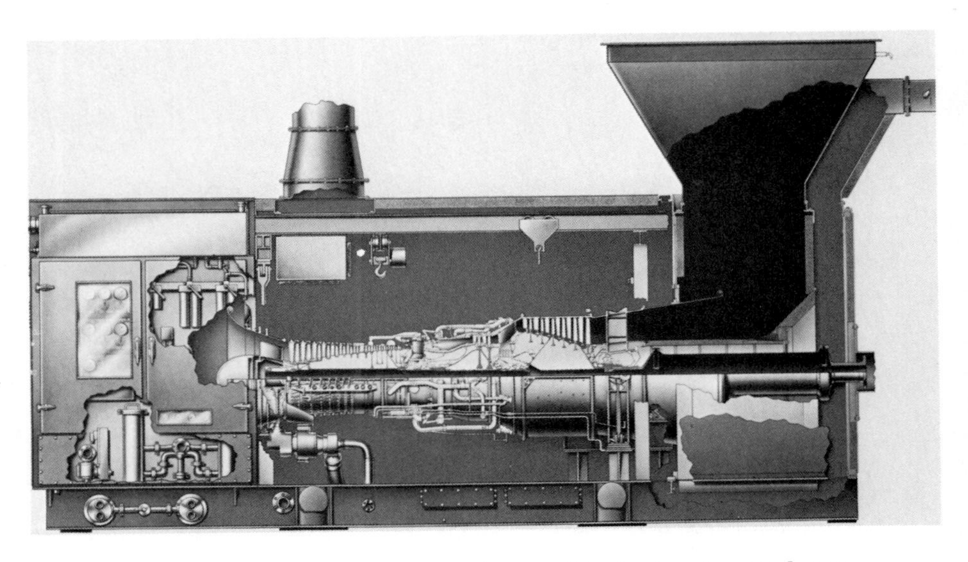

Figure 8.27 Industrial gas turbine with double tapered contoured coupling (courtesy of General Electric Schenectady, Gas Turbine Division).

Figure 8.28 Accessory drive multiple convoluted diaphragm coupling (courtesy of Zurn Industries, Inc., Mechanical Drives Division).

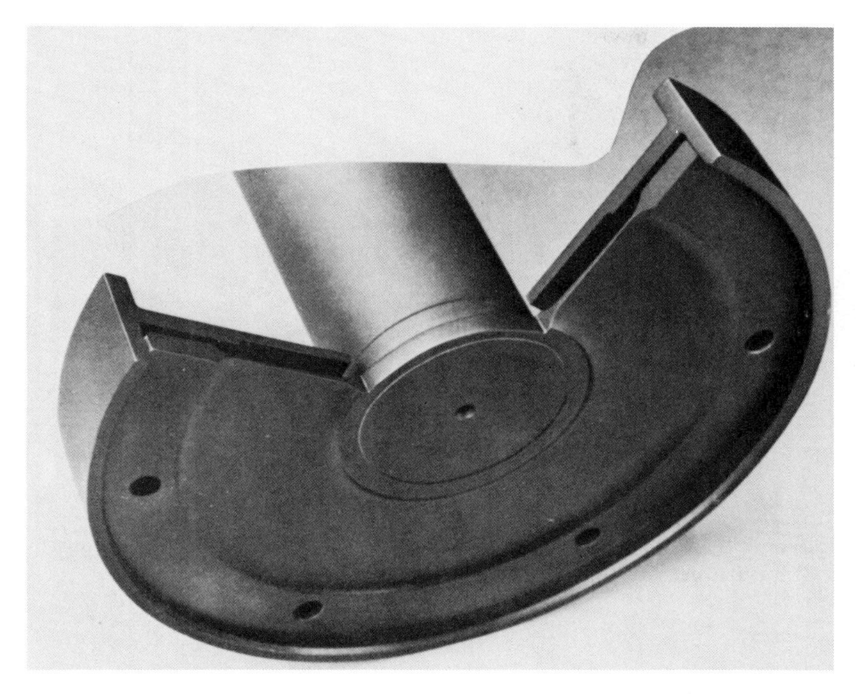

Figure 8.29 Tapered contoured diaphragm coupling (courtesy of Koppers Company, Inc., Engineered Metal Products Group).

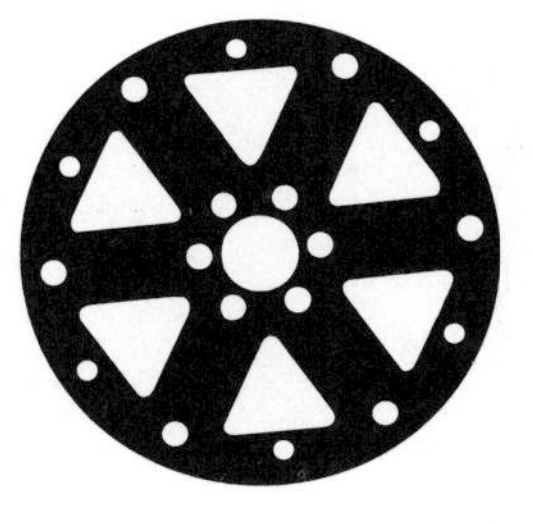

Figure 8.30 Multiple straight diaphragm coupling (courtesy of Flexibox International, Inc., Metastream Coupling).

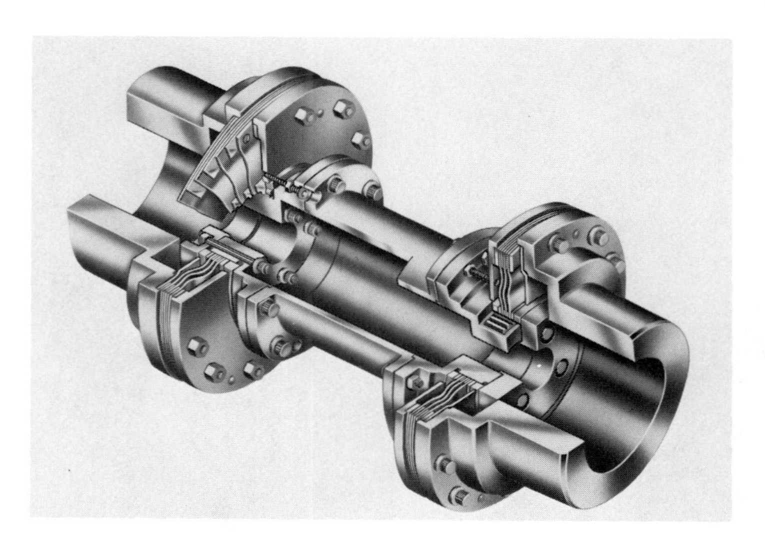

Figure 8.31 Multiple convoluted diaphragm coupling (courtesy of Zurn Industries, Inc., Mechanical Drives Division).

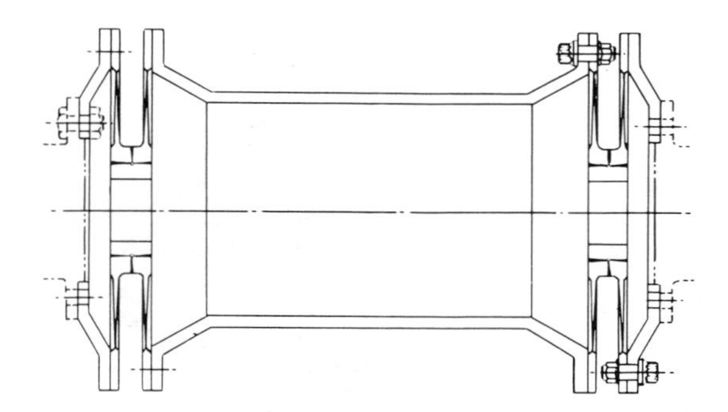

Figure 8.32 Contoured double flex coupling (courtesy of Bendix Fluid Power Division).

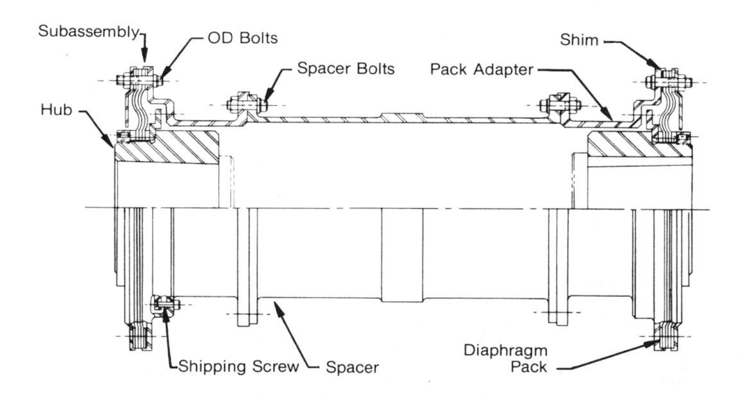

Figure 8.33 Reduced moment multiple convoluted diaphragm coupling (courtesy of Zurn Industries, Inc., Mechanical Drives Division).

Figure 8.34 Failure of a tapered contoured diaphragm coupling (from Michael Neale et al., *Proceedings of the International Conference on Flexible Couplings for High Power and Speeds*, 1977).

Figure 8.35 Failure of a multiple diaphragm coupling due to overtorque (from Michael Neale et al., *Proceedings of the International Conference on Flexible Couplings for High Power and Speeds*, 1977).

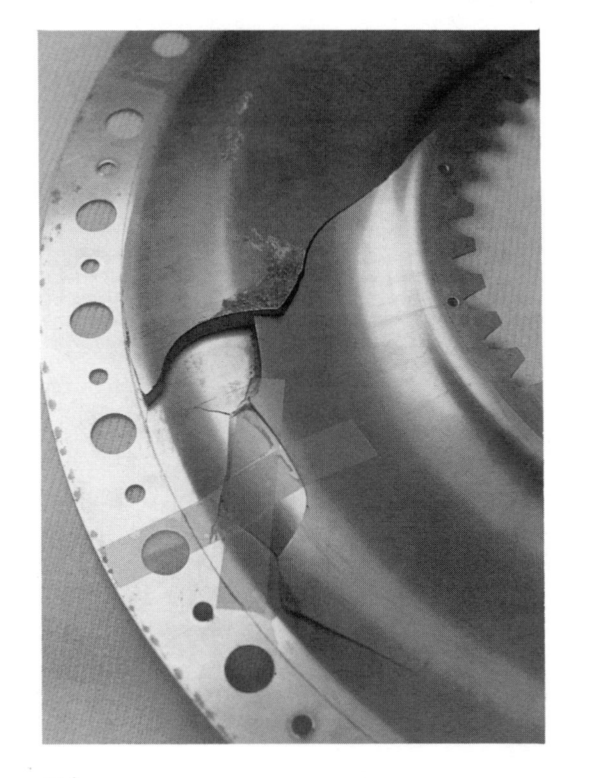

(A)

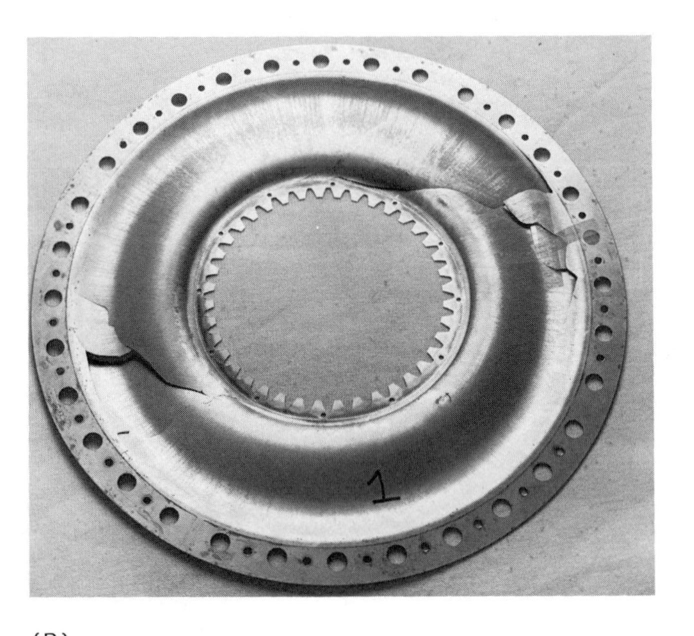

(B)

Figure 8.36 Stress corrosion fatigue failure of a diaphragm.

9
Miscellaneous Couplings

I. INTRODUCTION

Chapters 5 to 8 covered some of the most commonly used flexible couplings. There are many other types available, categorized as miscellaneous. This group of flexible couplings obtain their flexibility from a combination of the mechanisms covered in Chapters 6 to 8 or a unique mechanism. Some of those are:

1. The pin and bushing coupling (Figure 9.1)
2. The Schmidt coupling (Figure 9.2)
3. The spring coupling (Figures 9.3 and 9.4)
4. The slider block coupling (Figure 9.5)

In this chapter we describe their operating principles, functions, and variations available.

II. PIN AND BUSHING COUPLINGS (Figure 9.1A)

Resilient drive is provided by rubber-cushioned sleeve bearings. Hardened-steel drive studs are ground to a close tolerance. No lubrication is required, so the maintenance cost is low.

This type of coupling not only transmits torque and accommodates for misalignment and axial travel but also absorbs torsional vibration. The coupling provides for misalignment and axial travel by comparison of the bushing and sliding of the pin in the bushing. Pin and bushing couplings are used on hoists, conveyors, generators, compressors, roll-out tables, and motor fans (Figure 9.1B).

III. SCHMIDT COUPLINGS (Figure 9.2A)

Schmidt offset couplings belong in the family of torque rigid couplings and are designed to provide large parallel shaft offset at high accuracy with no side loads. This coupling design provides many features which are advantageous for applications in industries such as electronics, papermaking and converting, woodworking, textile operations, machine tools, printing, and steelmaking.

The Schmidt offset coupling offers flexibility in shaft displacement while maintaining undistrubed power transmission at constant angular velocity. The coupling does not add secondary forces to the drive. Schmidt offset couplings are available in parallel shaft ranging from 3/16- to 24-in. capacities from 5 to 400,000 in.-lb. The Schmidt coupling can be used on steel mill equipment and on conveyors (Figure 9.2B).

IV. SPRING COUPLINGS

A. Tangential Spring Couplings (Figure 9.3)

These couplings transmit torque through tangentailly arranged compression springs, need no lubrication, and are unaffected by most chemicals. The spring rate and inertia can be varied. The standard design permits maximum torsional deflections of approximately ±5°. Special designs are available for deflections between ±1° and ±8°.

The spring is available in sizes to transmit beyond 50,000 hp and up to 15 million lb-in. of torque. The spring coupling is also available with V-pulleys, flywheels, and special adapters.

Spring couplings protect connected equipment against expensive downtime by cushioning shock loads. These couplings will accommodate shaft misalignment and axial travel, both singly and in combination. When combined misalignments occur, permissible individual misalignments are reduced.

B. Spiral Spring Couplings (Figure 9.4)

The flexible portion of this coupling assembly consists of two units. Each unit has several spiral beams proportioned to have small resistance to bending due to torsional and shear loadings. The resistance to torsional deflection allows the coupling to absorb full rated torque with less than 2% windup. Its resistance to shear deflection allows two flex units to support a long drive shaft and maintain dynamic stability. The beams are shaped to maintain equal stress at the inside and outside diameters, which makes it possible to carry the same torque in either direction of rotation. These couplings are used for applications up to 100 hp.

V. SLIDER BLOCK COUPLINGS (Figure 9.5)

The slider block flexible coupling is a simple efficient unit. It performs all the required functions of a flexible coupling and compensates for angular misalignment up to ±1°. It is used for blind assembly or vertical applications.

The coupling is generally known as the "Oldham coupling." It transmits torque through an intermediate square floating member and compensates for all three types of misalignment by the combined sliding action between the closely fitted center member and the adjacent driving and driven jaw flanges.

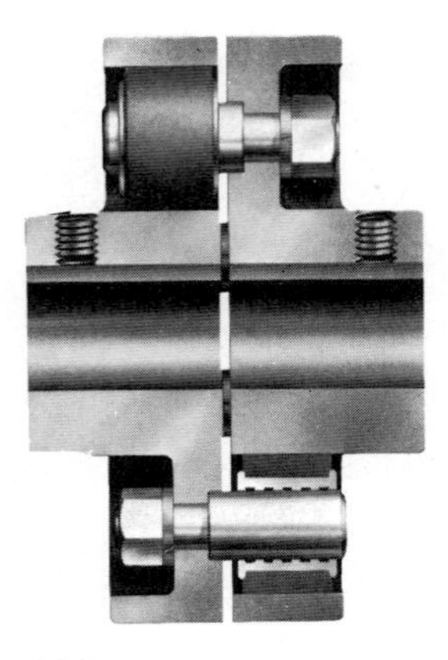

(A)

(B)

Figure 9.1 (A) Pin and bushing coupling; (B) an application of the pin and bushing coupling (courtesy of Renolds, Inc., Engineered Products Division).

(A)

(B)

Figure 9.2 (A) Schmidt coupling; (B) an application of the Schmidt coupling (courtesy of Schmidt Couplings, Inc.).

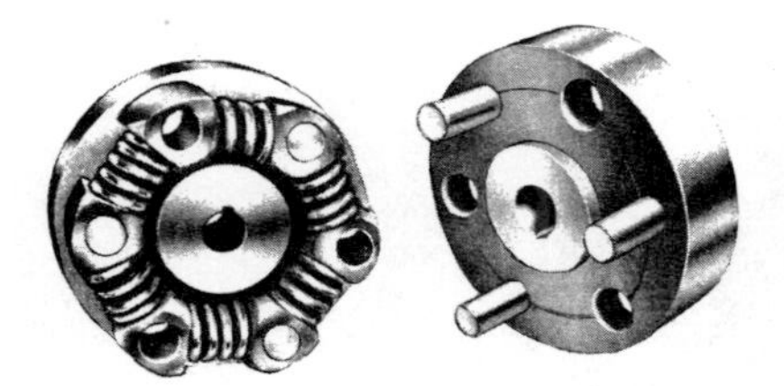

Figure 9.3 Tangential spring coupling (courtesy of Renolds, Inc., Engineered Products Division).

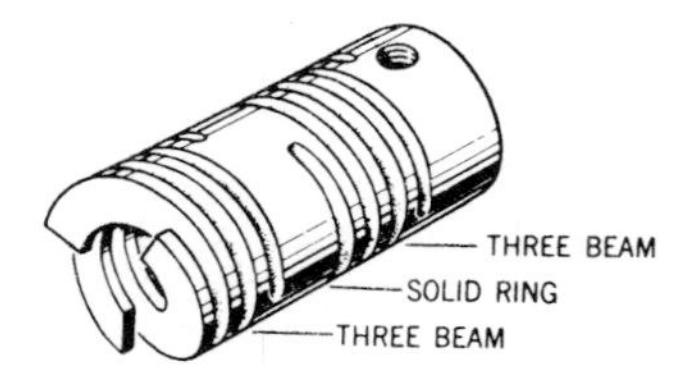

Figure 9.4 Spiral spring coupling (courtesy of Panamech Company).

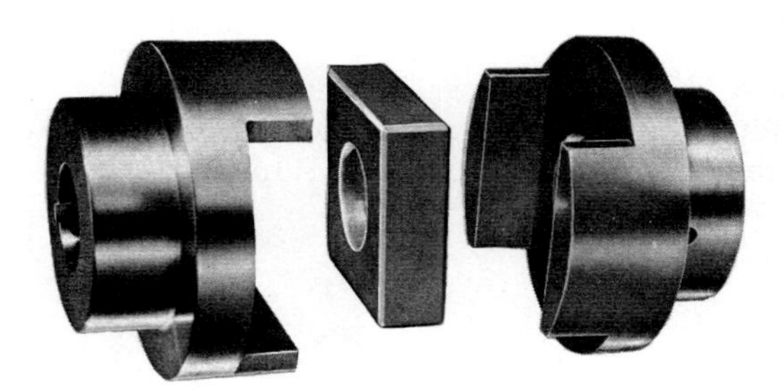

Figure 9.5 Slider block coupling (courtesy of Zurn Industries, Inc., Mechanical Drives Division).

Bibliography

American Gear Manufacturers Association (AGMA), AGMA 515, *Standard Balanced Classification for Flexible Couplings*, April 1977.
AGMA 510, *Nomenclature for Flexible Couplings.*
AGMA *Standards for Bores and Keyways for Flexible Couplings.*

American Petroleum Institute, *Special-Purpose Couplings for Refinery Service*, API Standard 671, First Edition, December 1969.

Barlett Hayward Company, "Fast's Self-Aligning Coupling," catalog, 1930.

Boylan, W., *Marine Application of Dental Couplings*, Paper 26, Society of Naval Architects and Marine Engineers, 1966.

Broersma, I. G., *Couplings and Bearings*, H. Stam N.V., 1968.

Calistrat, M., "Flexible Coupling Installation," for Koppers.

Calistrat, M., *Friction Between High Speed Gear Coupling Teeth*, ASME 80-C2/DET 5.

Calistrat, M., "Grease Separation Under Centrifugal Forces," Koppers Form 3924.

Calistrat, M., "Shaft Lubrication," *Journal of the American Society of Lubrication Engineers*, 1980, pp. 9–15.

Calistrat, M., "Wear and Lubrication," *Mechanical Engineering*, October 1975.

Calistrat, M., and Webb, S., "Sludge Accumulation in Continuously Lubricated Couplings," Koppers Form 1645.

Chase, T. W., "Maintenance of Gear-Type Couplings," *Iron and Steel Engineer*, October 1964, pp. 72–84.

Chironis, Nicholas, *Gear Design and Application*, McGraw–Hill, New York, 1967.

Condon, W., *Application of Universal Joints to Construction and Industrial Machinery*, Society of Automotive Engineers, September 1961.

Detroit Ball Bearing Co., *Coupling Alignment*.

Dodd, R., *Total Alignment*, Petroleum Publishing Co., 1975.

Eschleman, R., "Combating Vibration with Mechanical Couplings," *Machine Design*, September 1980.

Faires, V. M., *Design of Machine Elements*, 4th ed., Macmillan, New York, 1965.

Firth, D., "Couplings for Steel Mill Drives," *Iron and Steel Engineer*, August 1963.

Gensheimer, J., "How to Design Flexible Couplings," *Machine Design*, September 1961.

IRD Mechanalysis, *Audio-Visual Customer Training*, Instruction Manual, 1975.

Lord Corporation, *Understanding Torsional Vibration*, Design Monograph DM 1107, 1976.

Lovejoy, Inc., *Power Transmission in the Twentieth Century: An Overview*.

Lovejoy, Inc., *Preventing Unnecessary Failure of Power Transmission Components*, 1j-702.

MIL-C-23233—"For Couplings for Propulsion Units, Auxiliary Turbines and Line Shafts Naval Shipboard."

Mancuso, J., "Moment and Forces Imposed on Power Transmission Systems due to Misalignment of a Crowned Tooth Coupling," Master's thesis, Pennsylvania State University, June 1971.

Muster, J., and Flores, B., *Balancing Criteria and Their Relationship to Current American Practices*, ASME Paper 69 vibo-60.

Neale, M., et al., *"Proceedings of the International Conference on Flexible Couplings for High Power and Speeds,"* 1977.

Peterson, R., *Stress Concentration Factors*, Wiley-Interscience, New York, 1974.

Poole Foundry and Machinery Co., catalog, 1958.

Potgietier, F., "Cardan Universal Joints Applied to Steel Mill Drives," *Iron and Steel Engineer*, March 1969, pp. 73–78

Reynolds, catalog, 1979.

Roots, F., "Shaft Coupling," Patent 349,365, September 1886.

Rothfuss, N., *Design and Application of Flexible Diaphragm Couplings to Industrial-Marine Gas Turbines*, ASME 73-GT-75.

Rothfuss, N., "Design Criteria and Tables for Selecting High-Speed Power Couplings," *Product Engineering*, February 1963.

Schwerdlin, H., "Reaction Forces in Elastomeric Couplings," *Machine Design*, July 1979.

Society of Automotive Engineers, *Universal Joint and Driveshaft Design Manual*, AE-7.SAE, 1979.

Thomas Flexible Coupling Co., "Thomas Flexible Couplings," catalog, 1942.

Texaco, Inc., *Lubrication of Flexible Couplings*, Vols. 1 and 2.

Texaco, Inc., October–November 1981.

Wellman Bibby Co., Ltd., *The Bibby Resilient Coupling*.

Wood's Sons Company, *Home Study Course on Couplings*, 1979.

"Why a Universal Joint," *Power Transmission Design*, October 1961.

Wright, J., "Tracking Down the Cause of Coupling Failure," *Machine Design*, June 1977.

Zurn, F., *Crowned Tooth Gear Type Couplings*, AISI, 1957.

Zurn Industries, Inc., "The Evolution of Environmentalism," 1973.

Appendix A
Coupling Manufacturers

Company Name and Address	Type of Couplings Manufactured
Acme Chain, Incom International, Inc. 825 Main St. Holyoke, Mass. 01040	Chain
Acushnet Company, Rubber Div. P.O. Box E916 New Bedford, Mass. 02742	Elastomeric
Allen Couplings, Inc. 13109 Westchester Trail Chesterland, Ohio 44026	Disk
American Stock Gear, Div. of Perfection Gear Co. Darlington, South Carolina 20532	Elastomeric Variable-velocity U-joints
American Vulkan Corp. P.O. Drawer 673 Winter Haven, FL 33880	Elastomeric
Bendix, Fluid Power Div. 211 Seward Ave. Utica, NY 13503	Diaphragm

Company Name and Address	Type of Couplings Manufactured
Charles Bond Co. 1321 Windrim Ave. Philadelphia, PA 19141	Disk, Elastomeric, Variable-velocity U-joints
Boston Gear, Incon International Inc. 14 Hayward St. Quincy, Mass. 02171	Chain, Grid, Elastomeric
David Brown Gear Ind., Inc. 60 Emblem Court Agincourt, Ontario Canada M1S1B1	Rigid, Gear
Browning Mfg., Emerson Electric 1935 Browning Drive Maysville, KY 41056	Rigid, Gear, Chain, Grid, Disk, Elastomeric, Variable-velocity U-joints
Climax Metal Products Co. 30211 Lakeland Blvd. Cleveland, Ohio 44092	Rigid, Disk, Elastomeric
Coupling Corp. of America 2422 S. Queen St. York, PA 17402	Diaphragm
Curtis Jones 4 Birnie Ave. Springfield, MA 01107	Variable-velocity U-joints
Dana Industries, P.T. Div. 23601 Hoover Rd. P.O. Box 778 Warren, MI 48090	Rigid, Grid, Disk, Elastomeric, Constant-velocity U-joints, Variable-velocity U-joints
Dana Industries, Formsprag, P.T. Div. 801 East Industrial Ave. Mt. Pleasant, MI 48858	Disk
Dana Corp., Spicer Universal Joint Div. 1315 Directors Row P.O. Box 2229 Fort Wayne, IN 46801	Constant-velocity U-joints, Variable-velocity U-joints

Company Name and Address	Type of Couplings Manufactured
Dayco Corp. 333-7 W. First St. Dayton, OH 45402	Elastomeric
Deck Manufacturing Co. 51477 Bittersweet Rd. Graner, IN 46530	Gear
Dodge Div., Reliance Electric 500-TS Union St. Mishawake, IN 45644	Gear, Chain, Disk, Elastomeric
Falk Corp. Box 492 Milwaukee, WI 53201	Gear, Grid, Elastomeric
Fawick Corp. AirFlex 9919 Clinton Rd. Cleveland, OH 44111	Elastomeric
Fenner America Ltd. 400 E. Main St. Middletown, CT 06457	Disk, Elastomeric, Variable-velocity U-joint
Flender Corp. 950 Tollgate Rd. Elgin, IL 60120	Gear, Disk, Elastomeric
Flexibox International, Inc./Metastream 2417 Albright Houston, TX 77017	Disk
FMC Corp., Link Belt Drive Div. 2045 W. Hunting Park Ave. Philadelphia, PA 19140	Gear, Chain
Globe Couplings and Universal Joints 370 Commerce Dr. Fort Washington, PA 19034	Elastomeric, Variable-velocity U-joints
Guardian Industries, Inc. P.O. Box 478 Michigan City, IN 46360	Rigid, Gear, Elastomeric

Company Name and Address	Type of Couplings Manufactured
Hayes Mfg., Inc. 661 R.W. Harris Ind. Dr. P.O. Box 367 Monton, MI 49663	Elastomeric
Helical Products Co., Inc. 901 McCoy Lane Santa Maria, CA 93456	Other
Holset Engineering Co. Ltd. P.O. Box A9 Turnbridge Huddersfield, England HD1 GRD	Elastomeric
Hub City, Safeguard P.O. Drawer 1089 Aberdeen, SD 57401	Elastomeric, Variable-velocity U-joints
Kamatics Corp. Bloomfield, CT 06002	Disk
Koelling-Morgan 15 Belmont St. Worcester, MA 01605	Variable-velocity U-joints
Koppers Co., Inc. P.O. Box 1696 Baltimore, MD 21203	Rigid, Gear, Disk, Diaphragm, Elastomeric
Koyo International Inc. of America 29570 Clemens Rd. Westlake, OH 44145	Constant-velocity U-joints, Variable-velocity U-joints
Lord Corp., Industrial Products Div. 1635 W. 12 St. Erie, PA 16512	Elastomeric
Lovejoy, Inc. 2655 Wisconsin Ave. Downers Grove, IL 60515	Elastomeric, Constant-velocity U-joints, Variable-velocity U-joints, Other
Magnaloy Couping Co. P.O. Box 295 Alpena, MI 49707	Elastomeric

Company Name and Address	Type of Couplings Manufactured
Maurey Mfg. Co. 2907 S. Wabash Ave. Chicago, IL 60616	Elastomeric, Other
Metal Bellows Corp. 1075 Providence Hwy. Sharon, MA 02067	Other
Morse Industrial Corp. Ithaca, NY 14850	Chain, Elastomeric
Naugler Engineering Co. Off Dunham Rd. Beverly, MA 01915	Disk, Other
Parish Power Products, Inc. 3912 Funsten St. Toledo, OH 43612	Constant-velocity U-joints, Variable-velocity U-joints
Philadelphia Gear Corp. 181 S. Gulph Rd. King of Prussia, PA 19406	Gear, Other
Poole Co. 1700 Union Ave. Woodberry, Baltimore, MD 21211	Gear, Other
Renbrandt, Inc. 659 Massachusetts Ave. Boston, MA 02118	Rigid, Disk, Other
Renold Power Transmission Corp. Bourne St. Westfield, NY 14807	Rigid, Gear, Chain, Diaphragm, Elastomeric, Other
Rexnord, Inc., Coupling Div. P.O. Box 549 Warren, PA 16365	Chain, Disk, Elastomeric
Robertshaw Controls Co., Milford Div. 155 Hill St. Milford, CT 06460	Disk, Other

Company Name and Address	Type of Couplings Manufactured
Rockwell International Automotive Operation 2135 W. Maple Rd. Troy, MI 48084	Rigid, Variable-velocity U-joints
Schmidt Couplings, Inc. 2845 Harriot Ave. S. Minneapolis, MN 55408	Rigid, Other
Servometer Corp. 501 Little Falls Rd. Cedar Grove, NJ 07009	Disk, Other
Sier—Bath Bear Co., Inc. 9252 Kennedy Blvd. North Bergen, NJ 07047	Rigid, Gear
TGW Thyssen Getriebe, Kupplungswerke GmbH Oberhausenor Strasse 1 D-4330 Mulheim (Ruhr) P.O. Box 01 1507 West Germany	Disk, Other
TurboFlex Ltd. The Pine Tyler Green High Wycombe Buckinghamshire, England	Disk
Twin Disc, Inc. 1340 Racine St. Racine, WI 53403	Constant-velocity U-joints, Variable-velocity U-joints
H.S. Watson Co. 12061 E. Slauson Ave. Sante Fe Springs, CA 90670	Constant-velocity U-joints, Variable-velocity U-joints
Wellman Bibby Co. Ltd. Mill Street West Dowsbury West Yorkshire, England	Grid
Winsmith Div., UMC Industries, Inc. Springville, NY 14141	Elastomeric, Other

Company Name and Address	Type of Couplings Manufactured
T.B. Wood's Sons Co. 440 N. Fifth Ave. Chambersburg, PA 17201	Rigid, Elastomeric
Xtek, Inc. 211 Township Ave. Cincinnati, OH 45216	Gear
Zurn Industries, Inc., Mechanical Drives Div. 1801 Pittsburgh Ave. Erie, PA 16512	Rigid, Gear, Disk, Diaphragm, Elastomeric, Variable-velocity U-joints, Other

Appendix B

Typical Installation Instructions

I. INSTRUCTIONS FOR GEAR COUPLINGS

Alignment and Installation Instructions

Purpose: The purpose of aligning equipment is to avoid transmission of unwanted stresses to bearings, shafts, couplings, etc.

How: By providing minimum angularity and offset of shaft axis at normal operating conditions (Figs. 1 and 2).

Why: To increase life of bearings, couplings, shafts and seals. To get at the root of serious malfunctions involving shutdowns and costly repairs.

When:
1. During installation, before grouting.
2. Immediately after initial operation.
3. When final operating conditions and final temperature are attained.
4. Seasonally.
5. Whenever first symptoms of trouble occur — vibration, undue noise, sudden overheating of bearings.

Practical Considerations:
1. Verify shaft separation.
2. Locate rotor in running position (for example, on sleeve bearing motors).
3. Anticipate thermal changes.
4. Read instructions and review drawings.

Tools:
1. Dial indicator with attaching device.
2. Feeler gauges.
3. Inside micrometer.
4. Outside micrometer.
5. Snap gauges.
6. Straight edge.

Angular Misalignment Measurement:
1. Measure at 4 points the space between the shaft ends (Fig. 3).
2. Rotate both shafts 180° and repeat.
3. Perform calculations for angle.

Offset Misalignment Measurement:
1. Rotate shaft A (with dial indicator mounted) and note readings of shaft B offset (Fig. 4).
2. Or use straight edge and feeler gauge (Fig. 5).

CAUTION: Misalignment at installation should not exceed 1/3 of rated catalog misalignment.

CAUTION: Rotating equipment is potentially dangerous and could cause injury or damage if not properly protected. Follow applicable codes and regulations.

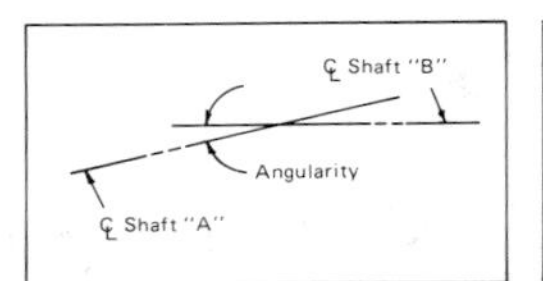

Figure 1 — Angularity is the acute angle formed at the intersection of the axes of the driving and the driven machine shafts. When shafts are exactly parallel, angular misalignment is zero; but vertical or horizontal displacement of axes may be present (See Fig. 2).

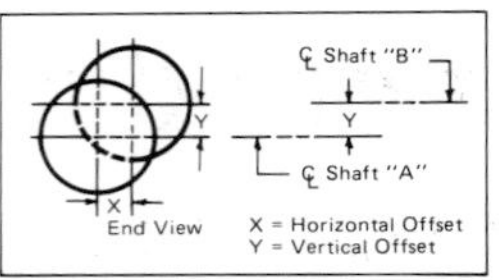

Figure 2 — Concentric alignment (also called offset alignment or parallel offset) is the relationship between the shaft axes in terms of vertical and horizontal displacements of the axis of one shaft from the axis of the other shaft.

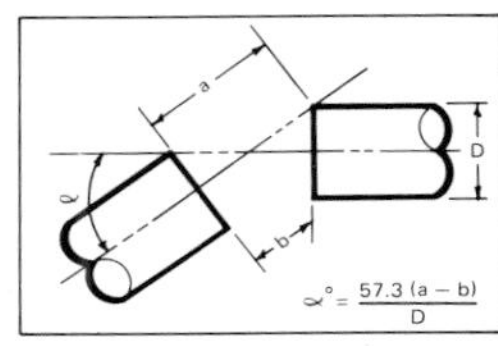

Figure 3 — To determine relative angular shaft-positions of driving and driven machines, measure at four points the space between the shaft ends. Choose the largest (a) and smallest dimension (b).

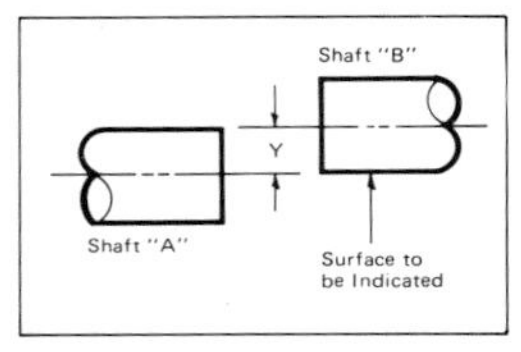

Figure 4 — To measure offsets with a dial indicator, attach the indicator to shaft "A", rotate shaft, and indicate to the periphery of shaft "B". To obtain actual displacements of shafts, divide dial indicator readings by 2.

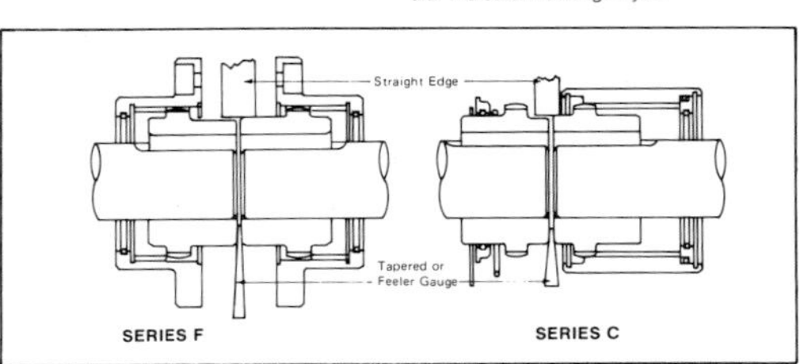

Figure 5 — Lay straight edge on one hub and measure gap between straight edge and other hub with feeler gauge. Measure at top, bottom, and both sides. Feeler gauge readings indicate actual displacements of shafts.

Installation and Lubrication Instructions

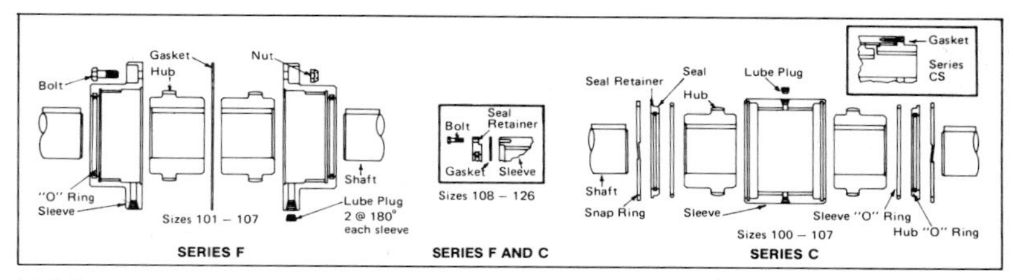

Installation
Disassemble coupling and clean all parts. Follow the appropriate 6 steps below and you are ready to go! Installed and lubricated in accordance with the instructions, your Amerigear coupling is prepared for a life of dependable, trouble-free service.

Series F Installation
Step 1. Lightly coat grease on "O" rings and insert "O" rings into grooves of sleeve (into grooves of seal retainer for sizes 108 and larger). Place sleeves (sizes 100-107) over shaft ends. For sizes 108 and larger, place only the seal retainer (with "O" rings inserted on shaft and check orientation to assure male pilot is toward shaft end. Care should be taken not to damage seal on shaft key seat.
Step 2. Check key fits and coat keys and keyways with oil resistant sealing compound (Permatex) to prevent leakage. Install size 101 through 107 hubs on shafts with long ends flush with shaft ends. Install size 100 hub on shaft with short end flush with shaft end. For shrink fits, apply heat to hubs uniformly, preferably submerged in oil not exceeding 350°F. Do not allow "O" ring seals to contact heated hubs. For sizes 108 and larger, place retainer gaskets and sleeves over hubs and onto shafts.
Step 3. Align shafts allowing clearance as per tabulation or in accordance with Dimension "D" from Engineering Data. Check gap with taper or feeler gauge at 90° points and align hubs with straight edge at 90° points.
Step 4. After thoroughly coating hub and sleeve teeth with lubricant, slip sleeves onto hubs, carefully engaging teeth (do not damage seal surface). Place sleeve gasket between sleeves and align bolt holes.
Step 5. Secure sleeves, using care to tighten fasteners uniformly. See tabulation "Flange Bolt Tightening Torque". For sizes 108 and larger bolt seal retainers to sleeves.
Step 6. Remove both Dryseal lube plugs and add grease in sufficient amount to overflow with lubricant holes in horizontal position. Install lube plugs using Permatex No. 2 for sealing and seat securely. For limited end float variation, Series FI and FP, lube both halves of coupling separately. For vertical couplings, remove one Dryseal plug in each coupling half and carefully insert thin probe under upper hub seal to assure complete filling.

Series C Installation
Step 1. For sizes 100-107 place snap ring, seal retainer (with "O" ring seated in retainer groove) and sleeve "O" ring on each shaft. For sizes 108 and larger, place seal retainer (with "O" ring inserted) and gasket over shaft and check orientation to assure male pilot is toward shaft end. For Series CS install gasket between rigid hub flange and sleeve.
Step 2. Check key fits and coat keys and keyways with oil resistant compound to prevent leakage. Install hubs on shafts with short ends flush with shaft ends. For shrink fits, apply heat to hubs uniformly, preferably submerged in oil not exceeding 350°F. Do not allow "O" rings to contact heated hubs.
Step 3. Slip sleeve over hub mounted on longest shaft.
Step 4. Align shafts allowing clearance as per tabulation or from Engineering Data, Dimension "D" Check gap with taper or feeler gauge at 90° intervals. Also align hubs with straight edge at 90° points.
Step 5. Pack hub and sleeve teeth with grease. Force grease into shaft gap. Lightly coat grease on "O" rings. Slide sleeve over hubs to center position. Remove Dryseal lube plugs* and add lubricant in sufficient amount to overflow capacity. For vertical application remove one Dryseal plug and carefully insert probe between upper hub seal to assure coupling is full of lubricant.
Step 6. For sizes 100-107 install sleeve "O" rings in sleeve counterbores — then press seal retainer assembly in place. Use fingertips or blunt tool. Seat snap rings in grooves, using a winding motion. Recheck to assure snap rings are positively seated. For sizes 108 and larger, bolt seal end plates to sleeve.

SIZE	HUB SEPARATION			FLANGE BOLT TIGHTENING TORQUE FT. LBS.*	
	F & C	FS	CS	F Exposed	F Shrouded
100	.125	.078	.125	10	10
101	.125	.078	.125	10	10
101¼	.125	.078	.125	10	10
101½	.125	.156	.125	29	32
102	.125	.156	.125	63	32
102½	.188	.188	.188	125	69
103	.188	.188	.188	125	69
103½	.250	.219	.250	210	133
104	.250	.312	.250	210	133
104½	.312	.344	.312	210	133
105	.312	.344	.312	313	232
105½	.312	.344	.312	313	232
106	.312	.406	.312	313	340
107	.375	.500	.375	440	476
108	.375	.500	–		
109	.500	.562	–		
110	.500	.625	–		
111	.500	.625	–		
112	.500	.625	–		
113	.750	.750	–		
114	.750	.750	–		
115	.750	.750	–		

*Tightening torque based on unlubricated threads, if threads are lubricated derate torque to 75% of above values.

Maintenance and Lubrication

LUBRICANTS

LUBRICANT MANUFACTURER	GENERAL	MOIST OR WET	HIGH TORQUE	150° - 300°F	CLASS III
American Lubricants Co.	Alubco Bison 1650	(same)	(same)	(same)	(same)
Atlantic Richfield Co.	Arco MP	Arco MP	Arco EP or Moly D	Dominion H2	Dominion H3
Amoco	Amolith #2	Amolith #2	Amolith #2	Rykon EP-2	Amoco CPLG Grease
Chevron USA, Inc.	Duralith EP-2	Duralith EP-2	Duralith EP-2	Duralith EP-2	NL Gear Compound 460
Cities Service Oil Co.	Citgo AP or HEP-2	AP or HEP-2	AP or HEP-2	AP or HEP-2	Citgo AP or EP Compound 130
Continental Oil Co.	Super Sta Grease	HD Calcium Grease	HD Calcium	HD Calcium	HD Calcium, Transmission Oil No. 140
Far Best	Molyvis ST-200	(same)	(same)	(same)	--
Fiske Bros. Refining Co.	Lubriplate 630-AA	Lubriplate 630-AA	Lubriplate 630-AA	Lubriplate 1200-2	Lubriplate No. 8
Gulf Oil Co.	Gulfcrown EP #2	Gulfcrown EP #2	Gulfcrown EP #2	Hi-Temp Grease	Precision No. 3
Exxon Co.	Pen-O-Lead EP 350	Rolubricant EP-300	Rolubricant EP-350	Unirex N2	Unirex N2 or Nuto No. 146
Kendall Refining Co.	Kenlube L-421 or Waverly Torque Lube A	L-421 or Torque Lube A	L-421 or Torque Lube A	L-421 or Torque Lube A	L-427
Mobil Oil Co.	Mobilux EP-0	Mobilux EP-0	Mobil Temp 78	Mobil Temp 78	Mobil No. 28
Pennzoil Co.	Pennlith 711 or 712	Pennlith 711 or 712	Pennlith 711 or 712	Pennlith 712 or Bearing Lube 706	Hi Speed Pennlith 712 or Bearing Lube 706
Suntech, Inc.	Sunaplex 991 EP or Prestige 741 EP	(same)	(same)	(same)	--
Syn-Tech	3913-G1	(same)	(same)	(same)	(same)
Tenneco Chemicals, Inc.	Anderol 786	(same)	(same)	(same)	(same)
Texaco, Inc.	Multifak EP-2	Multifak EP-2	Multifak EP-2	Thermatex EP-2	Thermatex EP-2
Union Oil Co. of Calif.	UNOBA EP-2	UNOBA EP-2	UNOBA EP-2	UNOBA EP-2	MP Gear Lube 140

For low temp. (-65°), Aeroshell #22 by Shell Oil Co., Anderol 793 by Tenneco Chemicals, Inc. & Mobil Grease #28 by Mobil Oil Co.

LUBRICANT QUANTITIES

Coupling Size	LUBRICATION			
	SERIES "F"		SERIES "C"	
	Wt. Lbs.	Vol. Qts.	Wt. Lbs.	Vol. Qts.
100	.020	.010	.015	.010
101	.045	.025	.036	.020
101¼	.06	.033	.045	.025
101½	.18	.10	.09	.05
102	.34	.19	.18	.10
102½	.45	.25	.30	.17
103	.79	.44	.45	.25
103½	1.0	.56	.57	.31
104	1.7	.94	1.1	.63
104½	2.0	1.1	1.2	.67
105	3.6	2.0	1.8	1.0
105½	5.0	2.7	2.3	1.2
106	5.9	3.2	2.7	1.5
107	10.9	6.0	6.0	3.3
108	16.6	9.2	8.0	4.4
109	20.3	11.2	10.9	6.0
110	25.4	14.0	13.8	7.6
111	38.4	21.2	22.5	12.4
112	43.5	24	25.4	14.0
113	52.2	28.8	31.2	17.2
114	68.9	38	42.8	23.6
115	77.5	42.8	49.3	27.2

Maintenance — The Amerigear Coupling requires a minimum of maintenance. Nevertheless, to ensure a trouble-free life a few checks and proper lubrication should be performed at regular intervals.

Zurn suggests that the maximum interval between checks and relube be one year. This is only a guide, and the actual interval should be in accordance with good operating practices for application.

To disassemble Series F remove flange fasteners, separate sleeves, slide sleeves over hubs, clean out old lubricant, and inspect seals and gear teeth. Reassemble, starting with Step 3 under Series F installation instructions on the previous page.

To disassemble Series C, remove one snap ring, slide sleeves off hubs, clean out old lubricant and inspect seals and gear teeth. Reassemble, starting at Step No. 4 under Series C installation instructions on the previous page.

If proper alignment of shafts is assured and it is not practical to disassemble coupling, remove both lube plugs and add grease in sufficient amount to overflow with lubricant holes in horizontal position. Recommended lubricants and quantities are listed on this page.

NOTE: Sizes 100 and 101 Series C are supplied without lube plugs — lubricate per Series C Step No. 5.

The lubricants listed above are recommended by the lubricant manufacturers for the indicated conditions. This list is solely for our customers' convenience and does not constitute an endorsement. The listing is not intended to be complete nor necessarily current due to continuous research and improvement by the various manufacturers.

Series F, FM, FA use quantities as recommended.
Series FS, FMS, FAS use one-half the quantities recommended.
Series C, CM, CA use quantities as shown.
Series CS, CMS, CAS use one-half the quantities recommended.
*Series F Class III use quantities as recommended for Series F but limited to the greases shown in Class III column above or the following oils:
Citgo EP Compound 130 by Cities Service; Nuto No. 146 by Exxon; Lubriplate No. 8 by Fiske Bros.; Transmission Oil No. 140 by Continental.

II. INSTRUCTIONS FOR DISK COUPLINGS

INSTALLATION AND ALIGNMENT
FLEXIBLE DISC COUPLINGS

INSTALLATION

Reasonable care in initial assembly and aligning will permit couplings to operate to full capacity in correcting future misalignments.

> **CAUTION**
> **All rotating power transmission producrs are potentially dangerous and must be properly guarded.**

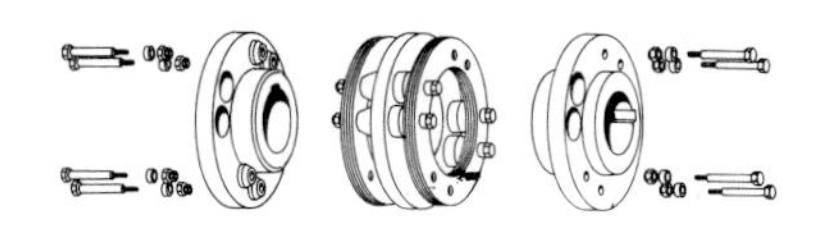

1. Coupling is shipped with center member assembled as shown above. If it is necessary to disassemble the coupling completely, tie a string or wire thru the bolt holes to maintain the dialing (alternating grain of the individual discs). Be careful to note the arrangement of parts.

2. Inspect both shafts and hub bores, making sure they are free from burrs. Be sure the keys fit the shafts and hubs properly.

3. Mount hubs on shafts. If hub is bored for an interference fit, they should be heated in water, oil or with soft open flame and quickly positioned on the shaft. Do not spot-heat hub as it may cause distortion.

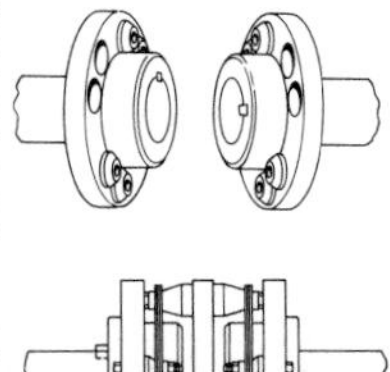

4. Move equipment to be connected into position, making sure that the distance between hubs is equal to the "C" dimension given in coupling dimensional tables ("L" dimension on floating shaft coupling types.)

5. Reassemble coupling, keeping the proper order of parts as noted in Step 1. Refer to drawings on coupling listing pages.

Note that coupling is in alignment.

NOTE: Floating shaft couplings should be mounted close to bearings of connected units to minimize overhang. Cooling tower and other right angle drives—dowel right angle gear boxes to their bases after couplings have been aligned. Right angle gear boxes tend to creep in a counter-rotational direction, which causes severe coupling misalignment and may result in premature failures.

ALIGNMENT

INDICATOR METHOD

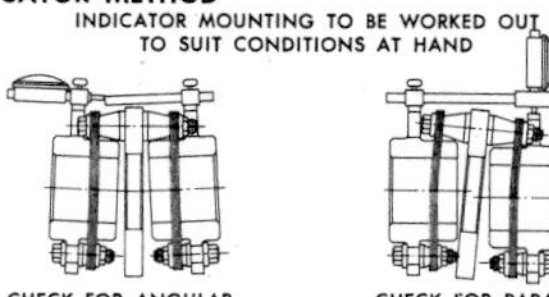

CHECK FOR ANGULAR MISALIGNMENT
Dial indicator measures maximum longitudinal variation in hub spacing through 360° rotation.

CHECK FOR PARALLEL MISALIGNMENT
Dial indicator measures displacement of one shaft center line from the other.

1. Attach dial indicator to hub, as with a hose clamp; rotate coupling 360° to locate point of minimum reading on dial; then rotate body or face of indicator so that zero reading lines up with pointer.
2. Rotate coupling 360°. Watch indicator for misalignment reading.
3. Driver and driven units will be lined up when dial indicator reading comes within maximum allowable variation of .002" for each inch of flange diameter; ie. a coupling with a 4-inch diameter flange has a maximum variation of .008 inches in indicator reading.
4. Reset pointer to zero and repeat above operations 1 and 2 when either driven unit or driver is moved during aligning trials.
5. Check for parallel misalignment as shown. Move or shim units so that parallel misalignment is at a minimum.
6. Coupling should be rotated several revolutions to make sure no "end-wise creep" in connected shafts is measured.
7. Tighten all locknuts.
8. Re-check and tighten all locknuts after several hours of operation.

CALIPER AND STRAIGHT EDGE METHOD

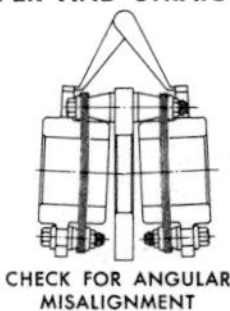

CHECK FOR ANGULAR MISALIGNMENT

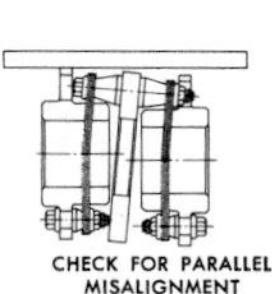

CHECK FOR PARALLEL MISALIGNMENT

1. Use calipers to check the gap between hubs. Gap should be the same at all points around the hub.
2. Place straight edge on the rims at the top and sides. When the coupling is in alignment the straight edge should rest evenly and both disc pack assemblies should be in a perfect plane at right angles to the straight edge.
3. Tighten all locknuts.
4. After several hours of operation recheck gap between hubs, and recheck tightness of all locknuts.
5. When operating at full speed, disc packs should have a clearly defined appearance—not blurred when viewed from top and side.

III. INSTRUCTIONS FOR ELASTOMERIC COUPLINGS

SURE-FLEX INSTALLATION INSTRUCTIONS

Sure-Flex flanges (outer metallic parts) and sleeves (inner elastomeric members) come in many sizes and types.
First, determine the size and type of components being used. Remove all components from their boxes, and loosely assemble the coupling on any convenient surface.
All rubber sleeves (EPDM and Neoprene) have the same ratings for a given size and may be used interchangeably. Hytrel sleeves, however, have completely different ratings. Rubber sleeves should not be substituted for Hytrel, or Hytrel for rubber.
(Do not attempt to install the wire ring on the two-piece E or N sleeve at this time.)

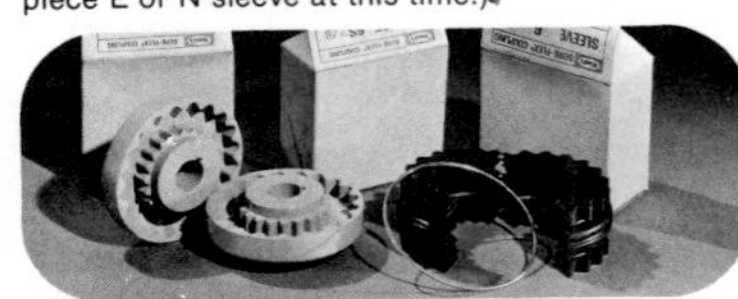

1 Inspect all coupling components and remove any protective coatings or lubricants from bores, mating surfaces and fasteners. Remove any existing burrs, etc. from the shafts.

2 Slide one coupling flange onto each shaft, using snug-fitting keys where required. With the Type B or C flanges, it may be necessary to expand the bore by wedging a screwdriver into the saw cut.

3 Position the flanges on the shafts to achieve the G dimension shown in table at right. It is usually best to have an equal length of shaft extending into each flange. Tighten one flange in its final position. Slide the other far enough away to install the sleeve. With a two-piece sleeve, do not move the wire ring to its final position; but allow it to hang loosely in the groove adjacent to the teeth.

4 Slide the loose flange on the shaft until the sleeve is completely seated in the teeth of each flange. (The "G" dimension is for reference and not critical.) Secure the flange to the shaft. (Note: When using sleeve bearing electric motors, the armature must be at its electrical center when determining sleeve engagement and flange location.)

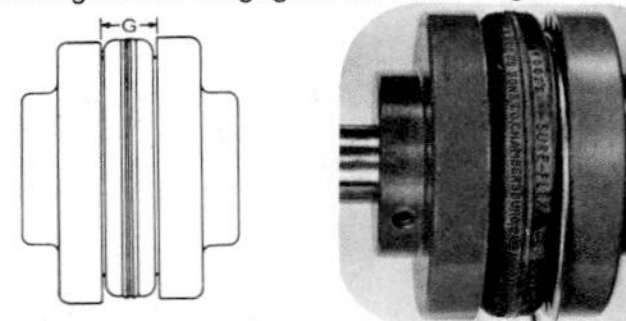

Different coupling sleeves require different degrees of alignment precision. Locate the alignment values for your sleeve size and type in table at right.

5 Check parallel alignment by placing a straightedge across the two coupling flanges and measuring the maximum offset at various points around the periphery of the coupling. DO NOT rotate the coupling. If the maximum offset exceeds the figure shown under "Parallel" in table below, realign the coupling.

6 Check angular alignment with a micrometer or caliper. Measure from the outside of one flange to the outside of the other at intervals around the periphery of the coupling. Determine the maximum and minimum dimensions. DO NOT rotate the coupling. The difference between the maximum and minimum must not exceed the figure given under "Angular" in table below. If a correction is necessary, be sure to recheck the parallel alignment.

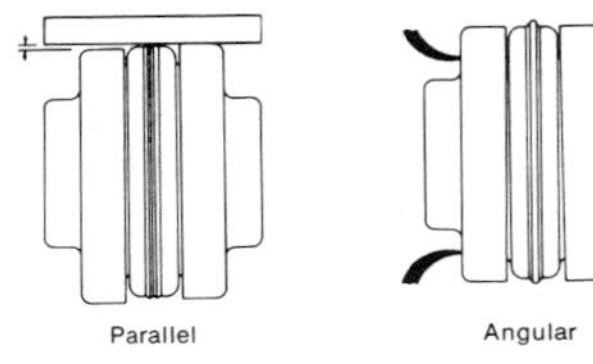

Parallel   Angular

MAXIMUM ALLOWABLE MISALIGNMENT

(Dimensions in inches)

Sleeve Size	G Dimension	Types JE & JN		Types E & N		Type H	
		Parallel	Angular	Parallel	Angular	Parallel	Angular
3	⅜	.010	.035	....	....	....	....
4	⅝	.010	.043	.010	.043	....	....
5	¾	.015	.056	.015	.056	....	....
6	⅞	.015	.070	.015	.070	.010	.016
7	1	.020	.081	.020	.081	.012	.020
8	1⅛	.020	.094	.020	.094	.015	.025
9	1 7/16	....	....	.025	.109	.017	.028
10	1⅝	....	....	.025	.128	.020	.032
11	1⅞	....	....	.032	.151	.022	.037
12	2 5/16	....	....	.032	.175	.025	.042
13	2 11/16	....	....	.040	.195	.030	.050
14	3¼	....	....	.045	.242	.035	.060
16	4¾	....	....	.062	.330	....	....

Note: Values shown above apply if the actual torque transmitted is more than ¼ the coupling rating. For lesser torque, reduce the above values by ½.

7 If the coupling employs the two-piece sleeve with the wire ring, force the ring into its groove in the center of the sleeve. It may be necessary to pry the ring into position with a blunt screwdriver.

8 Check safety codes, and install protective guards or shields as required.

Caution: Coupling sleeves may be thrown from the assembly when subjected to a severe shock load.

Index